CourseMate

Engaging. Trackable. Affordable.

CourseMate brings course concepts to life with interactive learning, study, and exam preparation tools that support FINITE.

INCLUDES:

Integrated eBook, interactive teaching and learning tools, and Engagement Tracker, a first-of-its-kind tool that monitors student engagement in the course.

ON THE
WEB

FINITE
Are you in?

ONLINE RESOURCES INCLUDED!

FOR INSTRUCTORS:
- First Day Class Instructions
- Solution Builder
- PowerPoint® Slides
- Instructor Prep Cards
- Student Review Cards
- Tech Cards

FOR STUDENTS:
- Quizzing
- Flashcards
- Glossary
- Interactive eBook
- Solution Videos
- Lecture Videos
- Additional Examples
- Extra Practice Exercises
- Answers & Solutions
- Technology Guides
- and More!

D1511459

Students sign in at **login.cengagebrain.com**

BROOKS/COLE
CENGAGE Learning™

FINITE
Geoffrey C. Berresford and Andrew M. Rockett

Editor in Chief: Michelle Julet

Publisher: Richard Stratton

Senior Development Editor: Laura Wheel

Editorial Assistant: Haeree Chang

Media Editor: Heleny Wong

Marketing Manager: Ashley Pickering

Marketing Coordinator: Erica O'Connell

Marketing Communications Manager: Mary Anne Payumo

Content Project Manager: Jill Clark

Executive Marketing Manager, 4LTR Press: Robin Lucas

Product Development Manager, 4LTR Press: Steve Joos

Project Manager, 4LTR Press, Kelli Strieby

Art Director: Jill Ort

Senior Manufacturing Buyer: Diane Gibbons

Rights Acquisition Specialist, Image: Mandy Groszko

Senior Rights Acquisition Specialist, Text: Katie Huha

Editorial Development and Composition: Lachina Publishing Services

Production Service: Lachina Publishing Services

Cover Designer: Hannah Wellman

Cover Image: © istockimages.com/Andresr

© 2012 Brooks/Cole, Cengage Learning

For product information and technology assistance, contact us at
Cengage Learning Customer & Sales Support, 1-800-354-9706

For permission to use material from this text or product, submit all requests online at **cengage.com/permissions**. Further permissions questions can be emailed to **permissionrequest@cengage.com**.

Library of Congress Control Number: 2010936957

Student Edition:

ISBN-13: 978-0-8400-6555-1

ISBN-10: 0-8400-6555-8

Brooks/Cole
20 Channel Center Street
Boston, MA 02210
USA

Cengage Learning is a leading provider of customized learning solutions with office locations around the globe, including Singapore, the United Kingdom, Australia, Mexico, Brazil and Japan. Locate your local office at **international.cengage.com/region**.

Cengage Learning products are represented in Canada by Nelson Education, Ltd.

For your course and learning solutions, visit **www.cengage.com**. Purchase any of our products at your local college store or at our preferred online store **www.cengagebrain.com**.

Printed in the United States of America

1 2 3 4 5 6 7 14 13 12 11 10

Brief Contents

finite

Remember

The portable cards at the back of the book are designed to help you with the course!

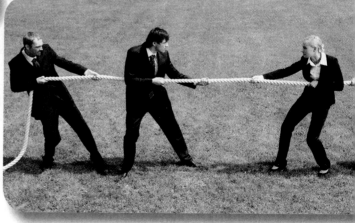

LEARN YOUR WAY

Chapter 1
Functions 2

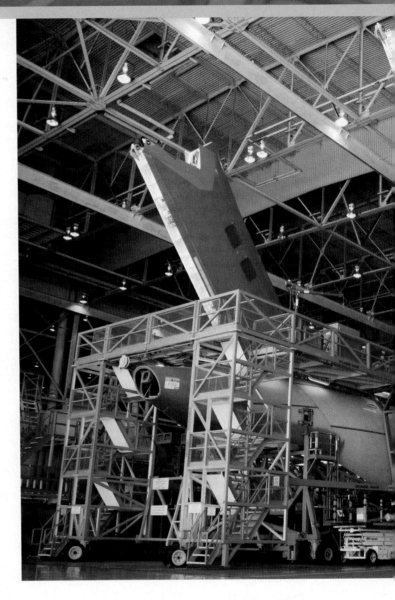

Contents

Chapter 3
Systems of Equations and Matrices 86

Contents

Contents

Chapter 6
Statistics 222

Contents

Contents

STUDY YOUR WAY

At no additional cost, you have access to online learning resources that include **tutorial videos, printable flashcards, quizzes,** and more!

Watch videos that offer step-by-step conceptual explanations and guidance for each chapter in the text.

Along with the printable flashcards and other online resources, you will have a multitude of ways to check your comprehension of key mathematical concepts.

You can find these resources at **login.cengagebrain.com.**

Functions

1.1   Real Numbers, Inequalities, and Lines

In this section we will study *linear* relationships between two quantities—that is, relationships that can be represented by *lines*. In later sections we will study *nonlinear* relationships, which can be represented by *curves*. Let's start by reviewing the fundamentals of real numbers and inequalities, sets and intervals, and the Cartesian plane. We'll then go over lines and slopes, exponents, and equations.

Real Numbers and Inequalities

In this book the word *number* means *real number*, a number that can be represented by a point on the number line (also called the *real line*).

The *order* of the real numbers is expressed by inequalities, with $a < b$ meaning "a is to the *left* of b" and $a > b$ meaning "a is to the *right* of b."

Inequalities

Inequality	In Words
$a < b$	a is less than (smaller than) b
$a \leq b$	a is less than or equal to b
$a > b$	a is greater than (larger than) b
$a \geq b$	a is greater than or equal to b

The inequalities $a < b$ and $a > b$ are called "strict" inequalities, and $a \le b$ and $a \ge b$ are called "nonstrict" inequalities.

➡ **EXAMPLE 1** Inequalities between Numbers

a. $3 \le 5$ **b.** $6 > -2$ **c.** $-10 < -5$

-10 is less than (smaller than) -5

Throughout this book are many **Practice Problems**, short questions designed to check your understanding of a topic before moving on to new material. Full solutions are given at the end of the section. Solve the following Practice Problem and then check your answer.

→ **Practice Problem 1**

Which number is smaller: $\dfrac{1}{100}$ or $-1,000,000$?

➤ **Solution on page 12**

Multiplying or dividing both sides of an inequality by a negative number reverses the direction of the inequality:

$$-3 < 2 \quad \text{but} \quad 3 > -2$$

Multiplying by -1

A *double* inequality, such as $a < x < b$, means that *both* the inequalities $a < x$ and $x < b$ hold. The inequality $a < x < b$ can be interpreted graphically as "x is between a and b."

$$a < x < b$$

Sets and Intervals

Braces { } are read "the set of all" and a vertical bar | is read "such that."

→ **EXAMPLE 2**   **Interpreting Sets**

The set of all

a. $\{x \mid x > 3\}$ means "the set of all x such that x is greater than 3."

Such that

b. $\{x \mid -2 < x < 5\}$ means "the set of all x such that x is between -2 and 5."

→ **Practice Problem 2**

a. Write in set notation "the set of all x such that x is greater than or equal to -7."

b. Express in words: $\{x \mid x < -1\}$.

➤ **Solutions on page 12**

The set $\{x \mid 2 \le x \le 5\}$ can be expressed in *interval* notation by enclosing the endpoints 2 and 5 in square brackets: [2, 5]. The *square* brackets indicate that the endpoints are *included*. The set $\{x \mid 2 < x < 5\}$ can be written (2, 5). The *parentheses* indicate that the endpoints 2 and 5 are *excluded*. An interval is *closed* if it includes both endpoints and is *open* if it includes neither endpoint. The four types of intervals are shown below: a *solid* dot • on the graph indicates that the point is *included* in the interval; a *hollow* dot ○ indicates that the point is *excluded*.

Finite Intervals

Interval Notation	Set Notation	Graph	Type
$[a, b]$	$\{x \mid a \leq x \leq b\}$		Closed
(a, b)	$\{x \mid a < x < b\}$		Open
$[a, b)$	$\{x \mid a \leq x < b\}$		Half-open or Half-closed
$(a, b]$	$\{x \mid a < x \leq b\}$		

An interval may extend infinitely far to the right (indicated by the symbol ∞ for "infinity") or infinitely far to the left (indicated by $-\infty$ for "negative infinity"). Note that ∞ and $-\infty$ are not numbers, but are merely symbols to indicate that the interval extends endlessly in one direction or the other. The infinite intervals in the next box are said to be *closed* or *open* depending on whether they *include* or *exclude* their single endpoint.

Infinite Intervals

Interval Notation	Set Notation	Graph	Type
$[a, \infty)$	$\{x \mid x \geq a\}$		Closed
(a, ∞)	$\{x \mid x > a\}$		Open
$(-\infty, a]$	$\{x \mid x \leq a\}$		Closed
$(-\infty, a)$	$\{x \mid x < a\}$		Open

→ **EXAMPLE 3**   Graphing Sets and Intervals

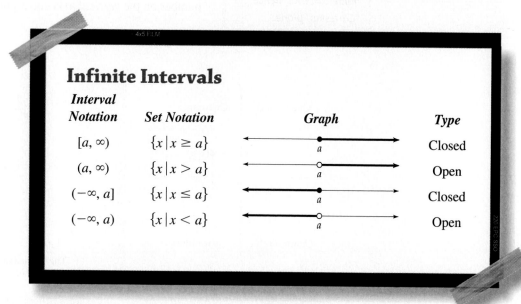

Interval Notation	Set Notation	Graph
$(-2, 5]$	$\{x \mid -2 < x \leq 5\}$	
$[3, \infty)$	$\{x \mid x \geq 3\}$	

We use *parentheses* rather than square brackets with ∞ and $-\infty$ since they are not actual numbers.

The interval $(-\infty, \infty)$ extends infinitely far in *both* directions (meaning the entire real line) and is also denoted by $\mathbb{R}$ (the set of all real numbers).

$$\mathbb{R} = (-\infty, \infty)$$
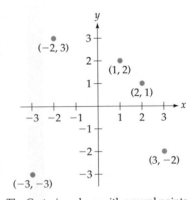

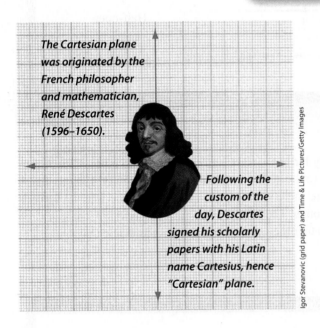

The Cartesian plane was originated by the French philosopher and mathematician, René Descartes (1596–1650).

Following the custom of the day, Descartes signed his scholarly papers with his Latin name Cartesius, hence "Cartesian" plane.

Igor Stevanovic (grid paper) and Time & Life Pictures/Getty Images

The Cartesian Plane

Two real lines or *axes*, one horizontal and one vertical, intersecting at their zero points, define the *Cartesian plane*. The axes divide the plane into four *quadrants*, I through IV, as shown below.

Any point in the Cartesian plane can be specified uniquely by an ordered pair of numbers (x, y); x, called the *abscissa* or *x-coordinate*, is the number on the *horizontal* axis corresponding to the point; y, called the *ordinate* or *y-coordinate*, is the number on the *vertical* axis corresponding to the point.

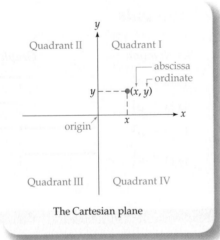

The Cartesian plane

The Cartesian plane with several points.
Order matters: $(1, 2)$ is not the same as $(2, 1)$.

Lines and Slopes

The symbol Δ (read "delta," the Greek letter D) means "the change in." For any two points (x_1, y_1) and (x_2, y_2) we define

$$\Delta x = x_2 - x_1 \qquad \text{The change in } x \text{ is the difference in the } x\text{-coordinates}$$

$$\Delta y = y_2 - y_1 \qquad \text{The change in } y \text{ is the difference in the } y\text{-coordinates}$$

Any two distinct points determine a line. A nonvertical line has a *slope* that measures the steepness of the line, defined as *the change in y divided by the change in x* for any two points on the line.

Slope of Line Through (x_1, y_1) and (x_2, y_2)

$$m = \frac{\Delta y}{\Delta x} = \frac{y_2 - y_1}{x_2 - x_1}$$

Slope is the change in y divided by the change in x $(x_2 \neq x_1)$

The changes Δy and Δx are often called, respectively, the "rise" and the "run," with the understanding that a negative "rise" means a "fall." Slope is then rise divided by run (or "rise over run").

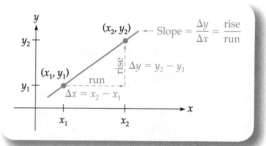

→ **EXAMPLE 4**   **Finding Slopes and Graphing Lines**

Find the slope of the line through each pair of points and graph the line.
a. $(1, 3), (2, 5)$ **b.** $(2, 4), (3, 1)$
c. $(-1, 3), (2, 3)$ **d.** $(2, -1), (2, 3)$

Solution
We use the slope formula $m = \dfrac{y_2 - y_1}{x_2 - x_1}$ for each pair $(x_1, y_1), (x_2, y_2)$.

a. For $(1, 3)$ and $(2, 5)$ the slope is

$$\frac{5 - 3}{2 - 1} = \frac{2}{1} = 2.$$

b. For $(2, 4)$ and $(3, 1)$ the slope is

$$\frac{1 - 4}{3 - 2} = \frac{-3}{1} = -3.$$

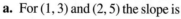

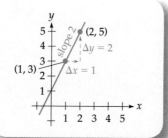

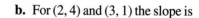

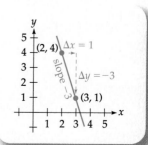

c. For $(-1, 3)$ and $(2, 3)$ the slope

is $\dfrac{3 - 3}{2 - (-1)} = \dfrac{0}{3} = 0.$

d. For $(2, -1)$ and $(2, 3)$ the slope

is *undefined*: $\dfrac{3 - (-1)}{2 - 2} = \dfrac{4}{0}.$

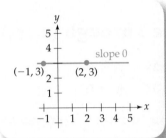

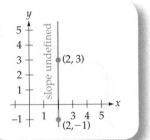

If $\Delta x = 1$ as in Examples 4a and 4b, then the slope is just the "rise," giving

$$\text{Slope} = \left(\begin{array}{c} \text{Amount that the line rises} \\ \text{when } x \text{ increases by 1} \end{array} \right)$$

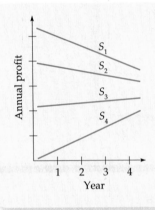

➜ Practice Problem 3

A company president is considering four different business strategies, called S_1, S_2, S_3, and S_4, each with different projected future profits. The graph on the left shows the annual projected profit for the first few years for each of the strategies. Which strategy will yield:

a. The highest projected profit in year 1?
b. The highest projected profit in the long run?

➤ **Solutions on page 12**

Equations of Lines

The point where a nonvertical line crosses the y-axis is called the *y-intercept* of the line. The y-intercept can be given either as the y-coordinate b or as the point $(0, b)$. Such a line can be expressed very simply in terms of its slope and y-intercept, representing the points by variable coordinates (or "variables") x and y.

Slope-Intercept Form of a Line

$y = mx + b$

$m = \text{slope}$
$b = y\text{-intercept}$

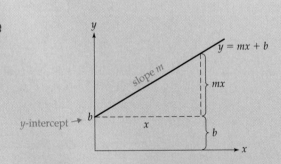

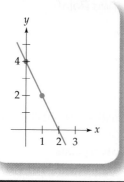

→ **EXAMPLE 5** **Using the Slope-Intercept Form**

Find an equation of the line with slope -2 and y-intercept 4, and graph it.

Solution

$$y = -2x + 4$$

$y = mx + b$ with
$m = -2$ and $b = 4$

We graph the line by first plotting the y-intercept $(0, 4)$. Using the slope $m = -2$, we plot another point 1 unit to the right and 2 units *down* from the y-intercept. We then draw the line through these two points, as shown on the left.

Point-Slope Form of a Line

$$y - y_1 = m(x - x_1)$$

$(x_1, y_1) =$ point on the line
$m =$ slope

This form comes directly from the slope formula $m = \dfrac{y_2 - y_1}{x_2 - x_1}$ by replacing x_2 and y_2 by x and y, and then multiplying each side by $(x - x_1)$. It is most useful when you know the slope of the line and a point on it.

→ **EXAMPLE 6** **Using the Point-Slope Form**

Find an equation of the line through $(6, -2)$ with slope $-\frac{1}{2}$.

Solution

$$y - (-2) = -\tfrac{1}{2}(x - 6)$$

$y - y_1 = m(x - x_1)$ with
$y_1 = -2$, $m = -\frac{1}{2}$,   and $x_1 = 6$

$$y + 2 = -\tfrac{1}{2}x + 3$$

Eliminating parentheses

$$y = -\tfrac{1}{2}x + 1$$

Subtracting 2 from each side

Alternatively, we could have found this equation using $y = mx + b$, replacing m by the given slope $-\frac{1}{2}$, and then substituting the given $x = 6$ and $y = -2$ to evaluate b.

→ EXAMPLE 7   Finding an Equation for a Line through Two Points

Find an equation for the line through the points $(4, 1)$ and $(7, -2)$.

Solution
The slope is not given, so we calculate it from the two points.

$$m = \frac{-2 - 1}{7 - 4} = \frac{-3}{3} = -1 \qquad\qquad m = \frac{y_2 - y_1}{x_2 - x_1} \quad \text{with } (4, 1) \text{ and } (7, -2)$$

Then we use the point-slope formula with this slope and either of the two points.

$$y - 1 = -1(x - 4) \qquad\qquad \begin{array}{l} y - y_1 = m(x - x_1) \quad \text{with} \\ \text{slope } -1 \text{ and point } (4, 1) \end{array}$$

$$y - 1 = -x + 4 \qquad\qquad \text{Eliminating parentheses}$$

$$y = -x + 5 \qquad\qquad \text{Adding 1 to each side}$$

→ Practice Problem 4

Find the slope-intercept form of the line through the points $(2, 1)$ and $(4, 7)$.

➤ Solution on page 12

Vertical and horizontal lines have particularly simple equations: a variable equaling a constant.

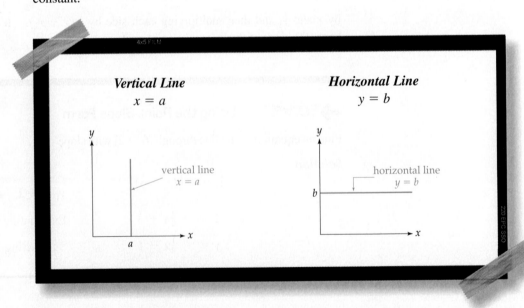

Vertical Line
$x = a$

Horizontal Line
$y = b$

→ EXAMPLE 8   Graphing Vertical and Horizontal Lines

Graph the lines $x = 2$ and $y = 6$.

Solution

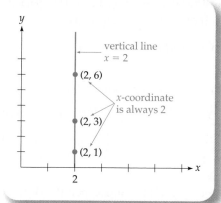

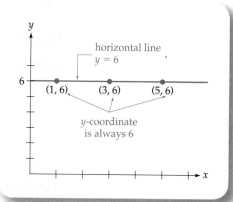

→ **EXAMPLE 9**   **Finding Equations of Vertical and Horizontal Lines**

a. Find an equation for the *vertical* line through the point $(3, 2)$.
b. Find an equation for the *horizontal* line through the point $(3, 2)$.

Solution

a. Vertical line $x = 3$ $x = a$,   with a being the
                                               x-coordinate from $(3, 2)$

b. Horizontal line $y = 2$ $y = b$,   with b being the
                                               y-coordinate from $(3, 2)$

→ **Practice Problem 5**

Find an equation for the vertical line through the point $(-2, 10)$.

➤ **Solution on page 12**

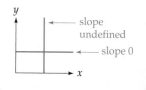

Distinguish carefully between slopes of vertical and horizontal lines:

    Vertical line: slope is *undefined*.
    Horizontal line: slope *is* defined and is *zero*.

There is one form that covers *all* lines, vertical and nonvertical.

General Linear Equation

$$ax + by = c$$

For constants a, b, and c,
with a and b not both zero

Any equation that can be written in this form is called a *linear equation*, and the variables are said to *depend linearly* on one another.

→ **EXAMPLE 10** **Finding the Slope and the y-Intercept from a Linear Equation**

Find the slope and y-intercept of the line $2x + 3y = 12$.

Solution

We write the line in slope-intercept form. Solving for y:

$$3y = -2x + 12 \qquad \text{Subtracting } 2x \text{ from both sides of } 2x + 3y = 12$$

$$y = -\frac{2}{3}x + 4 \qquad \text{Dividing each side by 3 gives the slope-intercept form } y = mx + b$$

Therefore, the slope is $-\frac{2}{3}$ and the y-intercept is $(0, 4)$.

→ **Practice Problem 6**

Find the slope and y-intercept of the line below $x - \dfrac{y}{3} = 2$.

➤ **Solution below**

→ **Solutions to Practice Problems**

1. $-1,000,000$ [the negative sign makes it less than (to the left of) the positive number $\frac{1}{100}$]

2. **a.** $\{x \,|\, x \geq -7\}$
   b. The set of all x such that x is less than -1

3. **a.** S_1
   b. S_4

4. $m = \dfrac{7 - 1}{4 - 2} = \dfrac{6}{2} = 3$ \qquad From points $(2, 1)$ and $(4, 7)$

   $y - 1 = 3(x - 2)$ \qquad Using the point-slope form with $(x_1, y_1) = (2, 1)$
   $y - 1 = 3x - 6$
   $y = 3x - 5$

5. $x = -2$

6. $x - \dfrac{y}{3} = 2$

   $-\dfrac{y}{3} = -x + 2$ \qquad Subtracting x from each side

   $y = 3x - 6$ \qquad Multiplying each side by -3

   Slope is $m = 3$ and y-intercept is $(0, -6)$.

→ 1.1 Exercises

Write each interval in set notation and graph it on the real line.

1. $[0, 6)$

2. $(-\infty, 2]$

Find the slope (if it is defined) of the line determined by each pair of points.

3. $(2, 3)$ and $(4, -1)$

4. $(2, -1)$ and $(2, 5)$

For each equation find the slope m and y-intercept $(0, b)$ (when they exist) and draw the graph.

5. $y = 3x - 4$

6. $y = 4$

7. $x = 4$

8. $2x - 3y = 12$

Write an equation of the line satisfying the following conditions. If possible, write your answer in the form $y = mx + b$.

9. Slope 5 and passing through the point $(-1, -2)$

10. Horizontal and passing through the point $(1.5, -4)$

11. Vertical and passing through the point $(1.5, -4)$

12. Passing through the points $(5, 3)$ and $(7, -1)$

13. **Energy Usage** A utility considers demand for electricity "low" if it is below 8 mkW (million kilowatts), "average" if it is at least 8 mkW but below 20 mkW, "high" if it is at least 20 mkW but below 40 mkW, and "critical" if it is 40 mkW or more. Express these demand levels in interval notation. [*Hint:* The interval for "low" is $[0, 8)$.]

14. **Corporate Profit** A company's profit increased linearly from $6 million at the end of year 1 to $14 million at the end of year 3.
    a. Use the two (year, profit) data points $(1, 6)$ and $(3, 14)$ to find the linear relationship $y = mx + b$ between $x =$ year and $y =$ profit.

b. Find the company's profit at the end of 2 years.

c. Predict the company's profit at the end of 5 years.

15. **Straight-Line Depreciation** Straight-line depreciation is a method for estimating the value of an asset (such as a piece of machinery) as it loses value ("depreciates") through use. Given the original *price* of an asset, its *useful lifetime*, and its *scrap value* (its value at the end of its useful lifetime), the value of the asset after t years is given by the formula

$$\text{Value} = (\text{price}) - \left(\frac{(\text{price}) - (\text{scrap value})}{(\text{useful lifetime})} \right) \cdot t$$

$$\text{for} \quad 0 \leq t \leq (\text{useful lifetime})$$

A farmer buys a harvester for $50,000 and estimates its useful life to be 20 years, after which its scrap value will be $6000.

a. Use the formula above to find a formula for the value V of the harvester after t years, for $0 \leq t \leq 20$.

b. Use your formula to find the value of the harvester after 5 years.

c. Graph the function found in part (a).

1.2 Exponents

Not all variables are related linearly. In this section we will discuss exponents, which will enable us to express many *nonlinear* relationships.

Positive Integer Exponents

Numbers may be expressed with exponents, as in $2^3 = 2 \cdot 2 \cdot 2 = 8$. More generally, for any positive integer n, x^n means the product of n x's.

$$\overbrace{x^n = x \cdot x \cdot \cdots \cdot x}^{n}$$

The number being raised to the power is called the *base* and the power is the *exponent*:

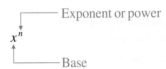

Exponent or power

x^n

Base

There are several *properties of exponents* for simplifying expressions. The first three are known, respectively, as the addition, subtraction, and multiplication properties of exponents.

Properties of Exponents

$x^m \cdot x^n = x^{m+n}$	To *multiply* powers of the same base, *add* the exponents
$\dfrac{x^m}{x^n} = x^{m-n}$	To *divide* powers of the same base, *subtract* the exponents (top exponent minus bottom exponent)
$(x^m)^n = x^{m \cdot n}$	To raise a power to a power, *multiply* the powers
$(xy)^n = x^n \cdot y^n$	To raise a product to a power, raise *each factor* to the power
$\left(\dfrac{x}{y}\right)^n = \dfrac{x^n}{y^n}$	To raise a fraction to a power, raise the numerator *and* denominator to the power

→ **EXAMPLE 1**   Simplifying Exponents

a. $x^2 \cdot x^3 = x^5$ (exponent $2 + 3$)

Since $x^2 \cdot x^3 = \overbrace{x \cdot x}^{2} \cdot \overbrace{x \cdot x \cdot x}^{3}$ (total 5)

b. $\dfrac{x^5}{x^3} = x^2$ (exponent $5 - 3$)

Since $\dfrac{x^5}{x^3} = \dfrac{\overset{2}{\overbrace{x \cdot x}} \cdot \cancel{x} \cdot \cancel{x} \cdot \cancel{x}}{\cancel{x} \cdot \cancel{x} \cdot \cancel{x}}$

c. $(x^3)^2 = x^6$ (exponent $3 \cdot 2$)

Since $(x^3)^2 = x^3 \cdot x^3$

d. $(2w)^3 = 2^3 w^3 = 8w^3$

Since $(2w)^3 = 2w \cdot 2w \cdot 2w = 2^3 w^3$

e. $\left(\dfrac{x}{5}\right)^3 = \dfrac{x^3}{5^3} = \dfrac{x^3}{125}$

Since $\left(\dfrac{x}{5}\right)^3 = \dfrac{x}{5} \cdot \dfrac{x}{5} \cdot \dfrac{x}{5}$

f. $\dfrac{[(x^2)^3]^4}{x^5 \cdot x^7 \cdot x} = \dfrac{x^{24}}{x^{13}} = x^{11}$ $\overbrace{2 \cdot 3 \cdot 4}\ \ \ 24 - 13$ $\underbrace{5 + 7 + 1}$ Combining all the rules

---

➡ **Practice Problem 1**

Simplify: **a.** $\dfrac{x^5 \cdot x}{x^2}$ **b.** $[(x^3)^2]^2$ ➤ **Solutions on page 21**

Remember: For exponents in the form $\ \ x^2 \cdot x^3 = x^5,\ \ \ add$ exponents.
                 For exponents in the form $\ \ (x^2)^3 = x^6,\ \ \ multiply$ exponents.

Zero and Negative Exponents

For any number x other than zero, we define

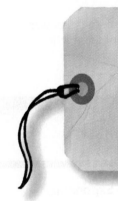

$x^0 = 1$ x to the power 0 is 1

$x^{-1} = \dfrac{1}{x}$ x to the power -1 is 1 divided by x

$x^{-2} = \dfrac{1}{x^2}$ x to the power -2 is 1 divided by x squared

$x^{-n} = \dfrac{1}{x^n}$ x to a negative power is 1 divided by x to the positive power

The definitions of x^0 and x^{-n} are motivated by the following calculations.

$$1 = \dfrac{x^2}{x^2} = x^{2-2} = x^0$$ The subtraction property of exponents leads to $x^0 = 1$

$$\dfrac{1}{x^n} = \dfrac{x^0}{x^n} = x^{0-n} = x^{-n}$$ $x^0 = 1$ and the subtraction property of exponents lead to $x^{-n} = \dfrac{1}{x^n}$

➡ **EXAMPLE 2**   **Simplifying Zero and Negative Exponents**

a. $5^0 = 1$ **b.** $7^{-1} = \dfrac{1}{7}$

c. $3^{-2} = \dfrac{1}{3^2} = \dfrac{1}{9}$ **d.** $(-2)^{-3} = \dfrac{1}{(-2)^3} = \dfrac{1}{-8} = -\dfrac{1}{8}$

e. 0^0 and 0^{-3} are undefined.

INVERT & MULTIPLY

A fraction to a negative power means *division* by the fraction, so we "invert and multiply."

$$\left(\frac{x}{y}\right)^{-1} = \frac{1}{\frac{x}{y}} = 1 \cdot \frac{y}{x} = \frac{y}{x}$$

Reciprocal of the original fraction

Therefore, for $x \neq 0$ and $y \neq 0$,

$$\left(\frac{x}{y}\right)^{-1} = \frac{y}{x}$$
A fraction to the power -1 is the reciprocal of the fraction

$$\left(\frac{x}{y}\right)^{-n} = \left(\frac{y}{x}\right)^{n}$$
A fraction to the negative power is the reciprocal of the fraction to the positive power

→ **EXAMPLE 3**   **Simplifying Fractions to Negative Exponents**

a. $\left(\frac{3}{2}\right)^{-1} = \frac{2}{3}$ **b.** $\left(\frac{1}{2}\right)^{-3} = \left(\frac{2}{1}\right)^{3} = \frac{2^3}{1^3} = 8$

Reciprocal of $\frac{3}{2}$

→ **Practice Problem 3**

Simplify: $\left(\frac{2}{3}\right)^{-2}$ ➤ **Solution on page 21**

Roots and Fractional Exponents

We may take the square root of any *nonnegative* number and the cube root of *any* number.

→ **EXAMPLE 4**   **Evaluating Roots**

a. $\sqrt{9} = 3$

b. $\sqrt{-9}$ is undefined.

Square roots of negative numbers are not defined

c. $\sqrt[3]{8} = 2$

d. $\sqrt[3]{-8} = -2$

Cube roots of negative numbers *are* defined

e. $\sqrt[3]{\dfrac{27}{8}} = \dfrac{\sqrt[3]{27}}{\sqrt[3]{8}} = \dfrac{3}{2}$

There are *two* square roots of 9, namely 3 and -3, but the radical sign $\sqrt{}$ means just the *positive* root (the "principal" square root).

$\sqrt[n]{a}$ means the principal nth root of a.

Principal means the positive root if there are two

In general, we may take *odd* roots of *any* number, but *even* roots only if the number is positive or zero.

→ **EXAMPLE 5**   **Evaluating Roots of Positive and Negative Numbers**

Odd roots of negative numbers *are* defined

a. $\sqrt[4]{81} = 3$

b. $\sqrt[5]{-32} = -2$

Since $(-2)^5 = -32$

Fractional Exponents

Fractional exponents are defined as follows:

$x^{\frac{1}{2}} = \sqrt{x}$ Power $\frac{1}{2}$ means the principal square root

$x^{\frac{1}{3}} = \sqrt[3]{x}$ Power $\frac{1}{3}$ means the cube root

$x^{\frac{1}{n}} = \sqrt[n]{x}$ Power $\frac{1}{n}$ means the principal nth root (for a positive integer n)

The definition of $x^{1/2}$ is motivated by the multiplication property of exponents:

$$(x^{\frac{1}{2}})^2 = x^{\frac{1}{2} \cdot 2} = x^1 = x$$

Taking square roots of each side of $(x^{\frac{1}{2}})^2 = x$ gives

$$x^{\frac{1}{2}} = \sqrt{x}$$

x to the half power means the square root of x

➡ EXAMPLE 6   Evaluating Fractional Exponents

a. $9^{1/2} = \sqrt{9} = 3$

b. $125^{1/3} = \sqrt[3]{125} = 5$

c. $81^{1/4} = \sqrt[4]{81} = 3$

d. $(-32)^{1/5} = \sqrt[5]{-32} = -2$

e. $\left(-\dfrac{27}{8}\right)^{1/3} = \sqrt[3]{-\dfrac{27}{8}} = -\dfrac{\sqrt[3]{27}}{\sqrt[3]{8}} = -\dfrac{3}{2}$

➡ Practice Problem 4

Evaluate: *a.* $(-27)^{\frac{1}{3}}$ *b.* $\left(\dfrac{16}{81}\right)^{\frac{1}{4}}$

➤ **Solutions on page 21**

To define $x^{\frac{m}{n}}$ for positive integers m and n, the exponent $\frac{m}{n}$ must be fully reduced (for example, $\frac{4}{6}$ must be reduced to $\frac{2}{3}$). Then

$$x^{\frac{m}{n}} = (x^{\frac{1}{n}})^m = (x^m)^{\frac{1}{n}}$$

Since in both cases the exponents multiply to $\frac{m}{n}$

Therefore, we define:

Fractional Exponents

$$x^{\frac{m}{n}} = \left(\sqrt[n]{x}\right)^m = \sqrt[n]{x^m}$$

$x^{\frac{m}{n}}$ means the mth power of the nth root, or equivalently, the nth root of the mth power

Both expressions, $\left(\sqrt[n]{x}\right)^m$ and $\sqrt[n]{x^m}$, will give the same answer. In either case the numerator determines the power and the denominator determines the root.

Power divided by root

→ EXAMPLE 7 Evaluating Fractional Exponents

a. $8^{2/3} = \sqrt[3]{8^2} = \sqrt[3]{64} = 4$

> same

First the power, then the root

b. $8^{2/3} = \left(\sqrt[3]{8}\right)^2 = (2)^2 = 4$

First the root, then the power

c. $25^{3/2} = \left(\sqrt{25}\right)^3 = (5)^3 = 125$

d. $\left(\dfrac{-27}{8}\right)^{2/3} = \left(\sqrt[3]{\dfrac{-27}{8}}\right)^2 = \left(\dfrac{-3}{2}\right)^2 = \dfrac{9}{4}$

→ Practice Problem 5

Evaluate: **a.** $16^{3/2}$ **b.** $(-8)^{2/3}$

➤ Solutions on page 21

→ EXAMPLE 8 Evaluating Negative Fractional Exponents

a. $8^{-2/3} = \dfrac{1}{8^{2/3}} = \dfrac{1}{\left(\sqrt[3]{8}\right)^2} = \dfrac{1}{2^2} = \dfrac{1}{4}$

A negative exponent means the reciprocal of the number to the positive exponent, which is then evaluated as before

b. $\left(\dfrac{9}{4}\right)^{-3/2} = \left(\dfrac{4}{9}\right)^{3/2} = \left(\sqrt{\dfrac{4}{9}}\right)^3 = \left(\dfrac{2}{3}\right)^3 = \dfrac{8}{27}$

Interpreting the power 3/2
Reciprocal to the positive exponent
Negative exponent

→ Practice Problem 6

Evaluate: **a.** $25^{-3/2}$ **b.** $\left(\dfrac{1}{4}\right)^{-1/2}$ **c.** $5^{1.3}$ [*Hint*: Use a calculator.]

➤ Solutions on page 21

Avoiding Pitfalls in Simplifying

The square root of a product is equal to the product of the square roots:

$$\sqrt{a \cdot b} = \sqrt{a} \cdot \sqrt{b} \qquad\qquad a > 0, \ b > 0$$

The corresponding statement for *sums*, however, is *not* true:

$$\sqrt{a + b} \quad \text{is } not \text{ equal to} \quad \sqrt{a} + \sqrt{b}$$

For example,

$$\underbrace{\sqrt{9 + 16}}_{\sqrt{25}} \neq \underbrace{\sqrt{9}}_{3} + \underbrace{\sqrt{16}}_{4}$$

The two sides are not equal: one is 5 and the other is 7

Therefore, do not "simplify" $\sqrt{x^2 + 9}$ into $x + 3$. The expression $\sqrt{x^2 + 9}$ *cannot be simplified.* Similarly,

$$(x + y)^2 \quad \text{is } not \text{ equal to} \quad x^2 + y^2$$

The expression $(x + y)^2$ means $(x + y)$ multiplied by itself:

$$(x + y)^2 = (x + y)(x + y) = x^2 + xy + yx + y^2 = x^2 + 2xy + y^2$$

$$(x + y)^2 = x^2 + 2xy + y^2$$

$(x + y)^2$ is the first number squared plus two times the product of the numbers plus the second number squared

Learning Curves in Airplane Production

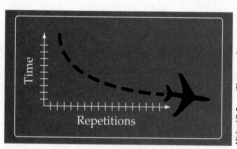

Michael D. Brown/Shutterstock

It is a truism that the more you practice a task, the faster you can do it. Successive repetitions generally take less time, following a "learning curve" like that on the left. Learning curves are used in industrial production. For example, it took 150,000 work-hours to build the first Boeing 707 airliner, while later planes $(n = 2, 3, \ldots, 300)$ took less time. (A work-hour is the amount of work that a person can do in 1 hour.)

$$\left(\begin{array}{c} \text{Time to build} \\ \text{plane number } n \end{array}\right) = 150n^{-0.322} \qquad \text{thousand work-hours}$$

The time for the 10th Boeing 707 is found by substituting $n = 10$:

$$\left(\begin{array}{c} \text{Time to build} \\ \text{plane 10} \end{array}\right) = 150(10)^{-0.322} \qquad 150n^{-0.322} \quad \text{with} \quad n = 10$$

$$\approx 71.46 \text{ thousand work-hours} \qquad \text{Using a calculator}$$

This equation shows that building the 10th Boeing 707 took about 71,460 work-hours, which is less than half of the 150,000 work-hours needed for the first. For the 100th 707:

$$\left(\begin{array}{c} \text{Time to build} \\ \text{plane 100} \end{array}\right) = 150(100)^{-0.322} \qquad 150n^{-0.322} \quad \text{with} \quad n = 100$$

$$\approx 34.05 \text{ thousand work-hours}$$

or about 34,050 work-hours, which is less than half the time needed to build the 10th plane. Such learning curves are used for determining the cost of a contract to build several planes.

Katrina Brown/Shutterstock

Notice that the learning curve graphed on the previous page decreases less steeply as the number of repetitions increases. So, although construction time continues to decrease, it does so more slowly for later planes. This behavior, called "diminishing returns," is typical of learning curves.

→ Solutions to Practice Problems

1. a. $\dfrac{x^5 \cdot x}{x^2} = \dfrac{x^6}{x^2} = x^4$

b. $[(x^3)^2]^2 = x^{3 \cdot 2 \cdot 2} = x^{12}$

2. a. $2^0 = 1$

b. $2^{-4} = \dfrac{1}{2^4} = \dfrac{1}{16}$

3. $\left(\dfrac{2}{3}\right)^{-2} = \left(\dfrac{3}{2}\right)^2 = \dfrac{9}{4}$

4. a. $(-27)^{1/3} = \sqrt[3]{-27} = -3$

b. $\left(\dfrac{16}{81}\right)^{1/4} = \sqrt[4]{\dfrac{16}{81}} = \dfrac{\sqrt[4]{16}}{\sqrt[4]{81}} = \dfrac{2}{3}$

5. a. $16^{3/2} = \left(\sqrt{16}\right)^3 = 4^3 = 64$

b. $(-8)^{2/3} = \left(\sqrt[3]{-8}\right)^2 = (-2)^2 = 4$

6. a. $25^{-3/2} = \dfrac{1}{25^{3/2}} = \dfrac{1}{\left(\sqrt{25}\right)^3} = \dfrac{1}{5^3} = \dfrac{1}{125}$

b. $\left(\dfrac{1}{4}\right)^{-1/2} = \left(\dfrac{4}{1}\right)^{1/2} = \sqrt{4} = 2$

c. $5^{1.3} \approx 8.103$

→ 1.2 Exercises

Evaluate each expression *without* using a calculator.

1. $(2^2 \cdot 2)^2$

2. $\left(\dfrac{1}{2}\right)^{-3}$

3. $25^{1/2}$

4. $25^{3/2}$

5. $\left(\dfrac{27}{125}\right)^{2/3}$

6. $4^{-3/2}$

Simplify.

7. $(x^3 \cdot x^2)^2$

8. $[z^2(z \cdot z^2)^2 z]^3$

9. $\dfrac{(ww^2)^3}{w^3 w}$

10. $\dfrac{(5xy^4)^2}{25x^3 y^3}$

11. **Dinosaurs** The study of size and shape is called "allometry," and many allometric relationships involve exponents that are fractions. For example, the body measurements of most four-legged animals, from mice to elephants, obey (approximately) the following power law:

$$\left(\begin{array}{c}\text{Average body}\\\text{thickness}\end{array}\right) = 0.4 \,(\text{hip-to-shoulder length})^{3/2}$$

where body thickness is measured vertically and all measurements are in feet. Assuming that this same relationship held for dinosaurs, find the average body thickness of Diplodocus, whose hip-to-shoulder length (measured from its skeleton) was 16 feet.

12. **The Rule of .6** Many chemical and refining companies use "the rule of point six" to estimate the cost of new equipment. According to this rule, if a piece of equipment (such as a storage tank) originally cost C dollars, then the cost of similar equipment that is x times as large will be approximately $x^{0.6}C$ dollars. For example, if the original equipment cost C dollars, then new equipment with twice the capacity of the old equipment $(x = 2)$ would cost $2^{0.6}C = 1.516C$ dollars—that is, about 1.5 times as much. Therefore, to increase capacity by 100% costs only about 50% more. Use the rule of .6 to find how costs change if a company wants to quadruple $(x = 4)$ its capacity.

1.3 Functions

In the previous section we saw that the time required to build a Boeing 707 airliner will vary, depending on the number that have already been built. Mathematical relationships such as this, in which one number depends on another, are called *functions*, and are central to the study of mathematics. In this section we define and give some applications of functions.

Functions

A *function* is a rule or procedure for finding, from a given number, a new number.* If the function is denoted by f and the given number by x, then the resulting number is written $f(x)$ (read "f of x") and is called the *value of the function f at x*. The set of numbers x for which a function f is defined is called the *domain* of f, and the set of all resulting function values $f(x)$ is called the *range* of f. For any x in the domain, $f(x)$ must be a *single* number.

---

* In this chapter the word *function* will mean *function of one variable*. In Chapter 4 we will discuss functions of more than one variable.

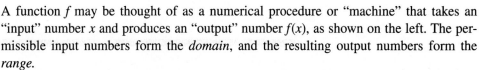

Function

A *function f* is a rule that assigns to each number x in a set a number $f(x)$. The set of all allowable values of x is called the *domain*, and the set of all values $f(x)$ for x in the domain is called the *range*.

For example, recording the temperature at a given location throughout a particular day would define a *temperature* function:

$$f(x) = \begin{pmatrix} \text{Temperature at} \\ \text{time } x \text{ hours} \end{pmatrix}$$

Domain would be [0, 24]

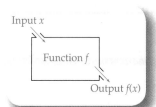

A function f may be thought of as a numerical procedure or "machine" that takes an "input" number x and produces an "output" number $f(x)$, as shown on the left. The permissible input numbers form the *domain*, and the resulting output numbers form the *range*.

We will be mostly concerned with functions that are defined by *formulas* for calculating $f(x)$ from x. If the domain of such a function is not stated, then it is always taken to be the *largest* set of numbers for which the function is defined, called the *natural domain* of the function. To *graph* a function f, we plot all points (x, y) such that x is in the domain and $y = f(x)$. We call x the *independent variable* and y the *dependent variable* because y *depends on* (is calculated from) x. The domain and range can be illustrated graphically.

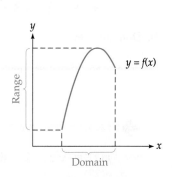

The domain of a function $y = f(x)$ is the set of all allowable x-values, and the range is the set of all corresponding y-values.

→ Practice Problem 1

Find the domain and range of the function graphed below.

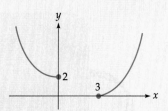

➤ **Solution on page 32**

→ **EXAMPLE 1** Finding the Domain

For the function $f(x) = \dfrac{1}{x-1}$, find:

a. $f(5)$ **b.** the domain

Solution

a. $f(5) = \dfrac{1}{5-1} = \dfrac{1}{4}$ $f(x) = \dfrac{1}{x-1}$ with $x = 5$

b. Domain $= \{x \mid x \neq 1\}$ $f(x) = \dfrac{1}{x-1}$ is defined for all x except $x = 1$

→ **EXAMPLE 2** Finding the Domain

For $f(x) = 2x^2 + 4x - 5$, determine:

a. $f(-3)$ **b.** the domain

Solution

a. $f(-3) = 2(-3)^2 + 4(-3) - 5$ $f(x) = 2x^2 + 4x - 5$ with

$= 18 - 12 - 5 = 1$ each x replaced by -3

b. Domain $= \mathbb{R}$ $2x^2 + 4x - 5$ is defined for *all* real numbers

---

→ **Practice Problem 2**

For $g(z) = \sqrt{z-2}$, determine:

  a. $g(27)$ *b.* the domain   ➤ Solutions on page 32

For each x in the domain of a function, there must be a *single* number $y = f(x)$, so the graph of a function cannot have two points (x, y) with the same x value but different y values. This requirement leads to the following *graphical* test for functions.

Vertical Line Test for Functions

A curve in the Cartesian plane is the graph of a *function* if and only if no vertical line intersects the curve at more than one point.

→ EXAMPLE 3 Using the Vertical Line Test

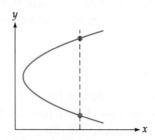

This is *not* the graph of a function of x because there is a vertical line (shown dashed) that intersects the curve twice.

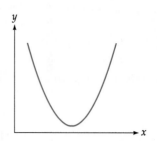

This *is* the graph of a function of x because no vertical line intersects the curve more than once.

A graph that has two or more points (x, y) with the same x-value but different y-values, such as the one on the left above, defines a *relation* rather than a function. We will be concerned exclusively with *functions*, and so we will use the terms *function*, *graph*, and *curve* interchangeably.

Functions can be classified into several types.

Linear Function

A *linear function* is a function that can be expressed in the form

$$f(x) = mx + b$$

with constants m and b. Its graph is a line with slope m and y-intercept b.

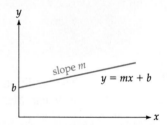

→ EXAMPLE 4 Finding a Company's Cost Function

An electronics company manufactures pocket calculators at a cost of $9 each, and the company's fixed costs (such as rent) amount to $400 per day. Find a function $C(x)$ that gives the total cost of producing x pocket calculators in a day.

Solution

Each calculator costs $9 to produce, so x calculators will cost $9x$ dollars, to which we must add the fixed costs of $400.

$$C(x) \;\; = \;\; 9x \;\; + \;\; 400$$

| Total cost | Unit cost | Number of units | Fixed cost |

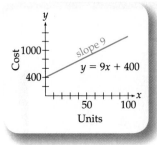

The graph of $C(x) = 9x + 400$ is a line with slope 9 and y-intercept 400, as shown on the left. Notice that the *slope* is the same as the *rate of change* of the cost (increasing at the rate of $9 per additional calculator), which is also the company's *marginal cost* (the cost of producing one more calculator is $9). For a linear cost function, the *slope*, the *rate of change*, and the *marginal cost* are always the same.

→ Practice Problem 3

A trucking company will deliver furniture for a charge of $25 plus 5% of the purchase price of the furniture. Find a function $D(x)$ that gives the delivery charge for a piece of furniture that costs x dollars.   ➤ **Solution on page 32**

A mathematical description of a real-world situation is called a *mathematical model*. For example, the cost function $C(x) = 9x + 400$ from Example 4 is a mathematical model for the cost of manufacturing calculators. In this model x, the number of calculators, should take only whole-number values (0, 1, 2, 3, ...), and the graph should consist of discrete dots rather than a continuous curve. Instead, we will find it easier to let x take *continuous* values, and round up or down as necessary at the end.

Quadratic Function

A *quadratic function* is a function that can be expressed in the form

$$f(x) = ax^2 + bx + c$$

with constants ("coefficients") $a \neq 0$, b, and c. Its graph is called a *parabola*.

The condition $a \neq 0$ keeps the function from becoming $f(x) = bx + c$, which would be linear. Many familiar curves are parabolas.

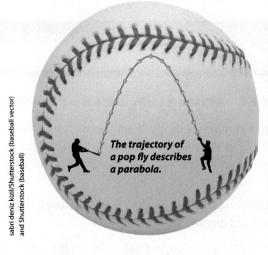

The trajectory of a pop fly describes a parabola.

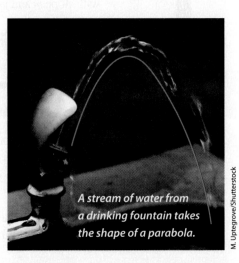

A stream of water from a drinking fountain takes the shape of a parabola.

The parabola $f(x) = ax^2 + bx + c$ opens *upward* if the constant a is *positive* and opens *downward* if the constant a is *negative*. The *vertex* of a parabola is its "central" point. The vertex is the *lowest* point on the parabola if it opens *up*, and the *highest* point if it opens *down*.

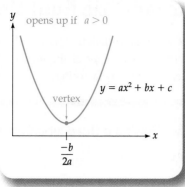

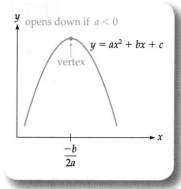

The x-coordinate of the vertex of a parabola may be found by the following formula. A derivation of this formula is given on page 31.

Vertex Formula for a Parabola

The vertex of the parabola $f(x) = ax^2 + bx + c$ has x-coordinate

$$x = \frac{-b}{2a}$$

➡ **EXAMPLE 5** **Graphing a Quadratic Function**

Graph the quadratic function $f(x) = 2x^2 - 8x + 9$.

Solution

The parabola opens *up* because x^2 has a positive coefficient. Using the vertex formula:

$$x_v = \frac{-b}{2a} = \frac{-(-8)}{2 \cdot 2} = \frac{8}{4} = 2$$

Substituting $x = 2$ into the original function gives $f(2) = 1$ for the y-value, so the vertex is $(2, 1)$. Then choosing the x-values on either side of the vertex, such as $x = 0$ and $x = 4$, and calculating the resulting y-values gives two more points, $(0, 9)$ and $(4, 9)$. Plotting these three points and sketching a parabola through them gives the graph below.

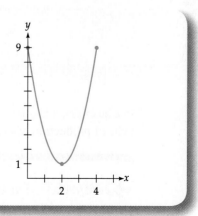

Solving Quadratic Equations

A value of x that solves an equation $f(x) = 0$ is called a *root* of the equation, or a *zero* of the function, or an *x-intercept* of the graph of $y = f(x)$. The roots of a quadratic equation can often be found by factoring.

➡ EXAMPLE 6   Solving a Quadratic Equation by Factoring

Solve $2x^2 - 4x = 6$.

Solution

$$2x^2 - 4x - 6 = 0 \qquad \text{Subtracting 6 from each side to get zero on the right}$$

$$2(x^2 - 2x - 3) = 0 \qquad \text{Factoring out a 2}$$

$$2(x - 3) \cdot (x + 1) = 0 \qquad \text{Factoring } x^2 - 2x - 3$$

$$\underbrace{}_{\substack{\text{Equals 0} \\ \text{at } x = 3}} \quad \underbrace{}_{\substack{\text{Equals 0} \\ \text{at } x = -1}} \qquad \text{Finding } x\text{-values that make each factor zero}$$

$$x = 3, \quad x = -1 \qquad \text{Solutions}$$

➡ Practice Problem 4

Solve by factoring or graphing: $9x - 3x^2 = -30$

➤ Solution on page 32

Quadratic equations can often be solved by the "Quadratic Formula." A derivation of this formula is given on page 32.

Quadratic Formula

The solutions to $ax^2 + bx + c = 0$ are

$$x = \frac{-b \pm \sqrt{b^2 - 4ac}}{2a}$$

The "plus or minus" $\pm$ sign means calculate *both* ways, first using the $+$ sign and then using the $-$ sign

In a business, it is often important to find a company's *break-even points*, the numbers of units of production where the company's costs are equal to its revenue.

➡ EXAMPLE 7   Finding Break-Even Points

A company that installs global positioning system (GPS) navigation devices in automobiles finds that if it installs x GPS devices per day, then its costs will be

Break Even
@ 20 or 120

lofoto/Shutterstock

$C(x) = 120x + 4800$ and its revenue will be $R(x) = -2x^2 + 400x$ (both in dollars). Find the company's break-even points.

Solution

We set the cost and revenue functions equal to each other and solve.

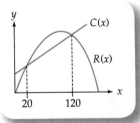

$$120x + 4800 = -2x^2 + 400x \qquad \text{Setting} \quad C(x) = R(x)$$

$$2x^2 - 280x + 4800 = 0 \qquad \text{Combining all terms on one side}$$

$$x = \frac{280 \pm \sqrt{(-280)^2 - 4 \cdot 2 \cdot 4800}}{2 \cdot 2} \qquad \begin{array}{l}\text{Quadratic Formula with} \quad a = 2, \\ b = -280, \quad \text{and} \quad c = 4800\end{array}$$

$$= \frac{280 \pm \sqrt{40{,}000}}{4} = \frac{280 \pm 200}{4} \qquad \begin{array}{l}\text{Working out the formula} \\ \text{on a calculator}\end{array}$$

$$= \frac{480}{4} \text{ or } \frac{80}{4} = 120 \text{ or } 20$$

The company will break even when it installs either 20 or 120 GPS systems (as shown in the graph on the left).

Although it is important for a company to know where its break-even points are, most companies want to do better than break even—they want to maximize their profit. Profit is defined as *revenue minus cost* (since profit is what is left over after subtracting expenses from income).

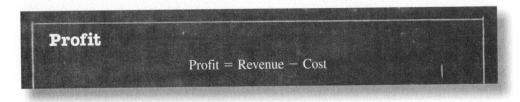

Profit

Profit = Revenue − Cost

➡ EXAMPLE 8   Maximizing Profit

For the GPS installer whose daily revenue and cost functions were given in Example 7, find the number of units that maximizes profit and the maximum profit.

Solution

The profit function is the revenue function minus the cost function.

$$P(x) = \underbrace{-2x^2 + 400x}_{R(x)} - \underbrace{(120x + 4800)}_{C(x)}$$

$P(x) = R(x) - C(x)$ with
$R(x) = -2x^2 + 400x$ and
$C(x) = 120x + 4800$

$$= -2x^2 + 280x - 4800 \qquad \text{Simplifying}$$

Since this function represents a parabola opening downward (because of the -2), it is maximized at its vertex, which is found using the vertex formula.

$$x = \frac{-280}{2(-2)} = \frac{-280}{-4} = 70 \qquad x = \frac{-b}{2a} \text{ with } a = -2 \text{ and } b = 280$$

Thus, profit is maximized when 70 GPS systems are installed. For the maximum profit, we substitute $x = 70$ into the profit function:

$$P(70) = -2(70)^2 + 280 \cdot 70 - 4800$$

$P(x) = -2x^2 + 280x - 4800$
with $x = 70$

$$= 5000 \qquad \text{Multiplying and combining}$$

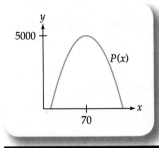

Therefore, the company will maximize its profit when it installs 70 GPS systems per day. Its maximum profit will be $5000 per day.

Why doesn't a company make more profit the more it sells? To increase its sales, a company must lower its prices, which eventually leads to lower profits. The relationship among the cost, revenue, and profit functions can be seen graphically as follows.

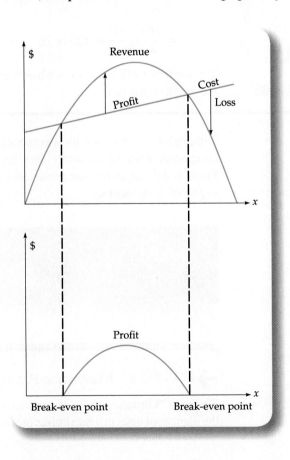

Not all quadratic equations have (real) solutions.

→ **EXAMPLE 9** **Using the Quadratic Formula**

Solve $\frac{1}{2}x^2 - 3x + 5 = 0$.

Solution

The Quadratic Formula with $a = \frac{1}{2}$, $b = -3$, and $c = 5$ gives

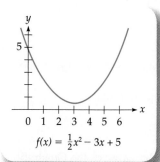

$f(x) = \frac{1}{2}x^2 - 3x + 5$

$$x = \frac{3 \pm \sqrt{9 - 4(\frac{1}{2})(5)}}{2(\frac{1}{2})} = \frac{3 \pm \sqrt{9 - 10}}{1} = 3 \pm \sqrt{-1} \qquad \text{Undefined}$$

Therefore, the equation $\frac{1}{2}x^2 - 3x + 5 = 0$ *has no real solutions* (because of the undefined $\sqrt{-1}$). The geometrical reason that there are no solutions can be seen in the graph on the left: The curve never reaches the *x*-axis, so the function never equals zero.

The quantity $b^2 - 4ac$, whose square root appears in the Quadratic Formula, is called the *discriminant*. If the discriminant is *positive* (as in Example 7), then the equation $ax^2 + bx + c = 0$ has *two* solutions (since the square root is added and subtracted). If the discriminant is *zero*, then there is only *one* root (since adding and subtracting zero gives the same answer). If the discriminant is *negative* (as in Example 9), then the equation has *no* real roots. Therefore, the discriminant being positive, zero, or negative corresponds to the parabola meeting the *x*-axis at 2, 1, or 0 points, as shown below.

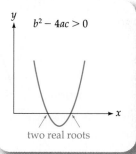

two real roots

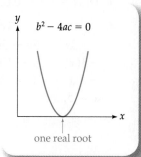

one real root

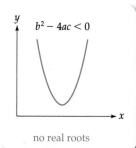

no real roots

Derivation of the Vertex Formula

Squaring, multiplying out, and simplifying the expression on the left shows that

$$a\left(x + \frac{b}{2a}\right)^2 - \frac{b^2}{4a} + c = ax^2 + bx + c$$

Therefore, if $a > 0$, then the above expression will be *minimized* when the expression in the parenthesis is zero (that is, when $x = \frac{-b}{2a}$). Correspondingly, if $a < 0$, then the above expression will be *maximized* when the expression in the parenthesis is zero (that is, when $x = \frac{-b}{2a}$). These maximum and minimum points will be the vertex, which in either case occurs at $x = \frac{-b}{2a}$.

Derivation of the Quadratic Formula

$ax^2 + bx + c = 0$	The quadratic set equal to zero
$ax^2 + bx = -c$	Subtracting c
$4a^2x^2 + 4abx = -4ac$	Multiplying by $4a$
$4a^2x^2 + 4abx + b^2 = b^2 - 4ac$	Adding b^2
$(2ax + b)^2 = b^2 - 4ac$	Since $4a^2x^2 + 4abx + b^2 = (2ax + b)^2$
$2ax + b = \pm\sqrt{b^2 - 4ac}$	Taking square roots
$2ax = -b \pm\sqrt{b^2 - 4ac}$	Subtracting b
$x = \dfrac{-b \pm\sqrt{b^2 - 4ac}}{2a}$	Dividing by $2a$ gives the Quadratic Formula

➜ Solutions to Practice Problems

1. Domain: $\{x \mid x \le 0 \text{ or } x \ge 3\}$;   Range: $\{y \mid y \ge 0\}$

2. a. $g(27) = \sqrt{27 - 2} = \sqrt{25} = 5$

  b. Domain: $\{z \mid z \ge 2\}$

3. $D(x) = 25 + 0.05x$

4.
$$9x - 3x^2 = -30$$
$$-3x^2 + 9x + 30 = 0$$
$$-3(x^2 - 3x - 10) = 0$$
$$-3(x - 5)(x + 2) = 0$$
$$x = 5, \; x = -2$$

or from:

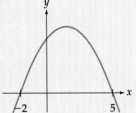

➜ 1.3 Exercises

Determine whether each graph defines a function of x.

1.

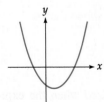

2.

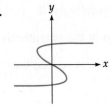

3. Find the domain and range of the function from its graph.

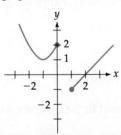

For each function in questions 4 through 6:

  a. Evaluate the given expression.

  b. Find the domain of the function.

4. $f(x) = \sqrt{x - 1}$; find $f(10)$

5. $h(z) = \dfrac{1}{z + 4}$; find $h(-5)$

6. $h(x) = x^{1/4}$; find $h(81)$

Graph the functions given in questions 7 through 10.

7. $f(x) = 3x - 2$

8. $f(x) = -x + 1$

9. $f(x) = 2x^2 + 4x - 16$

10. $f(x) = -3x^2 + 6x + 9$

Solve each equation in questions 11 through 14 by factoring or the Quadratic Formula.

11. $x^2 - 6x - 7 = 0$

12. $x^2 + 2x = 15$

13. $2x^2 + 40 = 18x$

14. $5x^2 - 50x = 0$

15. *Cost Functions* A lumberyard will deliver wood for $4 per board foot plus a delivery charge of $20. Find a function $C(x)$ for the cost of having x board feet of lumber delivered.

16. *Break-Even Points and Maximum Profit* A sporting goods store finds that if it sells x exercise machines per day, then its costs will be $C(x) = 100x + 3200$ and its revenue will be $R(x) = -2x^2 + 300x$ (both in dollars).

a. Find the store's break-even points.

b. Find the number of sales that will maximize profit, and find the maximum profit.

ATTENTION
NEED MORE PRACTICE? FIND MORE HERE:
CENGAGEBRAIN.COM

1.4 More About Functions

In this section we will define other useful types of functions and an important operation, the *composition* of two functions.

Polynomial Functions

A *polynomial function* (or simply a *polynomial*) is a function that can be written in the form

$$f(x) = a_n x^n + a_{n-1} x^{n-1} + \cdots + a_2 x^2 + a_1 x + a_0$$

where n is a nonnegative integer and $a_0, a_1, \ldots, a_n$ are (real) numbers, called *coefficients*. The *domain* of a polynomial is $\mathbb{R}$, the set of all (real) numbers. The *degree* of a polynomial is the highest power of the variable. The following are polynomials.

$$f(x) = 2x^8 - 3x^7 + 4x^5 - 5$$ A polynomial of degree 8 (since the highest power of x is 8)

$$f(x) = -4x^2 - \tfrac{1}{3}x + 19$$ A polynomial of degree 2 (a quadratic function)

$$f(x) = x - 1$$ A polynomial of degree 1 (a linear function)

$$f(x) = 6$$ A polynomial of degree 0 (a constant function)

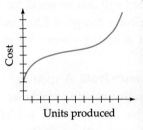

Cost

Units produced

A cost function may increase at different rates at different production levels.

Polynomials are used to model many situations in which change occurs at different rates. For example, the polynomial in the graph on the left might represent the total cost of manufacturing x units of a product. At first, costs rise quite steeply as a result of high start-up expenses, then they rise more slowly as the economies of mass production come into play, and finally they rise more steeply as new production facilities need to be built.

Polynomial equations can often be solved by factoring (just as with quadratic equations).

➜ EXAMPLE 1   Solving a Polynomial Equation by Factoring

Solve $3x^4 - 6x^3 = 24x^2$.

Solution

$$3x^4 - 6x^3 - 24x^2 = 0 \qquad \text{Rewritten with all the terms on the left side}$$

$$3x^2(x^2 - 2x - 8) = 0 \qquad \text{Factoring out } 3x^2$$

$$\underbrace{3x^2}_{\substack{\text{Equals} \\ \text{zero at} \\ x = 0}} \underbrace{(x - 4)}_{\substack{\text{Equals} \\ \text{zero at} \\ x = 4}} \underbrace{(x + 2)}_{\substack{\text{Equals} \\ \text{zero at} \\ x = -2}} = 0 \qquad \text{Factoring further}$$

Finding the zeros of each factor

$$x = 0, \quad x = 4, \quad x = -2 \qquad \text{Solutions}$$

As in this example, if a positive power of x can be factored out of a polynomial, then $x = 0$ is one of the roots.

➜ Practice Problem 1

Solve $2x^3 - 4x^2 = 48x$.

➤ Solution on page 40

Rational Functions

The word *ratio* means fraction or quotient, and the quotient of two polynomials is called a *rational function*. The following are rational functions.

$$f(x) = \frac{4x^3 + 3x^2}{x^2 - 2x + 1} \qquad g(x) = \frac{1}{x^2 + 1}$$

A rational function is a polynomial divided by a polynomial

The domain of a rational function is the set of all numbers for which the denominator is not zero. For example, the domain of the left function $f(x)$ is $\{x \mid x \neq 1\}$ (since $x = 1$ makes the denominator zero), and the domain of the right function $g(x)$ is $\mathbb{R}$ (since $x^2 + 1$ is never zero).

→ Practice Problem 2

What is the domain of the rational function $f(x) = \dfrac{18}{(x+2)(x-4)}$?

➤ Solution on page 40

Simplifying a rational function by canceling a common factor from the numerator and the denominator can change the domain of the function, so the "simplified" and "original" versions may not be equal (since they have different domains). For example, the rational function on the left below is not defined at $x = 1$, but the simplified version on the right *is* defined at $x = 1$, so the two functions are technically not equal.

$$\underbrace{\frac{x^2 - 1}{x - 1}}_{\substack{\text{Not defined at } x = 1, \\ \text{so the domain is } \{x \mid x \neq 1\}}} = \frac{(x+1)(x-1)}{x-1} \neq \underbrace{x + 1}_{\substack{\textit{Is} \text{ defined at } x = 1, \\ \text{so the domain is } \mathbb{R}}}$$

The functions *are*, however, equal at every x-value *except* $x = 1$, and the graphs (shown below) are the same except that the rational function omits the point at $x = 1$.

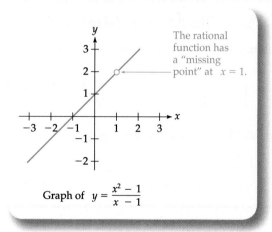

The rational function has a "missing point" at $x = 1$.

Graph of $y = \dfrac{x^2 - 1}{x - 1}$

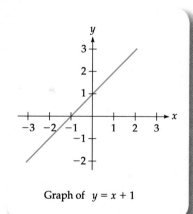

Graph of $y = x + 1$

Piecewise Linear Functions

The rule for calculating the values of a function may be given in several parts. If each part is linear, the function is called a *piecewise linear function*, and its graph consists of "pieces" of straight lines.

→ EXAMPLE 2   **Graphing a Piecewise Linear Function**

Graph

$$f(x) = \begin{cases} 5 - 2x & \text{if } x \geq 2 \\ x + 3 & \text{if } x < 2 \end{cases}$$

This notation means to use the top formula for $x \geq 2$ and the bottom formula for $x < 2$

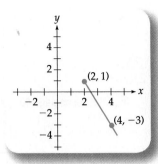

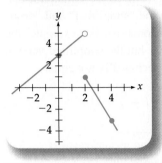

Solution
We graph one "piece" at a time.

Step 1: To graph the first part, $f(x) = 5 - 2x$ if $x \geq 2$, we use the "endpoint" $x = 2$ and also $x = 4$ (or any other x-value satisfying $x \geq 2$). The points are $(2, 1)$ and $(4, -3)$, with the y-coordinates calculated from $f(x) = 5 - 2x$. Draw the line through these two points, but only for $x \geq 2$ (from $x = 2$ to the *right*), as shown on the left.

Step 2: For the second part, $f(x) = x + 3$ if $x < 2$, the restriction $x < 2$ means that the line ends just *before* $x = 2$. We mark this "missing point" $(2, 5)$ by a "hollow dot" ∘ to indicate that it is *not* included in the graph [the y-coordinate comes from $f(x) = x + 3$]. For a second point, choose $x = 0$ (or any other $x < 2$), giving $(0, 3)$. Draw the line through these two points, but only for $x < 2$ (to the *left* of $x = 2$), completing the graph of the function.

An important piecewise linear function is the *absolute value* function.

➡ EXAMPLE 3   The Absolute Value Function

The absolute value function is $f(x) = |x|$ defined as

$$f(x) = \begin{cases} x & \text{if } x \geq 0 \\ -x & \text{if } x < 0 \end{cases}$$
The second line, for *negative x*, attaches a *second* negative sign to make the result *positive*

For example, when applied to either 3 or -3, the function gives *positive* 3:

$f(3) = 3$ Using the top formula (since $3 \geq 0$)

$f(-3) = -(-3) = 3$ Using the bottom formula (since $-3 < 0$)

To graph the absolute value function, we may proceed as in Example 2 or we may simply observe that for $x \geq 0$, the function gives $y = x$ (a half-line from the origin with slope 1), and for $x < 0$, it gives $y = -x$ (a half-line on the other side of the origin with slope -1), as shown in the following graph.

Absolute Value Function

$$f(x) = |x| = \begin{cases} x & \text{if } x \geq 0 \\ -x & \text{if } x < 0 \end{cases}$$

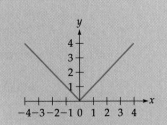

The absolute value function $f(x) = |x|$ has a "corner" at the origin.

Examples 2 and 3 show that the "pieces" of a piecewise linear function may or may not be connected.

➡ **EXAMPLE 4** Graphing an Income Tax Function

Federal income taxes are "progressive," meaning that they take a higher percentage of higher incomes. For example, the 2010 federal income tax for a single taxpayer whose taxable income was no more than $82,400 was determined by a three-part rule: 10% of income up to $8,375, plus 15% of any amount over $8,375 up to $34,000, plus 25% of any amount over $34,000 up to $82,400. For an income of x dollars, the tax $f(x)$ may be expressed as follows:

$$f(x) = \begin{cases} 0.10x & \text{if } 0 \le x \le 8{,}375 \\ 837.50 + 0.15\,(x - 8{,}375) & \text{if } 8{,}375 < x \le 34{,}000 \\ 4{,}681.25 + 0.25\,(x - 34{,}000) & \text{if } 34{,}000 < x \le 82{,}400 \end{cases}$$

Graphing this expression by the same technique as before leads to the following graph. The slopes 0.10, 0.15, and 0.25 are called the *marginal tax rates*.

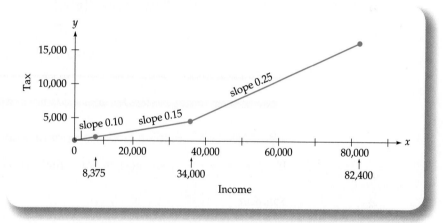

Composite Functions

Just as we substitute a *number* into a function, we may substitute a *function* into a function. For two functions f and g, evaluating f at $g(x)$ gives $f(g(x))$, called the *composition of f with g evaluated at x.*

Composite Functions

The *composition* of f with g evaluated at x is $f(g(x))$.

The *domain* of $f(g(x))$ is the set of all numbers x in the domain of g such that $g(x)$ is in the domain of f. If we think of the functions f and g as "numerical machines," then the composition $f(g(x))$ may be thought of as a *combined* machine in which the output of g is connected to the input of f.

---

* The composition $f(g(x))$ may also be written $(f \circ g)(x)$, although we will not use this notation.

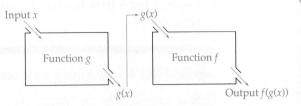

Input x → $g(x)$

Function g

$g(x)$

Function f

Output $f(g(x))$

A "machine" for generating the composition of f with g. A number x is fed into the function g, and the output $g(x)$ is then fed into the function f, resulting in $f(g(x))$.

➡ EXAMPLE 5   Finding a Composite Function

If $f(x) = x^7$ and $g(x) = x^3 - 2x$, find the composition $f(g(x))$.

Solution

$$f(g(x)) \;=\; \underbrace{[g(x)]^7}_{\substack{f(x) = x^7 \text{ with } x \\ \text{replaced by } g(x)}} \;=\; \underbrace{(x^3 - 2x)^7}_{\substack{\text{Using} \\ g(x) = x^3 - 2x}}$$

➡ EXAMPLE 6   Finding Both Composite Functions

If $f(x) = \dfrac{x+8}{x-1}$ and $g(x) = \sqrt{x}$, find $f(g(x))$ and $g(f(x))$.

Solution

$$f(g(x)) = \frac{g(x) + 8}{g(x) - 1} = \frac{\sqrt{x} + 8}{\sqrt{x} - 1} \qquad\qquad f(x) = \frac{x+8}{x-1} \text{ with } x$$
$$\text{replaced by } g(x) = \sqrt{x}$$

$$g(f(x)) = \sqrt{f(x)} = \sqrt{\frac{x+8}{x-1}} \qquad\qquad g(x) = \sqrt{x} \text{ with } x$$
$$\text{replaced by } f(x) = \frac{x+8}{x-1}$$

The order of composition is important: $f(g(x))$ is *not* the same as $g(f(x))$. To illustrate, we evaluate the above $f(g(x))$ and $g(f(x))$ at $x = 4$:

$$f(g(4)) = \frac{\sqrt{4} + 8}{\sqrt{4} - 1} = \frac{2 + 8}{2 - 1} = \frac{10}{1} = 10 \qquad\qquad f(g(x)) = \frac{\sqrt{x} + 8}{\sqrt{x} - 1} \text{ at } x = 4$$

Different answers

$$g(f(4)) = \sqrt{\frac{4+8}{4-1}} = \sqrt{\frac{12}{3}} = \sqrt{4} = 2 \qquad\qquad g(f(x)) = \sqrt{\frac{x+8}{x-1}} \text{ at } x = 4$$

➜ **Practice Problem 3**

If $f(x) = x^2 + 1$ and $g(x) = \sqrt[3]{x}$, find: *a.* $f(g(x))$ *b.* $g(f(x))$

➤ **Solutions on page 40**

➜ **EXAMPLE 7** Predicting Water Usage

A planning commission estimates that if a city's population is p thousand people, its daily water usage will be $W(p) = 30p^{1.2}$ thousand gallons. The commission further predicts that the population in t years will be $p(t) = 60 + 2t$ thousand people. Express the water usage W as a function of t, the number of years from now, and find the water usage 10 years from now.

Solution

Water usage W as a function of t is the *composition* of $W(p)$ with $p(t)$:

$$W(p(t)) = 30[p(t)]^{1.2} = 30(60 + 2t)^{1.2}$$

$W = 30p^{1.2}$ with p replaced by $p(t) = 60 + 2t$

To find water usage in 10 years, we evaluate $W(p(t))$ at $t = 10$:

$$W(p(10)) = 30(60 + 2 \cdot 10)^{1.2}$$

$30(60 + 2t)^{1.2}$ with $t = 10$

$$= 30(80)^{1.2} \approx 5765$$

Using a calculator

Thousand gallons

Therefore, in 10 years the city will need about 5,765,000 gallons of water per day.

Shifts of Graphs

Sometimes the composition of two functions is just a shift of an original graph. This shift occurs when one of the functions is simply the addition or subtraction of a constant. The following diagrams show the graph of $y = x^2$ together with various shifts and the functions that generate them.

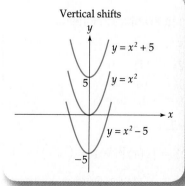

Vertical shifts

$y = x^2 + 5$

$y = x^2$

$y = x^2 - 5$

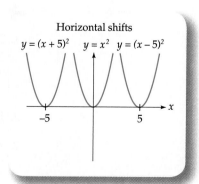

Horizontal shifts

$y = (x + 5)^2$ $y = x^2$ $y = (x - 5)^2$

In general, adding to or subtracting from the *x-value* means a *horizontal* shift, while adding to or subtracting from the *function* means a *vertical* shift. These same ideas hold for *any* function: given the graph of $y = f(x)$, adding a positive number a to the function $f(x)$ or to the variable x or subtracting a positive number a from the function $f(x)$ or from the variable x shifts the graph as follows:

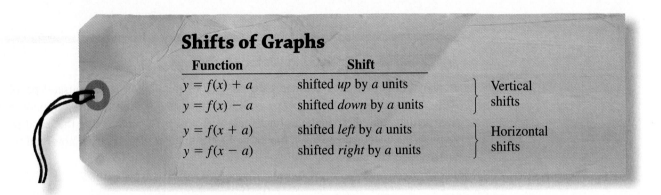

Shifts of Graphs

Function	Shift	
$y = f(x) + a$	shifted *up* by a units	Vertical shifts
$y = f(x) - a$	shifted *down* by a units	
$y = f(x + a)$	shifted *left* by a units	Horizontal shifts
$y = f(x - a)$	shifted *right* by a units	

Of course, a graph can be shifted both horizontally *and* vertically:

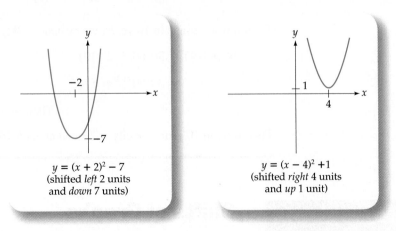

$y = (x + 2)^2 - 7$
(shifted *left* 2 units
and *down* 7 units)

$y = (x - 4)^2 + 1$
(shifted *right* 4 units
and *up* 1 unit)

Such double shifts can be applied to *any* function $y = f(x)$: the graph of $y = f(x + a) + b$ is shifted *left* a units and *up* b units (with the understanding that a *negative* a or b means that the direction is reversed).

Be careful: Remember that adding a *positive* number to x means a *left* shift.

→ **Solutions to Practice Problems**

1. $2x^3 - 4x^2 - 48x = 0$
$2x(x^2 - 2x - 24) = 0$
$2x(x + 4)(x - 6) = 0$
$x = 0, \quad x = -4, \quad x = 6$

2. $\{x \mid x \neq -2, x \neq 4\}$

3. a. $f(g(x)) = [g(x)]^2 + 1 = \left(\sqrt[3]{x}\right)^2 + 1 \quad$ or $\quad x^{2/3} + 1$

   b. $g(f(x)) = \sqrt[3]{f(x)} = \sqrt[3]{x^2 + 1} \quad$ or $\quad (x^2 + 1)^{1/3}$

➔ 1.4 Exercises

For each function in Exercises 1 and 2:

   a. Evaluate the given expression.

   b. Find the domain of the function.

1. $f(x) = \dfrac{x^2}{x-1}$;   find $f(-1)$

2. $g(x) = |x+2|$;   find $g(-5)$

Solve each equation by factoring.

3. $x^5 + 2x^4 - 3x^3 = 0$

4. $2x^3 + 18x = 12x^2$

Graph each function.

5. $f(x) = |x-3| - 3$

6. $f(x) = \begin{cases} 2x - 7 & \text{if } x \geq 4 \\ 2 - x & \text{if } x < 4 \end{cases}$

For each pair $f(x)$ and $g(x)$, find   **a.** $(f(g(x))$,   **b.** $g(f(x))$.

7. $f(x) = x^5$; $g(x) = 7x - 1$

8. $f(x) = \dfrac{1}{x}$; $g(x) = x^2 + 1$

9. *Income Tax* The following function expresses an income tax that is 10% for incomes up to $5000 and

otherwise is $500 plus 30% of income in excess of $5000.

$$f(x) = \begin{cases} 0.10x & \text{if } 0 \leq x \leq 5000 \\ 500 + 0.30(x - 5000) & \text{if } x > 5000 \end{cases}$$

   a. Calculate the tax on an income of $3000.

   b. Calculate the tax on an income of $5000.

   c. Calculate the tax on an income of $10,000.

   d. Graph the function.

10. *Insurance Reserves* An insurance company keeps reserves (money to pay claims) of $R(v) = 2v^{0.3}$, where v is the value of all its policies, and the value of its policies is predicted to be $v(t) = 60 + 3t$, where t is the number of years from now. (Both R and v are in millions of dollars.) Express the reserves R as a function of t and evaluate the function at $t = 10$.

1.5   Exponential Functions

We now introduce exponential functions, showing how they are used to model the processes of growth and decay.

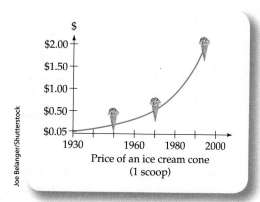

Price of an ice cream cone
(1 scoop)

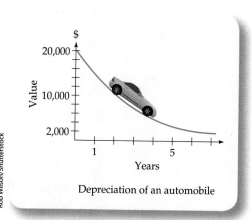

Depreciation of an automobile

We will also define the very important mathematical constant e.

Exponential Functions

A function that has a variable in an exponent, such as $f(x) = 2^x$, is called an *exponential function*.

$$f(x) = 2^{\overset{\text{Exponent}}{x}}$$

Base

The table below shows some values of the exponential function $f(x) = 2^x$, and its graph (based on these points) is shown on the right.

x	y = 2^x
−3	$2^{-3} = \frac{1}{8}$
−2	$2^{-2} = \frac{1}{4}$
−1	$2^{-1} = \frac{1}{2}$
0	$2^0 = 1$
1	$2^1 = 2$
2	$2^2 = 4$
3	$2^3 = 8$

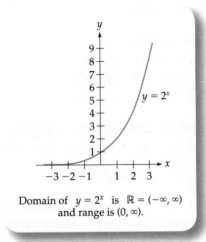

Domain of $y = 2^x$ is $\mathbb{R} = (-\infty, \infty)$ and range is $(0, \infty)$.

Clearly, the graph of the exponential function 2^x is quite different from the parabola x^2.

The exponential function $f(x) = \left(\frac{1}{2}\right)^x$ has base $\frac{1}{2}$. The following table shows some of its values, and its graph is shown to the right of the table. Notice that it is the mirror image of the curve $y = 2^x$.

x	y = (1/2)^x
−3	$\left(\frac{1}{2}\right)^{-3} = 8$
−2	$\left(\frac{1}{2}\right)^{-2} = 4$
−1	$\left(\frac{1}{2}\right)^{-1} = 2$
0	$\left(\frac{1}{2}\right)^0 = 1$
1	$\left(\frac{1}{2}\right)^1 = \frac{1}{2}$
2	$\left(\frac{1}{2}\right)^2 = \frac{1}{4}$
3	$\left(\frac{1}{2}\right)^3 = \frac{1}{8}$

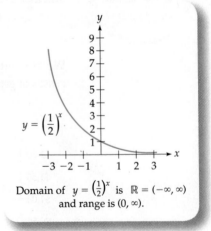

Domain of $y = \left(\frac{1}{2}\right)^x$ is $\mathbb{R} = (-\infty, \infty)$ and range is $(0, \infty)$.

We can define an exponential function $f(x) = a^x$ for any positive base a. We always take the base to be positive, so for the rest of this section *the letter a will stand for a positive constant.*

Exponential functions with bases $a > 1$ are used to model *growth*, as in populations or savings accounts, and exponential functions with bases $a < 1$ are used to model *decay*, as in depreciation. (For base $a = 1$, the graph is a horizontal line, since $1^x = 1$ for all x.)

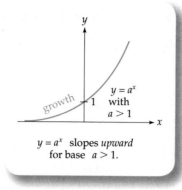

$y = a^x$ slopes *upward*
for base $a > 1$.

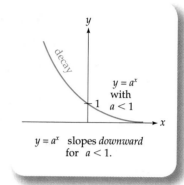

$y = a^x$ slopes *downward*
for $a < 1$.

Compound Interest

Money invested at compound interest grows exponentially. (The word *compound* means that the interest is added to the account, earning more interest.) Banks always state *annual* interest rates, but the compounding may be done more frequently. For example, if a bank offers 8% compounded quarterly, then each quarter you get 2% (one quarter, or one fourth, of the annual 8%), so that 2% of your money is added to the account each quarter. If you begin with P dollars (the "principal"), then at the end of the first quarter you would have P dollars plus 2% of P dollars:

$$\begin{pmatrix} \text{Value after} \\ \text{1 quarter} \end{pmatrix} = P + 0.02P = P \cdot (1 + 0.02)$$

Notice that increasing a quantity by 2% is the same as multiplying it by $(1 + 0.02)$. Since a year has four quarters, t years will have $4t$ quarters. Therefore, to find the value of your account after t years, we simply multiply the principal by $(1 + 0.02)$ a total of $4t$ times, obtaining

$$\begin{pmatrix} \text{Value after} \\ t \text{ years} \end{pmatrix} = \overbrace{P \cdot (1 + 0.02) \cdot (1 + 0.02) \cdots \cdots (1 + 0.02)}^{4t \text{ times}}$$

$$= P \cdot (1 + 0.02)^{4t}$$

The 8%, which gave the $\frac{0.08}{4} = 0.02$ quarterly rate, can be replaced by any interest rate r (written in decimal form), and the 4 can be replaced by any number m of compounding periods per year, leading to the following general formula.

Compound Interest

For P dollars invested at interest rate r compounded m times a year,

$$\begin{pmatrix} \text{Value after} \\ t \text{ years} \end{pmatrix} = P \cdot \left(1 + \frac{r}{m}\right)^{mt}$$

r = annual rate
m = periods per year
t = number of years

For example, for monthly compounding we would use $m = 12$ and for daily compounding $m = 365$ (the number of days in the year).

→ EXAMPLE 1   Finding a Value Under Compound Interest

Find the value of $4000 invested for 2 years at 12% compounded quarterly.

Solution

$$4000 \cdot \left(1 + \frac{0.12}{4}\right)^{4 \cdot 2} = 4000(1 + 0.03)^8$$

$$\underbrace{}_{0.03}$$

$$= 4000 \cdot 1.03^8 \approx 5067.08$$

$P \cdot \left(1 + \frac{r}{m}\right)^{mt}$

with $P = 4000$, $r = 0.12$, $m = 4$, and $t = 2$

Using a calculator

The value after two years will be $5067.08.

We may interpret the formula $P(1 + 0.03)^8$ intuitively as follows: multiplying the principal by $(1 + 0.03)$ means that you keep the original amount (the 1) plus some interest (the 0.03), and the exponent 8 means that this is done a total of eight times.

→ Practice Problem 1

Find the value of $2000 invested for 3 years at 24% compounded monthly.

➤ Solution on page 49

Much more will be said about compound interest in Section 2.2.

Depreciation by a Fixed Percentage

Depreciation by a fixed percentage means that a piece of equipment loses a fixed percentage (say 30%) of its value each year. Losing a percentage of value is like compound interest but with a *negative* interest rate. Therefore, we use the compound interest formula with $m = 1$ (since depreciation is annual) and with r being *negative*.

→ EXAMPLE 2   Depreciating an Asset

A car worth $15,000 depreciates in value by 40% each year. How much is it worth after 3 years?

Solution
The car loses 40% of its value each year, which is equivalent to an annual interest rate of *negative* 40%. The compound interest formula gives

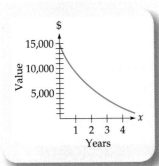

$$15,000(1 - 0.40)^3 = 15,000(0.60)^3 = \$3240$$

$$\underbrace{}_{\text{Using a calculator}}$$

$P(1 + r/m)^{mt}$ with $P = 15,000$, $r = -0.40$, $m = 1$, and $t = 3$

The exponential function $f(x) = 15,000(0.60)^x$, giving the value of the car after x years of depreciation, is graphed on the left.

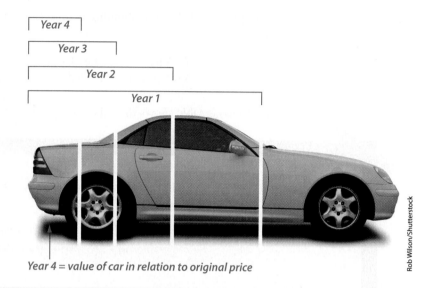

Year 4

Year 3

Year 2

Year 1

Year 4 = value of car in relation to original price

Rob Wilson/Shutterstock

→ **Practice Problem 2**

A printing press, originally worth $50,000, loses 20% of its value each year. What is its value after 4 years?   **Solution on page 49**

The graph on the previous page shows that depreciation by a fixed percentage is quite different from "straight-line" depreciation. Under straight-line depreciation the same *dollar* value is lost each year, while under fixed-percentage depreciation the same *percentage* of value is lost each year, resulting in larger dollar losses in the early years and smaller dollar losses in later years. Depreciation by a fixed percentage (also called the "declining balance" method) is one type of "accelerated" depreciation. The method of depreciation that one uses depends on how one chooses to estimate value and in practice is often determined by tax laws.

The Number *e*

Imagine that a bank offers 100% interest and that you deposit $1 for 1 year. Let us see how the value changes under different types of compounding.

For *annual* compounding, your $1 would in 1 year grow to $2 (the original $1 plus $1 interest).

For *quarterly* compounding, we use the compound interest formula with $P = 1$, $r = 1$ (for 100%), $m = 4$, and $t = 1$:

$$1\left(1 + \frac{1}{4}\right)^{1 \cdot 4} = (1 + 0.25)^4 = (1.25)^4 \approx 2.44 \qquad P(1 + r/m)^{mt}$$

or 2 dollars and 44 cents, an improvement of 44 cents over annual compounding.

For *daily* compounding, the value after 1 year would be

$$\left(1 + \frac{1}{365}\right)^{365} \approx 2.71$$

$m = 365$ periods

$\dfrac{r}{m} = \dfrac{100\%}{365} = \dfrac{1}{365}$

an increase of 27 cents over quarterly compounding. Clearly, if the interest rate, the principal, and the amount of time stay the same, the value increases because the compounding is done more frequently.

In general, if the compounding is done m times a year, the value of \$1 after 1 year will be

$$\begin{pmatrix} \text{Value of \$1 after 1 year at 100\%} \\ \text{interest compounded } m \text{ times a year} \end{pmatrix} = \left(1 + \frac{1}{m}\right)^m$$

The following table shows the value of $\left(1 + \frac{1}{m}\right)^m$ for various values of m.

Value of \$1 at 100% Interest Compounded m Times a Year for 1 Year

m	$\left(1 + \dfrac{1}{m}\right)^m$	Answer (rounded)	
1	$\left(1 + \dfrac{1}{1}\right)^1$	$= 2.00000$	Annual compounding
4	$\left(1 + \dfrac{1}{4}\right)^4$	≈ 2.44141	Quarterly compounding
365	$\left(1 + \dfrac{1}{365}\right)^{365}$	≈ 2.71457	Daily compounding
10,000	$\left(1 + \dfrac{1}{10,000}\right)^{10,000}$	≈ 2.71815	
100,000	$\left(1 + \dfrac{1}{100,000}\right)^{100,000}$	≈ 2.71827	
1,000,000	$\left(1 + \dfrac{1}{1,000,000}\right)^{1,000,000}$	≈ 2.71828	Answers agree to five decimal places
10,000,000	$\left(1 + \dfrac{1}{10,000,000}\right)^{10,000,000}$	≈ 2.71828	

Notice that as m increases, the values in the right-hand column seem to be settling down to a definite value, approximately 2.71828. That is, as m becomes infinitely large, the limit of $\left(1 + \dfrac{1}{m}\right)^m$ is approximately 2.71828. This particular number is very important in mathematics and is given the name e (just as 3.14159… is given the name π). In the following definition, we use the letter n to state the definition in its traditional form.

The Constant e

$$e = \lim_{n \to \infty} \left(1 + \frac{1}{n}\right)^n = 2.71828\ldots$$

$n \to \infty$ is read "n approaches infinity" (the dots mean that the decimal expansion goes on forever)

The same e appears in probability and statistics in the formula for the "bell-shaped" or "normal" curve (see page 239). Its value has been calculated to many thousands of decimal places, and its value to 15 decimal places is $e \approx 2.718281828459045$.

Continuous Compounding of Interest

This kind of compound interest, the limit as the compounding frequency approaches infinity, is called *continuous* compounding. We have shown that $1 at 100% interest compounded continuously for 1 year would be worth precisely e dollars (about $2.72). The formula for continuous compound interest at other rates is as follows (a justification for the formula is given at the end of this section).

Continuous Compounding

For P dollars invested at interest rate r compounded continuously,

$$\left(\begin{array}{c}\text{Value after}\\ t \text{ years}\end{array}\right) = Pe^{rt}$$

→ **EXAMPLE 3** **Finding Value with Continuous Compounding**

Find the value of $1000 at 8% interest compounded continuously for 20 years.

Solution

We use the formula Pe^{rt} with $P = 1000$, $r = 0.08$, and $t = 20$.

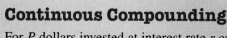

$$Pe^{rt} = 1000 \cdot e^{(0.08)(20)} = 1000 \cdot e^{1.6} \approx \$4953.03$$

$e^{1.6}$ is usually found using the [2nd] and [ln] keys

      P r t 4.95303

Intuitive Meaning of Continuous Compounding

Under quarterly compounding, your money is, in a sense, earning interest throughout the quarter, but the interest is not added to your account until the end of the quarter. Under continuous compounding, the interest is added to your account *as it is earned*, with no delay. The extra earnings in continuous compounding come from this "instant crediting" of interest because your interest starts earning more interest immediately.

The Function $y = e^x$

The number e gives us a new exponential function $f(x) = e^x$. This function is used extensively in business, economics, and all areas of science. The table on the next page shows the values of e^x for various values of x. These values lead to the graph of $f(x) = e^x$ shown to the right of the table.

x	$y = e^x$
−3	$e^{-3} \approx 0.05$
−2	$e^{-2} \approx 0.14$
−1	$e^{-1} \approx 0.37$
0	$e^0 = 1$
1	$e^1 \approx 2.72$
2	$e^2 \approx 7.39$
3	$e^3 \approx 20.09$

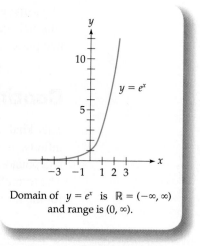

Domain of $y = e^x$ is $\mathbb{R} = (-\infty, \infty)$ and range is $(0, \infty)$.

Notice that e^x is never zero and is positive for all values of x, even when x is negative. We restate this important observation as follows:

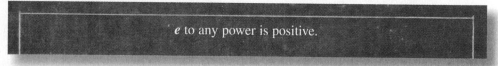

e to any power is positive.

The following graph shows the function $f(x) = e^{kx}$ for various values of the constant k. For positive values of k the curve rises, and for negative values of k the curve falls (as you move to the right). For higher values of k the curve rises more steeply. Each curve crosses the y-axis at 1.

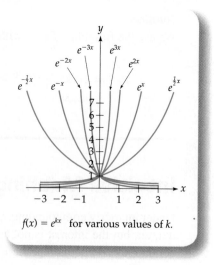

$f(x) = e^{kx}$ for various values of k.

Exponential Growth

All exponential growth, whether continuous or discrete, has one common characteristic: the amount of growth is proportional to the size. For example, the interest that a bank account earns is proportional to the size of the account, and the growth of a population is proportional to the size of the population. This characteristic is in contrast, for example, to a person's height, which does not increase exponentially. That is, exponential growth occurs in those situations in which a quantity grows *in proportion to its size*.

Justification of the Formula for Continuous Compounding

The compound interest formula Pe^{rt} is derived as follows: P dollars invested for t years at interest rate r compounded m times a year yields

$$P\left(1 + \frac{r}{m}\right)^{mt}$$

From the formula on page 43

Define $n = m/r$ so that $m = rn$. Replacing m by rn and letting m (and therefore n) approach ∞, this becomes

$$P\left(1 + \frac{r}{rn}\right)^{rnt} = P\left[\left(1 + \frac{1}{n}\right)^{n}\right]^{rt} \rightarrow Pe^{rt}$$

Letting $n \rightarrow \infty$

$$\underbrace{\phantom{\left(1 + \frac{1}{n}\right)^{n}}}$$
Approaches
e as $n \rightarrow \infty$

The limit shown on the right is the continuous compounding formula Pe^{rt}.

→ **Solutions to Practice Problems**

1. $2000\left(1 + \dfrac{0.24}{12}\right)^{12 \cdot 3} = 2000(1 + 0.02)^{36} = 2000(1.02)^{36} \approx \$4079.77$

2. $50{,}000(1 - 0.20)^{4} = 50{,}000(0.8)^{4} = 50{,}000(0.4096) = \$20{,}480$

→ **1.5 Exercises**

Graph each function.

1. $y = 3^x$

2. $y = \left(\frac{1}{3}\right)^x$

3. *Interest* Find the value of $1000 deposited in a mutual fund at 10% interest for 8 years compounded:

   a. annually **b.** quarterly **c.** continuously

4. *Millionaires* Your rich uncle, wanting to make you a millionaire, deposited $5810 at the time of your birth in a trust fund paying 8% compounded quarterly. Will he succeed by the time you retire at age 65?

5. *Zero-Coupon Bonds* A *zero-coupon bond* is a bond that makes no payments (coupons) until it matures, at which time it pays its "face value" of $1000. (You buy it for much less than $1000.) A bond trader prices a zero-coupon bond at $560 and the amount grows at interest rate 5.8% compounded continuously. Will it reach its "par" value of $1000 in 10 years? (Continuous compounding is frequently used in bond trading.)

6. *Options Trading* An "option" is an offer to buy or sell an asset at some time in the future. The Black–Scholes formula provides an accurate way to price options and involves continuous compounding. If an option is now priced at $2000 and its price grows at interest rate 5.5% compounded continuously, what will be its price in 8 years?

7. *Automobile Depreciation* A $20,000 automobile depreciates by 35% each year. Find its value after:

   a. 4 years **b.** 6 months

8. *Population* According to the United Nations Fund for Population Activities, the population of the world x years after the year 2000 will be $5.89e^{0.0175x}$ billion people (for $0 \leq x \leq 20$). Use this formula to predict the world population in the year 2020.

9. *Nuclear Meltdown* According to the Nuclear Regulatory Commission, the probability of a "severe core meltdown accident" at a nuclear reactor in the United States within the next n years is $1 - (0.9997)^{100n}$. Find the probability of a meltdown:

   a. within 25 years **b.** within 40 years

10.  *Temperature* A covered mug of coffee originally at 200 degrees Fahrenheit, if left for t hours in a room whose temperature is 70 degrees, will cool to a temperature of $70 + 130e^{-1.8t}$ degrees. Find the temperature of the coffee after:

a. 15 minutes   *b.* half an hour

1.6 Logarithmic Functions

In this section we will introduce *logarithmic* functions, concentrating on *common* (base 10) and *natural* (base e) logarithms. We will then use common and natural logarithms to solve problems about depreciation and carbon-14 dating. Logarithms will be particularly useful in Chapter 2.

Common Logarithms

The word *logarithm* (abbreviated "log") means *power* or *exponent*. The number being raised to the power is called the *base* and is written as a subscript. For example, the expression

$$\log_{10} 1000$$

Read: log (base 10) of 1000

means the *exponent* to which we have to raise 10 to get 1000. Since $10^3 = 1000$, the exponent is 3, so the *logarithm* is 3.

$$\log_{10} 1000 = 3 \qquad \text{Since } 10^3 = 1000$$

Logarithms with base 10 are called *common logarithms*. For common logarithms we often omit the subscript, with base 10 understood.

Common Logarithms

$\log x = y$ is equivalent to $10^y = x$ $\log x$ means $\log_{10} x$ (base 10)

Since $10^y = x$ is positive for every value of y, $\log x$ is defined only for *positive* values of x. The common logarithm of a number can often be found by expressing the number as a power of 10 and then taking the exponent.

→ **EXAMPLE 1** Finding a Common Logarithm

Evaluate $\log 100$.

Solution

$$\log 100 = y \qquad \text{is equivalent to} \qquad 10^y = 100 \qquad y = 2 \quad \text{works, so} \\ 2 \text{ is the logarithm}$$

The logarithm *y* is the exponent that solves

Therefore,

$$\log 100 = 2 \qquad\qquad\qquad \text{Since} \quad 10^2 = 100$$

→ EXAMPLE 2   **Finding a Common Logarithm**

Evaluate $\log \dfrac{1}{10}$.

Solution

$$\log \frac{1}{10} = y \qquad \text{is equivalent to} \qquad 10^y = \frac{1}{10} \qquad y = -1 \quad \text{works, so} \\ -1 \text{ is the logarithm}$$

The logarithm *y* is the exponent that solves

Therefore,

$$\log \frac{1}{10} = -1 \qquad\qquad \text{Since} \quad 10^{-1} = \frac{1}{10}$$

→ Practice Problem 1

Evaluate $\log 10,000$.

➤ Solution on page 58

Properties of Common Logarithms

Because logarithms are exponents, the properties of exponents can be restated as properties of logarithms. For positive numbers *M* and *N* and any number *P*:

Properties of Common Logarithms

1. $\log 1 = 0$ — The log of 1 is 0 (since $10^0 = 1$)

2. $\log 10 = 1$ — The log of 10 is 1 (since $10^1 = 10$)

3. $\log 10^x = x$ — The log of 10 to a power is just the power (since $10^x = 10^x$)

4. $10^{\log x} = x$ — 10 raised to the log of a number is just the number $(x > 0)$

5. $\log (M \cdot N) = \log M + \log N$ — The log of a product is the sum of the logs

Continues

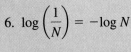

6. $\log\left(\dfrac{1}{N}\right) = -\log N$ The log of 1 divided by a number is minus the log of the number

7. $\log\left(\dfrac{M}{N}\right) = \log M - \log N$ The log of a quotient is the difference of the logs

8. $\log(M^P) = P \cdot \log M$ The log of a number to a power is the power multiplied by the log

The first two properties are simply special cases of the third (with $x = 0$ and $x = 1$). Because logs are exponents, the third property simply says that the exponent of 10 that gives 10^x is x, which is obvious when you think about it. Because $\log x$ is the power of 10 that gives x, raising 10 to that power must give x, which is the fourth property. Properties 5–8 are just restatements of the properties of exponents given on page 14. Property 8 will be particularly useful in applications and can be summarized: *Logarithms bring down exponents.*

➡ **EXAMPLE 3**   **Using the Properties of Common Logarithms**

a. $\log 10^7 = 7$ Property 3

b. $10^{\log 13} = 13$ Property 4

c. $\log(3 \cdot 4) = \log 3 + \log 4$ Property 5

d. $\log\left(\dfrac{1}{4}\right) = -\log 4$ Property 6

e. $\log\left(\dfrac{3}{4}\right) = \log 3 - \log 4$ Property 7

f. $\log(5^3) = 3 \log 5$ Property 8: $\log(5^3) = 3 \log 5$

➡ **EXAMPLE 4**   **Finding When a Car Depreciates by Half**

A car depreciates by 20% per year. When will it be worth only half its original value?

Solution

We know from page 44 that depreciation follows the compound interest formula but with a *negative* interest rate (since it *loses* value). Because we are not told the price of the car, we let P stand for the price. The compound interest formula with negative interest rate $r = -0.20$, set equal to $0.5P$ (representing *half* the original price) gives

$$P(1 - 0.20)^t = 0.5P \qquad\qquad \begin{array}{l} P(1 + r/m)^{mt} \text{ with} \\ r = -0.20 \text{ and } m = 1 \end{array}$$

$$(0.80)^t = 0.5 \qquad\qquad \begin{array}{l} \text{Canceling } Ps \\ \text{and simplifying} \end{array}$$

$$\log(0.80)^t = \log 0.5 \qquad\qquad \text{Taking log of each side}$$

$$t \cdot \log 0.80 = \log 0.5$$

Bringing down the power
(property 8)

$$t = \frac{\log 0.5}{\log 0.80}$$

Dividing by log 0.80

$$t \approx 3.1$$

Using a calculator

The car will be worth half its value in about 3.1 years.

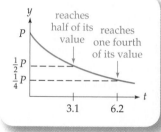

Incidentally, in another 3.1 years the car will again fall to half its value, thereby reaching *one fourth* its original value in 6.2 years. This halving of value will continue every 3.1 years.

Graphs of Logarithmic and Exponential Functions

If a point (x, y) lies on the graph of $y = \log x$ or, equivalently, $x = 10^y$, then, reversing x and y, the point (y, x) lies on the graph of $y = 10^x$. That is, the curves $y = \log x$ and $y = 10^x$ are related by having their x- and y-coordinates *reversed*, so the curves are *mirror images* of each other across the line $y = x$.

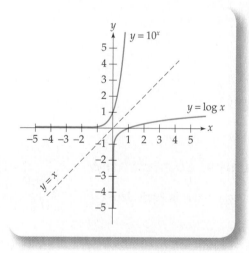

$f(x) = 10^x$ has domain $(-\infty, \infty)$ and range $(0, \infty)$

$f(x) = \log x$ has domain $(0, \infty)$ and range $(-\infty, \infty)$

This graphical relationship illustrates that the functions $\log x$ and 10^x "undo" or "reverse" each other, as shown by properties 3 and 4 on page 51. Such functions are called *inverse functions*:

$$\log x \quad \text{and} \quad 10^x \quad \text{are inverse functions}$$

Logarithms to Other Bases

We may calculate logarithms to bases other than 10. In fact, *any* positive number other than 1 may be used as a base for logs, using the following definition.

Base a Logarithms

$$\log_a x = y \quad \text{is equivalent to} \quad a^y = x \qquad (x > 0)$$

For example,

$$\log_2 8 = 3 \qquad\qquad \text{Since } 2^3 = 8$$

and

$$\log_9 3 = \tfrac{1}{2} \qquad\qquad \text{Since } 9^{1/2} = \sqrt{9} = 3$$

Natural Logarithms

We will use only one other base, the number e (approximately 2.718) that we defined on page 46. Logarithms to the base e are called *natural* or *Napierian* logarithms. The natural logarithm of a positive number x is written $\ln x$ ("n" for "natural") and may be found using the $\boxed{\text{ln}}$ key on a calculator.

John Napier (1550–1617), the Scottish mathematician who developed Napierian logarithms, is also the father of the decimal point.

Hulton Archive/Getty Images (Napier) and
Petr Vaclavek/Shutterstock (speech bubble)

Natural Logarithms

$$\ln x = y \quad \text{is equivalent to} \quad e^y = x \qquad \ln x \quad \text{means} \quad \log_e x \text{ (base } e\text{)}$$

→ **Practice Problem 2**

Use a calculator to find $\ln 8.34$.

➤ **Solution on page 58**

The properties of natural logarithms are similar to those of common logarithms but with ln instead of log and e instead of 10. For positive numbers M and N and any number P:

Properties of Natural Logarithms

1. $\ln 1 = 0$ The natural log of 1 is 0 (since $e^0 = 1$)

2. $\ln e = 1$ The natural log of e is 1 (since $e^1 = e$)

3. $\ln e^x = x$ The natural log of e to a power is just the power (since $e^x = e^x$)

4. $e^{\ln x} = x$ e raised to the natural log of a number is just the number $(x > 0)$

5. $\ln (M \cdot N) = \ln M + \ln N$ The natural log of a product is the sum of the logs

6. $\ln \left(\dfrac{1}{N}\right) = -\ln N$ The natural log of 1 divided by a number is minus the log of the number

7. $\ln \left(\dfrac{M}{N}\right) = \ln M - \ln N$ The natural log of a quotient is the difference of the logs

8. $\ln (M^P) = P \cdot \ln M$ The natural log of a number to a power is the power multiplied by the log

As with common logs, the first two properties are special cases of the third (with $x = 0$ and $x = 1$). The third and fourth properties have interpretations analogous to their "common" counterparts (for example, the third says that the exponent of e that gives e^x is x). As before, property 8 can be summarized: *Logarithms bring down exponents.*

➡ **EXAMPLE 5**   **Using the Properties of Natural Logarithms**

a. $\ln e^7 = 7$ Property 3

b. $e^{\ln 13} = 13$ Property 4

c. $\ln (3 \cdot 4) = \ln 3 + \ln 4$ Property 5

d. $\ln \left(\dfrac{1}{4}\right) = -\ln 4$ Property 6

e. $\ln \left(\dfrac{3}{4}\right) = \ln 3 - \ln 4$ Property 7

f. $\ln (5^3) = 3 \ln 5$ Property 8: $\ln (5^3) = 3 \ln 5$

Properties 3 and 4 show that $y = \ln x$ and $y = e^x$ are *inverse functions*, so their graphs are reflections of each other in the diagonal line $y = x$.

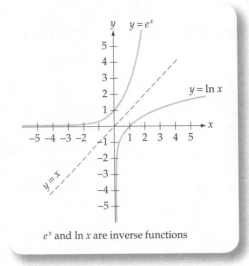

e^x and ln x are inverse functions

$f(x) = e^x$ has domain $(-\infty, \infty)$ and range $(0, \infty)$

$f(x) = \ln x$ has domain $(0, \infty)$ and range $(-\infty, \infty)$

The properties of natural logarithms are helpful for simplifying functions.

➡ **EXAMPLE 6** **Simplifying a Function**

$$f(x) = \ln (2x) - \ln 2$$

$$= \ln 2 + \ln x - \ln 2 \qquad \text{Since} \quad \ln (2x) = \ln 2 + \ln x \quad \text{by property 5}$$

$$= \ln x \qquad \text{Canceling}$$

➡ **EXAMPLE 7** **Simplifying a Function**

$$f(x) = \ln \left(\frac{x}{e}\right) + 1$$

$$= \ln x - \ln e + 1 \qquad \text{Since} \quad \ln (x/e) = \ln x - \ln e \quad \text{by property 7}$$

$$= \ln x - 1 + 1 \qquad \text{Since} \quad \ln e = 1 \quad \text{by property 2}$$

$$= \ln x \qquad \text{Canceling}$$

➡ **EXAMPLE 8** **Simplifying a Function**

$$f(x) = \ln (x^5) - \ln (x^3) = 5 \ln x - 3 \ln x \qquad \text{Bringing down exponents by property 8}$$

$$= 2 \ln x \qquad \text{Combining}$$

Carbon-14 Dating

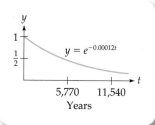

© Shutterstock

All living things absorb small amounts of radioactive carbon-14 from the atmosphere. When they die, the carbon-14 stops being absorbed and decays exponentially into ordinary carbon. Therefore, the proportion of carbon-14 still present in ancient organic remains can be used to estimate how old they are. The proportion of the original carbon-14 that will be present after t years is

$$\left(\begin{array}{l}\text{Proportion of carbon-14}\\ \text{remaining after } t \text{ years}\end{array}\right) = e^{-0.00012t}$$

→ EXAMPLE 9   Dating by Carbon-14

The Dead Sea Scrolls, discovered in a cave near the Dead Sea in what was then Jordan, are among the earliest documents of Western civilization. Estimate the age of the Dead Sea Scrolls if the animal skins on which some were written contain 78% of their original carbon-14.

Solution

The proportion of carbon-14 remaining after t years is $e^{-0.00012t}$. We equate this formula to the actual proportion (expressed as a decimal):

$$e^{-0.00012t} = 0.78 \qquad \text{Equating the proportions}$$

$$\ln e^{-0.00012t} = \ln 0.78 \qquad \text{Taking natural logs}$$

$$-0.00012t = \ln 0.78 \qquad \begin{array}{l}\ln e^{-0.00012t} = -0.00012t\\ \text{by property 3}\end{array}$$

$$t = \frac{\ln 0.78}{-0.00012} \approx \frac{-0.24846}{-0.00012} \approx 2071 \qquad \begin{array}{l}\text{Solving for } t \text{ and using a}\\ \text{calculator}\end{array}$$

Therefore, the Dead Sea Scrolls are approximately 2070 years old.

Both here and in Example 4 we solved for the variable in the exponent by taking logarithms, but here we used *natural* logarithms and there we used *common* logarithms. Is there a difference? There really isn't—you can use logarithms to *any* base to solve for exponents, and the final answer will be the same. When e is involved, however, it is usually easier to use *natural* logs.

→ **1.6 Exercises**

Find each logarithm *without* using a calculator.

1. *a.* $\log 100{,}000$ *b.* $\log \frac{1}{100}$ *c.* $\log \sqrt{10}$

2. *a.* $\ln e^5$ *b.* $\ln \frac{1}{e}$ *c.* $\ln \sqrt[3]{e}$

3. *a.* $\log_4 16$ *b.* $\log_4 \frac{1}{4}$ *c.* $\log_4 2$

Use the properties of natural logarithms to simplify each function.

4. $f(x) = \ln(9x) - \ln 9$

5. $f(x) = \ln\left(\frac{x}{4}\right) + \ln 4$

6. $f(x) = \ln(e^{5x}) - 2x - \ln 1$

7. ***Depreciation*** A car depreciates by 30% per year. When will it be worth only:

    a. half of its original value?

    b. one fourth of its original value?

8. ***Depreciation*** An industrial printing press depreciates by 15% per year. When will it be worth two thirds of its original value?

9. ***Carbon-14 Dating*** The proportion of carbon-14 still present in a sample after *t* years is $e^{-0.00012t}$. Use this formula to estimate the age of the cave paintings discovered in 1994 in the Ardèche region of France if the carbon with which they were drawn contains only 2.3% of its original carbon-14. These images are the oldest known paintings in the world.

10. ***Potassium-40 Dating*** The radioactive isotope potassium-40 is used to date very old remains. The proportion of potassium-40 that remains after *t* million years is $e^{-0.00054t}$. The most nearly complete skeleton of an early human ancestor ever found was discovered in Kenya in 1984. Use the formula to estimate the age of the remains if rocks near them contained 99.91% of their original potassium-40.

TURN TO A TRUSTED RESOURCE

With the Student Solutions Manual, now available online, you will have all the learning resources you need for your course in one convenient place, at no additional charge!

Solutions to all odd-numbered exercises, organized by chapter, are easily found online.

Simply visit **login.cengagebrain.com** to access this trusted resource.

Chapter 2

Mathematics of Finance

Tischenko Irina/Shutterstock

2.1 Simple Interest

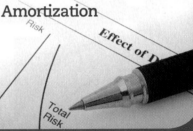

In the modern credit world, the old adage "time is money" has become a fact of economic life. When you open a savings account or take out a car loan, you directly experience the "time value of money." This chapter covers the financial properties of a loan between a lender and a borrower and the calculation of the interest and payments. We begin with simple interest.

Simple Interest Formula

The *principal* of a loan is the amount of money borrowed from the lender, the *term* is the time the borrower has the money, and the *interest* is the additional money paid by the borrower for the use of the lender's money. The interest is called *simple interest* if it is calculated as a fixed percentage of the principal and is paid at the end of the term. The *interest rate* of the loan is the dollars of interest per \$100 of principal per year (or "per annum"). It is usually stated as a percentage but is always written as a decimal in calculations. Simple interest is calculated as follows:

Simple Interest Formula

The interest I on a loan of P dollars at simple interest rate r for t years is

$$I = Prt$$

P = principal
r = rate
t = term

→ **EXAMPLE 1**   Finding Simple Interest

Find the interest on a loan of $2 million at 7.2% simple interest for 4 months.

Solution

Writing the interest rate in decimal form and changing the term to years, we find that

$$I = 2{,}000{,}000 \cdot 0.072 \cdot \frac{4}{12} = 48{,}000$$

$I = Prt$ with
$P = 2{,}000{,}000,$
$r = 0.072,$ and $t = 4/12$

The interest is $48,000.

Be careful! When using a calculator, *round only your final answer*. For example, if you use $t = 0.33$ in the previous example instead of $\frac{4}{12}$ or $\frac{1}{3}$, you would get the value $2{,}000{,}000 \cdot 0.072 \cdot 0.33 = \$47{,}520$, which is wrong by $480.

→ **Practice Problem 1**

Find the interest on a loan of $50,000 at 19.8% simple interest for 3 months.

> **Solution on page 66**

If you have values for any three variables in the interest formula $I = Prt$, you can solve for the fourth one.

→ **EXAMPLE 2** Finding the Simple Interest Rate

What is the interest rate of a loan charging $18 simple interest on a principal of $150 after 2 years?

Solution

We solve $I = Prt$ for interest rate r:

$$r = \frac{I}{Pt} \qquad\qquad I = Prt \quad \text{divided by } Pt$$

Then

$$r = \frac{18}{150 \cdot 2} = \frac{18}{300} = 0.06 \qquad\qquad \begin{array}{l} \text{Substituting} \quad I = 18, \\ P = 150, \quad \text{and} \quad t = 2 \end{array}$$

The interest rate is 6%.

Total Amount Due on a Loan

When the term of a loan is over, the borrower repays the principal and interest, so the total amount due is

$$\underbrace{P}_{\text{Principal}} + \underbrace{Prt}_{\text{Interest}} = \underbrace{P(1 + rt)}_{\text{Total amount}} \qquad\qquad \begin{array}{l}\text{Factoring out the} \\ \text{common term}\end{array}$$

which gives the following formula:

Total Amount Due for Simple Interest

The total amount A due at the end of a loan of P dollars at simple interest rate r for t years is

$$A = P(1 + rt)$$

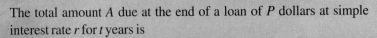

The amount due may also be regarded as the *accumulated value* of an investment or the *future value* of the principal. As before, this formula may be used as it is or it may be solved for any one of the other variables, as the next few examples will show.

➜ EXAMPLE 3 Finding the Total Amount Due

What is the total amount due on a loan of $3000 at 6% simple interest for 4 years?

Solution

$$A = 3000(1 + 0.06 \cdot 4)$$
$$= 3000 \cdot 1.24 = 3720$$

$A = P(1 + rt)$ with
$P = 3000,\quad r = 0.06,$
and $t = 4$

The total due on the loan is $3720.

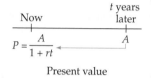

t years
later

Now

$P = \dfrac{A}{1 + rt}$ A

Present value

The $3000 loan in the above example grew to $3720 in 4 years. The $3720 is sometimes called the *future value* of the original $3000. Reversing our viewpoint, we say that the principal of $3000 is the *present value* of the later $3720. To find a formula for the present value, we solve the "total amount due" formula (on the previous page) for P by dividing by $1 + rt$:

$$P = \frac{A}{1 + rt}$$

Present value of the
future amount A

➜ EXAMPLE 4 Finding a Present Value

How much should be invested now at 8.6% simple interest if $10,000 is needed in 6 years?

Solution

The amount to invest now means the *present value* of $10,000 in 6 years.
Using the above formula:

$$P = \frac{10,000}{1 + 0.086 \cdot 6} = \frac{10,000}{1.516} \approx 6596.31$$

$P = A/(1 + rt)$
with $A = 10,000,$
$r = 0.086,$ and $t = 6$

The amount required is $6596.31.

We emphasize that the *present value* is the amount that will grow to the required sum in the given time period (at the stated interest rate). For this reason, it gives the actual value *now* of an amount to be paid later.

Amounts like this one, to be paid at some time in the future, occur in many situations, from personal credit card balances to government bonds. Such "debts" are frequently bought and sold by banks and companies, and the present value gives the current value of such future payments. For example, at the interest rate stated in the preceding example, a payment of $10,000 to be received in 6 years is worth exactly $6596.31 now and so should be bought or sold for this price.

➜ Practice Problem 2

Find the present value of a "promissory note" that will pay $6000 in 4 years at 5% simple interest.

➤ Solution on page 66

We found the present value by deriving a formula for it. We could instead have substituted the given numbers into the "total amount due" formula and then solved for P (which we may think of as standing for *principal* or *present value*). We will do the next example in this way, substituting numbers into the original formula and then solving for the remaining variable. You can use either method to solve these problems.

→ **EXAMPLE 5** **Finding the Term of a Simple Interest Loan**

What is the term of a loan of $2000 at 4% simple interest if the amount due is $2400?

Solution
Substituting the given numbers for the appropriate variables in the "total amount due" formula (page 62) gives

$$2400 = 2000(1 + 0.04t)$$ $A = P(1 + rt)$ with $A = 2400$, $P = 2000$,   and $r = 0.04$

$$1.2 = 1 + 0.04t$$ Dividing by 2000

$$0.2 = 0.04t$$ Subtracting 1

$$t = \frac{0.2}{0.04} = 5$$ Dividing by 0.04 and reversing sides

The term is 5 years.

Discounted Loans and Effective Interest Rates

In a *discounted loan* the lender deducts the interest from the amount the borrower receives at the start. A discounted loan is better for the lender because getting the money early (so it can earn interest somewhere else) is always better than later. Since the borrower actually is receiving a smaller amount, we may recalculate the interest rate as a "standard" loan on this smaller amount. The resulting rate is called the *effective* simple interest rate of the loan.

→ **EXAMPLE 6** **Finding the Effective Simple Interest Rate**

Determine the effective simple interest rate on a discounted loan of $1000 at 6% simple interest for 2 years.

Solution
The interest is $\$1000 \cdot 0.06 \cdot 2 = \$120$,   so as a simple interest loan, the borrower receives only $P = 1000 - 120 = \$880$ and agrees to pay back $1000 at the end of 2 years. If we write r_s for the simple interest rate of this loan, the "total amount due" formula on page 62 gives

$$1000 = 880(1 + r_s \cdot 2)$$ $A = P(1 + rt)$ with $A = 1000$, $P = 880$,   and $t = 2$

Solving for r_s,

$$r_s = \frac{1}{2}\left(\frac{1000}{880} - 1\right) \approx 0.0682$$

Dividing by 880, subtracting 1, and dividing by 2

The effective simple interest rate on this discounted loan is 6.82%.

Notice that the effective rate of 6.82% is significantly greater than the stated (or "nominal") rate of 6%. In general, the nominal rate in a discounted loan may be used to conceal the true cost to the borrower (and, similarly, the true benefit to the lender). That is why effective rates are so important.

Using the same method as in Example 6, we can find a formula for the effective rate. For a discounted loan of amount A at interest rate r for term t, the borrower receives $A - Art = A(1 - rt)$ dollars. Therefore, the "total amount due" formula from page 62, but with the simple interest rate replaced by the effective rate r_s, gives

$$A = \underbrace{A(1 - rt)}_{\text{Principal}}(1 + r_s t)$$

$A = P(1 + rt)$ with $A(1 - rt)$ for P

Solving for r_s:

$$\frac{1}{1 - rt} = 1 + r_s t$$

Canceling the As and dividing by $(1 - rt)$

$$r_s t = \frac{1}{1 - rt} - 1 = \frac{1 - (1 - rt)}{1 - rt} = \frac{rt}{1 - rt}$$

Reversing sides, subtracting 1, and simplifying

$$r_s = \frac{1}{t}\frac{rt}{1 - rt} = \frac{r}{1 - rt}$$

Dividing by t and then canceling

which gives the following formula for r_s.

Effective Simple Interest Rate for a Discounted Loan

For a discounted loan at interest rate r for t years, the effective simple interest rate r_s is

$$r_s = \frac{r}{1 - rt}$$

Notice that the effective rate r_s will be larger than r (since the denominator is less than 1) and that it depends only on the rate and term of the loan, and not on its amount. For the loan in Example 6, our formula gives the answer that we found:

$$r_s = \frac{0.06}{1 - 0.06 \cdot 2} = \frac{0.06}{0.88} \approx 0.0682$$

$r_s = \frac{r}{1 - rt}$ with $r = 0.06$ and $t = 2$

→ Practice Problem 3

What is the effective simple interest rate of a discounted loan at 6% interest for 3 years? for 5 years?

➤ **Solution below**

→ Solutions to Practice Problems

1. $I = 50{,}000 \cdot 0.198 \cdot \frac{3}{12} = 2475$

2. $P = \frac{6000}{1 + 4 \cdot 0.05} = \frac{6000}{1.2} = 5000$

3. $r_s = 0.06/(1 - 0.06 \cdot 3) \approx 0.0732$
   The effective simple interest rate is 7.32%.
   For 5 years, $r_s = 0.06/(1 - 0.06 \cdot 5) \approx 0.0857$

The effective simple interest rate is 8.57%. Can you think of an intuitive reason for the effective rate to be higher if the term is longer? [*Hint*: Think of how much earlier the lender gets the money or of how much less the borrower really gets.]

→ 2.1 Exercises

Find the simple interest on each loan.

1. $1500 at 7% for 10 years.

2. $825 at 6.58% for 5 years 6 months.

Find the total amount due for each simple interest loan.

3. $1500 at 7% for 10 years.

4. $6100 at 5.7% for 4 years 9 months.

5. *Interest Rate* Find the interest rate on a loan charging $704 simple interest on a principal of $2750 after 8 years.

6. *Principal* Find the principal of a loan at 4.2% if the simple interest after 5 years 6 months is $1155.

7. *Term* Find the term of a loan of $175 at 4.5% if the simple interest is $63.

8. *Present Value* How much should be invested now at 5.2% simple interest if $8670 is needed in 3 years?

2.2 Compound Interest

When a simple interest loan is not paid off at the end of its term but is taken out as a new loan, the borrower owes the principal plus the interest. The interest on the new loan is called *compound interest* because it combines interest on both the original principal and on the unpaid interest. For example, if you borrow $8000 at 5% simple interest for 1 year, the formula on page 62 says that the amount due will be $\$8000(1 + 0.05) = \$8400$. If you do not pay it off, but instead borrow this sum for the next year, the amount due after 2 years will be $\$8000(1 + 0.05)(1 + 0.05) = \$8000(1 + 0.05)^2 = \$8820$. After a third year, the amount due would become $\$8000(1 + 0.05)^3 = \$9261$. For each subsequent year, this amount would be multiplied by another $(1 + 0.05)$, which is 1 plus the interest rate. Compounding clearly increases the amount of interest paid because of the interest on the interest.

Compound Interest Formula

In general, the amount A due on a compound interest loan of P dollars at interest rate r per year compounded annually for t years is found by repeatedly multiplying the principal P by $(1 + r)$, once for each year:

$$A = \underbrace{P(1 + r)(1 + r) \cdots (1 + r)}_{t \text{ multiplications by } (1 + r)} = P(1 + r)^t$$

Notice how the formulas for the amounts due under simple and compound interest differ:

Simple	$P(1 + rt)$
Compound	$P(1 + r)^t$

The formula for *compound* interest has t in the *exponent*, which is why it grows so much faster (eventually).

Banks always state *annual* interest rates, but the compounding can be done more than once a year. Some standard compounding periods are as follows:

Compounding Frequency	Periods per Year
Annually	1
Semiannually	2
Quarterly	4
Monthly	12
Daily	365

Robyn Mackenzie/Shutterstock

Given the same annual interest rate, more frequent compounding is better because your money starts earning interest sooner. For example, if a bank offers 8% compounded quarterly, then the interest is calculated on a *quarterly* basis: each quarter you get 2% (one quarter of the 8%), and that is done every *quarter*, so the exponent is the number of quarters, which is *four times* the number of years. In general, if the annual interest rate is r with m compoundings per year, then the interest rate per period is r/m and the number of compounding periods in t years is mt. The following is the general formula for compound interest:

Compound Interest Formula

The amount A due on a loan of P dollars at yearly interest rate r compounded m times per year for t years is

$$A = P\left(1 + \frac{r}{m}\right)^{mt}$$

r = annual rate
m = periods per year
t = term (in years)

The stated yearly interest rate r is called the *nominal rate* of the loan.

➡ EXAMPLE 1   Finding the Amount Due on a Compound Interest Loan

Find the amount due on a loan of $1500 at 4.8% compounded monthly for 2 years.

Solution
The nominal rate of 4.8% expressed as a decimal is 0.048, and monthly compounding means that $m = 12$, so we have

$$A = 1500\left(1 + \frac{0.048}{12}\right)^{12 \cdot 2}$$

$A = P(1 + r/m)^{mt}$
with $P = 1500$, $r = 0.048$, $m = 12$, and $t = 2$

$$= 1500(1.004)^{24} \approx 1650.82$$

The amount due is $1650.82.

➡ Practice Problem 1

Find the amount that is due on a loan of $3500 at 5.1% compounded quarterly for 3 years.
➤ **Solution on page 73**

A loan is really an amount that a bank *invests* in a borrower, so any question about a loan can be rephrased as a question about an *investment*. Example 1 could have asked for the value of an investment of $1500 that grows by 4.8% compounded monthly for 2 years, and the answer would still be $1650.82. In this section we will speak interchangeably of *loans* and *investments* since they are the same but from different viewpoints, depending whether you are the borrower or the lender.

In Example 1, the amount $1650.82 may be called the *future value* of the original $1500; conversely, the $1500 is the *present value* of the later amount $1650.82. As before, the present value gives the value *today* of a payment that will be received at some time in the future. To find a formula for the present value, we solve the compound interest formula (above) for P, using the familiar $1/x^n = x^{-n}$ rule of exponents.

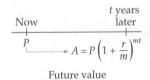

t years later

Now

$P \longrightarrow A = P\left(1 + \frac{r}{m}\right)^{mt}$

Future value

t years later

Now

$P = \dfrac{A}{\left(1 + \frac{r}{m}\right)^{mt}} \longleftarrow A$

Present value

$$P = \frac{A}{\left(1 + \dfrac{r}{m}\right)^{mt}} = A\left(1 + \frac{r}{m}\right)^{-mt}$$ Present value

This formula should be intuitively reasonable: To find *future* value you *multiply* by $(1 + r/m)^{mt}$, so to find *present* value you should *divide* by the same quantity.

→ **EXAMPLE 2** **Finding the Present Value**

How much should be invested now at 4.3% compounded weekly if $10,000 is needed in 6 years?

Solution

$$P = 10{,}000\left(1 + \frac{0.043}{52}\right)^{-52 \cdot 6} \approx 7726.78$$

$P = A(1 + r/m)^{-mt}$ with $A = 10{,}000$, $r = 0.043$, $m = 52$, and $t = 6$

The amount required is $7726.78.

Growth Times

We solved the preceding example by using the formula for present value. Instead, we could have substituted the given numbers into the compound interest formula and *then* solved for the remaining variable. We will do the next example in this way, finding a "growth time," the time for a loan to reach a given value. We will use the rule of logarithms $\log(x^n) = n \log x$ to "bring down the power." Logarithms with *any* base may be used in such calculations.

→ **EXAMPLE 3** **Finding the Term of a Compound Interest Loan**

What is the term of a loan of $2000 at 6% compounded monthly that will have an amount due of $2400?

Solution

We substitute the given numbers into the compound interest formula (see page 68):

$$2400 = 2000\left(1 + \frac{0.06}{12}\right)^{12 \cdot t}$$

$A = P(1 + r/m)^{mt}$ with $A = 2400$, $P = 2000$, $r = 0.06$, and $m = 12$

$$1.2 = 1.005^{12t}$$

Dividing by 2000 and simplifying

$$\log 1.2 = \log 1.005^{12t}$$

Taking logarithms

$$\log 1.2 = 12t \log 1.005$$

Using $\log(x^n) = n \log x$ to bring down the power

$$12t = \frac{\log 1.2}{\log 1.005} \approx 36.555$$

Dividing by $\log 1.005$, reversing sides, and calculating logarithms

Months

This amount of time, $12t = 36.555$, is in *months* (t is years, but $12t$ is *months*). We round *up* to the nearest month (because a shorter term will not reach the needed amount), so the term needed is 37 months, or 3 years 1 month.

Hint: Instead of solving for *t*, if the compounding is monthly, stop when you find 12*t* (the number of months); if quarterly, stop when you find 4*t* (the number of quarters); if weekly, stop when you find 52*t* (the number of weeks); and so on. Then round *up* to the next whole number of periods to ensure that the needed amount will actually be reached.

Notice that the actual amounts $2400 and $2000 did not matter in this calculation, only their *ratio*   2400/2400 = 1.2.   Thus, a loan of $2000 grows to $2400 in the same amount of time that a loan of $20 *million* would grow to $24 million (at the stated interest rate) since the ratios are the same.

Therefore, in Example 3 we could have omitted the dollar amounts and simply asked for the term of a loan that would *multiply the principal by 1.2*. Furthermore, multiplying by 1.2 means *increasing* by 0.2, or 20%, so we could have asked for the term of a loan that would *increase the principal by 20%*. All three formulations are equivalent.

Be careful! Distinguish carefully between the *increase* and the *multiplier*. *Increasing by 20% means multiplying by 1.2* (the "1" keeps the original amount and the ".2" increases it by 20%). For example, to *increase* an amount by 35%, you would *multiply* it by 1.35; to increase an amount by 75%, you would multiply by 1.75; and to increase it by 100%, you would multiply by 2.

➡ Practice Problem 2

What is the multiplier that corresponds to increasing the amount by 50%?

➤ Solution on page 73

➡ EXAMPLE 4   Finding the Term to Increase the Principal

What is the term of a loan at 6% compounded weekly that will double the principal?

Solution
Doubling means multiplying the amount by 2. Using *P* for the unknown principal, we must solve

$$2P = P\left(1 + \frac{0.06}{52}\right)^{52 \cdot t} \qquad \text{*P* (for principal) on both sides}$$

$$2 = \left(1 + \frac{0.06}{52}\right)^{52t} \qquad \text{Canceling the *P*s and simplifying}$$

$$\log 2 = 52t \log (1 + 0.06/52) \qquad \text{Taking logarithms and bringing down the power}$$

$$52t = \frac{\log 2}{\log (1 + 0.06/52)} \approx 601.1 \qquad \text{Dividing by   log (1 + 0.06/52), reversing sides, and calculating logarithms}$$

Weeks

Rounding *up*, the term needed is 602 weeks, or 11 years 30 weeks.

→ Practice Problem 3

What is the term of a loan at 6% compounded quarterly that increases the principal by 50%?
> **Solution on page 73**

Rule of 72

The *doubling time* of an investment or a loan is the number of years it takes for the value to double. Doubling times are often estimated by using the *rule of 72*:

$$\left(\begin{array}{c}\text{Doubling}\\\text{time}\end{array}\right) \approx \frac{72}{r \times 100}$$ Divide 72 by the rate times 100

The rule of 72, however, gives only an *approximation* for the doubling time, but it is often quite accurate. (It follows from solving $2P = Pe^{rt}$ for t and then approximating the numerator by 0.72.) For example, for a loan at 6% compounded quarterly the rule of 72 gives

$$\left(\begin{array}{c}\text{Doubling}\\\text{time}\end{array}\right) \approx \frac{72}{0.06 \times 100} = \frac{72}{6} = 12 \text{ years}$$ Rule of 72 estimate

In Example 4 we found the correct doubling time to be 11 years 30 weeks, so the rule of 72 is not far off.

Effective Rates

Suppose you are considering two certificates of deposit (CDs), one offering 6.2% compounded semiannually and the other offering 6.15% compounded monthly. Which should you choose, the one with the higher rate or the one with more frequent compounding? More generally, how can we compare any two different interest rates with different compoundings? We do so by calculating the actual percentage increase that each will generate in a year. This number is called the *annual percentage yield* or *APY*. More formally, the APY is also called the *effective rate of interest*, denoted r_e, which is the simple interest rate that will return the same amount on a 1-year loan. Therefore, r_e can be found by solving

$$P(1 + r_e) = P\left(1 + \frac{r}{m}\right)^m$$ Finding the "simple" rate that gives the "compound" rate for $t = 1$

Canceling the Ps and subtracting the 1 on the left gives the following formula:

Effective Rate for a Compound Interest Loan

For a compound interest loan at interest rate r compounded m times per year, the effective rate r_e is

$$r_e = \left(1 + \frac{r}{m}\right)^m - 1$$ $r_e = $ APY

The effective rate can be interpreted as the amount of interest that would be earned by $1 in 1 year.

→ EXAMPLE 5 Using Effective Rates to Compare Investments

A self-employed carpenter setting up a Keogh retirement plan must choose between a CD with the First & Federal Bank at 6.2% compounded semiannually or a CD with the Chicago Nationswide Bank at 6.15% compounded monthly. Which is better?

Solution

The effective rate for 6.2% compounded semiannually (First & Federal) is

$$r_e = \left(1 + \frac{0.062}{2}\right)^2 - 1 \approx 0.0630 \qquad r_e \approx 6.30\%$$

For 6.15% compounded monthly (Chicago Nationswide), it is

$$r_e = \left(1 + \frac{0.0615}{12}\right)^{12} - 1 \approx 0.0633 \qquad r_e \approx 6.33\%$$

The CD at the Chicago Nationswide Bank is better.

Sometimes the interest rate is not given, and only the amounts at the beginning and end of the loan are known. How can we find the interest rate (called the *effective* rate of interest)?

The effective rate of return for an investment of P dollars that returns an amount A after t years can be found from the compound interest formula (page 68) with $m = 1$:

$$A = P(1 + r_e)^t$$

$A = P(1 + r/m)^{mt}$ with
$m = 1$ and $r = r_e$

An example is a *zero-coupon bond*, which pays its *face value* on maturity and sells now for a lower price. This kind of bond makes no payments before maturity and so has no interest "coupons."

→ EXAMPLE 6 Effective Rate for a Zero-Coupon Bond

What is the effective rate of return of a $10,000 zero-coupon bond maturing in 6 years and offered now for sale at $7635?

Solution

We solve

$$10{,}000 = 7635(1 + r_e)^6$$

$A = P(1 + r_e)^t$ with
$A = 10{,}000$, $P = 7635$,
and $t = 6$

$$\frac{10{,}000}{7635} = (1 + r_e)^6$$

Dividing by 7635

$$\left(\frac{10{,}000}{7635}\right)^{1/6} = 1 + r_e$$

Raising to power $\frac{1}{6}$

$$r_e = \left(\frac{10{,}000}{7635}\right)^{1/6} - 1 \approx 0.0460$$

Solving for r_e

The effective rate is 4.6%.

→ 2.2 Exercises

Determine the amount due on each compound interest loan.

1. $15,000 at 3% for 10 years if the interest is compounded:

   a. annually.   *b.* quarterly.

2. $12,000 at 4.1% for 4 years 6 months if the interest is compounded:

   a. semiannually.   *b.* monthly.

Calculate the present value of each compound interest loan.

3. $25,000 after 7 years at 4% if the interest is compounded:

   a. annually.   *b.* quarterly.

4. $11,500 after 4 years 3 months at 3.2% if the interest is compounded:

   a. semiannually.   *b.* monthly.

Find the term of each compound interest loan.

5. 3.9% compounded quarterly to obtain $8400 from a principal of $2000.

6. 4.5% compounded monthly to increase the principal by 65%.

Use the "rule of 72" to estimate the doubling time (in years) for each interest rate, and then calculate it exactly.

7. 9% compounded annually.

8. 7.9% compounded weekly.

Find the effective rate of each compound interest rate or investment.

9. 18% compounded monthly. [*Note*: This rate is a typical credit card interest rate, often stated as 1.5% per month.]

10. A $50,000 zero-coupon bond maturing in 8 years and selling now for $41,035.

11. *Bond Funds* Since 2007, the Reynolds Blue Chip fund returned 13.9% compounded monthly. How much would a $5000 investment in this fund have been worth after 3 years?

12. *Bond Funds* You have just received $125,000 from the estate of a long-lost rich uncle. If you invest all your inheritance in a tax-free bond fund earning 6.9% compounded quarterly, how long do you have to wait to become a millionaire?

13. *Rate Comparisons* The Second Peoples National Bank offers a long-term certificate of deposit earning 6.43% compounded monthly. Your broker locates a $20,000 zero-coupon bond rated AA by Standard & Poor's for $7965 and maturing in 14 years. Which investment will give the greater rate of return?

14. *Lottery Winnings* You have just won $100,000 from a lottery. If you invest all this amount in a tax-free money market fund earning 7% compounded weekly, how long do you have to wait to become a millionaire?

②.③ Annuities

An *annuity* is a scheduled sequence of payments. With some annuities, such as retirement pensions, you receive the payments, while with others, such as car loans and home mortgages, you pay someone else. In this section we will consider an *ordinary annuity*, which is an annuity with equal payments made at the end of regular intervals, with the interest compounded at the end of each. For example, a car loan for 4 years with monthly payments of $200 at 6% compounded monthly is an ordinary annuity.

A First Example

While having a new car is nice, when the first payment of the above loan comes due at the end of the first month, you may wish that instead you were saving that $200 for yourself. If you were, then 4 years later your first $200 at 6% compounded monthly would have grown to

$$200\left(1 + \frac{0.06}{12}\right)^{47} = 200 \cdot 1.005^{47}$$

$$\underbrace{1 + 0.005 = 1.005}$$

The exponent is 47 rather than 48 because the first payment was made at the end of the first month and so earns interest for only 47 months. Similarly, your second payment would grow to $200 \cdot 1.005^{46}$, your third to $200 \cdot 1.005^{45}$, and so on down to your last payment of 200 (at the end of the last month, so earning no interest).

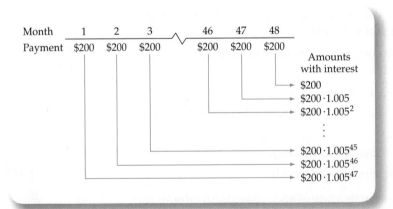

With patience and care we could evaluate and add up the numbers in the right side of this diagram, but there is an easier way. We will develop a formula for such sums and then use it to find the total amount of this annuity.

74 *Chapter 2: Mathematics of Finance*

Geometric Series

The numbers in the right-hand side of the diagram are found by successively multiplying by 1.005. In general, a sum of numbers each of which is a fixed multiple of the one before it is called a *geometric series*. The simplest geometric series is of the form

$$S = 1 + x + x^2 + \cdots + x^{n-1}$$

Multiplying by x gives the next number, and S is the sum of the series

If we multiply the sum by x (which increases the exponent of each term) and subtract the original sum, we obtain

$$x \cdot S = \quad x + x^2 + \cdots + x^{n-1} + x^n \qquad \text{x times S}$$

$$S = 1 + x + x^2 + \cdots + x^{n-1} \qquad \text{S (lining up "like" terms)}$$

$$\underbrace{x \cdot S - S}_{S(x - 1)} = -1 \qquad\qquad\qquad + x^n \qquad \text{Subtracting (most terms cancel)}$$

This last line can be written

$$S(x - 1) = x^n - 1$$

Factoring on the left, reversing the order on the right

so that

$$S = \frac{x^n - 1}{x - 1}$$

Dividing both sides by $x - 1$ (for $x \neq 1$)

Since S was defined as the sum $S = 1 + x + x^2 + \cdots + x^{n-1}$, we have a formula for this sum:

$$1 + x + x^2 + \cdots + x^{n-1} = \frac{x^n - 1}{x - 1}$$

Multiplying both sides by a gives a formula for the sum of a geometric series:

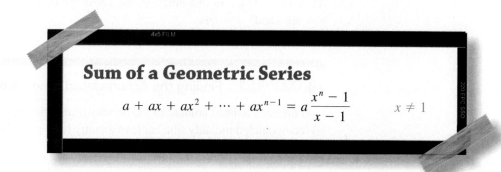

Sum of a Geometric Series

$$a + ax + ax^2 + \cdots + ax^{n-1} = a\,\frac{x^n - 1}{x - 1} \qquad x \neq 1$$

For instance, the geometric series $3 + 6 + 12 + 24 + 48$, in which $a = 3$, $x = 2$, and the number of terms is $n = 5$, has the value

$$3 \cdot \frac{2^5 - 1}{2 - 1} = 93$$

$a\,\frac{x^n - 1}{x - 1}$ with $a = 3$, $x = 2$, and $n = 5$

as you should check by adding up the five numbers.

Accumulated Amount Formula

For the payments at 6% compounded monthly for 4 years on page 74, we added terms of the form $200(1 + 0.06/12)^k$ with the exponent k taking values from 0 to $47 = 12 \cdot 4 - 1$. More generally, for payments of P dollars m times a year for t years at interest rate r, we would need to sum the geometric series

$$P + P\left(1 + \frac{r}{m}\right) + P\left(1 + \frac{r}{m}\right)^2 + \cdots + P\left(1 + \frac{r}{m}\right)^{mt-1}$$

Using the geometric series sum formula on the previous page, the sum of this series is

$$P\frac{\left(1 + \dfrac{r}{m}\right)^{mt} - 1}{\left(1 + \dfrac{r}{m}\right) - 1} \qquad a\frac{x^n - 1}{x - 1} \quad \text{with} \quad a = P,$$
$$x = (1 + r/m), \quad \text{and} \quad n = mt$$

Simplifying the denominator to r/m gives the following very useful formula:

Accumulated Amount of an Annuity

For an ordinary annuity with payments of P dollars m times a year for t years at interest rate r compounded at the end of each payment period, the accumulated amount (payments plus interest) will be

$$A = P\frac{(1 + r/m)^{mt} - 1}{r/m}$$

Although we will not use it, the archaic symbol $s_{\overline{n}i}$ (pronounced "s sub n angle i") sometimes still occurs in business textbooks and tables to denote the value of $((1 + i)^n - 1)/i$. In this notation, the above formula takes the form $A = Ps_{\overline{n}i}$ with $n = mt$ and $i = r/m$.

➡ EXAMPLE 1 Finding the Accumulated Amount of an Annuity

Find the accumulated amount of the annuity of monthly payments for 4 years of $200 at 6% compounded monthly (the car loan annuity from page 74).

Solution

$$A = 200\frac{(1 + 0.06/12)^{12 \cdot 4} - 1}{0.06/12} \approx 10{,}819.57 \qquad \begin{array}{l} P\dfrac{(1 + r/m)^{mt} - 1}{r/m} \\ \text{with} \quad P = 200, \quad r = 0.06, \\ m = 12, \quad \text{and} \quad t = 4 \end{array}$$

Therefore, regular $200 payments at the end of each month for 4 years into a savings account earning 6% compounded monthly will total $10,819.57.

The amount of this annuity is significantly more than just the sum of 48 payments of $200, which would total $9600. The difference, 10,819.57 − 9600 = $1219.57, is the total interest earned by the payments.

> **Practice Problem 1**
>
> What is the final balance of a retirement account earning 5% interest compounded weekly if $40 is deposited at the end of every week for 35 years? How much of this final balance is interest?
>
> ➤ Solution on page 80

Sinking Funds

A *sinking fund* is a regular savings plan designed to provide a given amount after a certain number of years. For example, suppose you decide to save up for a $18,000 car instead of taking out a loan. What amount should you save each month to have the $18,000 in 3 years? If you just put your money in a shoe box, then you will need to set aside $18,000 ÷ (12 × 3) = $500 each month. It would make more sense to put your money into a savings account each month and let it earn compound interest. It is the same as setting up an annuity with a given accumulated amount and asking what the regular payment should be. Solving the accumulated amount formula (on the previous page) for P by dividing by the fraction following P gives the following formula:

Sinking Fund Payment

The payment P to make m times per year for t years at interest rate r compounded at each payment to accumulate amount A is

$$P = A\frac{r/m}{(1 + r/m)^{mt} - 1}$$

This formula may also be written $P = A/S_{\overline{n}i}$ with $n = mt$ and $i = r/m$.

> ➤ **EXAMPLE 2** Finding a Sinking Fund Payment

What amount should be deposited at the end of each month for 3 years in a savings account earning 4.5% interest compounded monthly to accumulate $18,000 to buy a new car?

Solution

$$P = 18,000\frac{0.045/12}{(1 + 0.045/12)^{12\cdot3} - 1} \approx 467.945$$

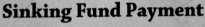

$P = A\dfrac{r/m}{(1 + r/m)^{mt} - 1}$
with $A = 18,000$,
$r = 0.045$, $m = 12$,
and $t = 3$

Rounding *up* to the next penny, $467.95 should be saved each month. (If we rounded down to $467.94, then the savings account would accumulate slightly less than the amount needed.)

Notice that this amount is significantly lower than the $500 that would be needed *without* compound interest.

This example could also have been done using the accumulated amount formula (page 76) by substituting $A = 18{,}000$, $r = 0.045$, $m = 12$, and $t = 3$, and then solving for the remaining variable to find $P = \$467.95$.

> ### → Practice Problem 2
>
> For wage earners paid every other week instead of monthly, Example 2 becomes: "What amount should be deposited biweekly for 3 years in a savings account earning 4.5% interest compounded biweekly to accumulate $18,000 for a new car?" Find this amount.
>
> ➤ **Solution on page 80**

How Long Will It Take?

Continuing with our sinking fund example, suppose you could save only $200 each month. How long would it take to accumulate the $18,000?

We solve the accumulated amount formula (page 76) for the number of periods mt.

$$\frac{A}{P}\frac{r}{m} + 1 = \left(1 + \frac{r}{m}\right)^{mt}$$

$$A = P\,\frac{(1 + r/m)^{mt} - 1}{r/m}$$

dividing by P, multiplying by r/m, and adding 1

$$mt \log\left(1 + \frac{r}{m}\right) = \log\left(\frac{A}{P}\frac{r}{m} + 1\right)$$

Switching sides and taking logs of both to bring down the exponent

Dividing by $\log(1 + r/m)$, we have:

Number of Periods

$$mt = \frac{\log\left(\dfrac{A}{P}\dfrac{r}{m} + 1\right)}{\log\left(1 + \dfrac{r}{m}\right)} \quad \text{periods}$$

Nino Cavalier/Shutterstock (fortune cookie)
and r. martens/Shutterstock (clocks)

$t = ?$

As usual, we round *up* to the next whole number of compounding periods and then express the answer as years plus any extra periods. Notice that the time t depends only on the ratio A/P between the accumulated amount A and the regular payment P.

→ **EXAMPLE 3** Finding the Term for a Sinking Fund

How long will it take to accumulate \$18,000 by depositing \$200 at the end of each month in a savings account earning 4.5% interest compounded monthly?

Solution

$$12t = \frac{\log\left(\dfrac{18{,}000}{200}\dfrac{0.045}{12} + 1\right)}{\log\left(1 + \dfrac{0.045}{12}\right)} \approx 77.7 \qquad mt = \frac{\log\left(\dfrac{A}{P}\dfrac{r}{m} + 1\right)}{\log\left(1 + \dfrac{r}{m}\right)}$$

with $A = 18{,}000$, $P = 200$, $r = 0.045$, and $m = 12$

The time is 77.7 months, which we round *up* to 78 months. Therefore, it will take 6 years 6 months to accumulate the \$18,000.

This example could also have been done using the accumulated amount formula (page 76) by substituting $A = 18{,}000$, $P = 200$, $r = 0.045$, and $m = 12$, and then solving (using logarithms) for the remaining variable to find that $12t$ is 78 months.

→ **Practice Problem 3**

How long will it take to accumulate \$18,000 by depositing \$250 each month in a savings account earning 4.5% interest compounded monthly?

➤ **Solution on next page**

In the previous section we found the eventual value of an investment, the initial amount necessary to reach that later amount, and the time it takes, just as we did in this section. What's the difference between the two sections? *There* we were considering investments of a *single* payment of P dollars, while *here* we are considering *multiple and regular payments* of P dollars.

Timmary/Shutterstock

Solutions to Practice Problems

1. $A = 40 \dfrac{(1 + 0.05/52)^{52 \cdot 35} - 1}{0.05/52} \approx 197{,}590.27.$ The final balance is
$197,590.27. Because the deposits total $40 \cdot 52 \cdot 35 = \$72{,}800,$ the final balance contains $124,790.27 interest.

2. $P = 18{,}000 \dfrac{0.045/26}{(1 + 0.045/26)^{26 \cdot 3} - 1} \approx 215.742.$ $215.75 should be
deposited every other week. (Compare double this amount, $431.50, to the $467.95 monthly payment found in Example 2 on page 77.)

3. $12t = \dfrac{\log\left(\dfrac{18{,}000}{250} \dfrac{0.045}{12} + 1\right)}{\log\left(1 + \dfrac{0.045}{12}\right)} \approx 63.9$ months,   which rounds up to

64 months. It will take 5 years 4 months.

→2.3 Exercises

In the following ordinary annuities, the interest is compounded with each payment, and the payment is made at the end of the compounding period.

Find the accumulated amount of each annuity.

1. $1500 annually at 7% for 10 years.
2. $1000 monthly at 6.9% for 20 years.

Find the required payment for each sinking fund.

3. Monthly deposits earning 5% to accumulate $5000 after 10 years.
4. Yearly deposits earning 12.3% to accumulate $8500 after 12 years.

Find the amount of time needed for each sinking fund to reach the given accumulated amount.

5. $1500 yearly at 8% to accumulate $100,000.
6. $235 monthly at 5.9% to accumulate $25,000.
7. *Retirement Savings* An individual retirement account, or IRA, earns tax-deferred interest and allows the owner to invest up to $5000 each year. Joe and Jill both will make IRA deposits for 30 years (from age 35 to 65) into stock mutual funds yielding 9.8%. Joe deposits $5000 once each year, while Jill has $96.15 (which is 5000/52) withheld from her weekly paycheck and deposited automatically. How much will each have at age 65?

8. *Mutual Funds* How much must you invest each month in a mutual fund yielding 13.4% compounded monthly to become a millionaire in 10 years?

9. *Lifetime Savings* The Oseola McCarty Scholarship Fund at the University of Southern Mississippi was established by a $150,000 gift from an 87-year-old woman who had dropped out of sixth grade and worked for most of her life as a washerwoman. How much would she have had to save each week in a bank account earning 3.9% compounded weekly to have $150,000 after 75 years?

10. *Home Buying* You and your new spouse each bring home $1500 each month after taxes and other payroll deductions. By living frugally, you intend to live on just one paycheck and save the other in a mutual fund yielding 7.86% compounded monthly. How long will it take to have enough for a 20% down payment on a $165,000 condo in the city?

2.4 Amortization

In this section we will find the *present value of an annuity*, or the total value *now* of all the future payments. Such calculations are important for buying and selling annuities and for using them to pay off debts (amortization). For example, on page 76 we found the accumulated amount for monthly payments of $200 for 4 years at 6% compounded monthly. How much would a loan company give *right now* for that promise of future payments?

Present Value of an Annuity

We know that for an ordinary annuity with payments of P dollars m times per year for t years at interest rate r compounded at each payment, the accumulated amount A is

$$A = P\,\frac{(1 + r/m)^{mt} - 1}{r/m}$$ From page 76

On page 68 we saw that to find the present value of an amount you simply divide by $(1 + r/m)^{mt}$.   Therefore, the present value of this amount (denoted PV to distinguish it from P for payment) is

$$PV = P\,\frac{(1 + r/m)^{mt} - 1}{(r/m)(1 + r/m)^{mt}}$$ Dividing by $(1 + r/m)^{mt}$

$$= P\,\frac{1 - (1 + r/m)^{-mt}}{r/m}$$ Dividing numerator and denominator by $(1 + r/m)^{mt}$

which gives the following formula:

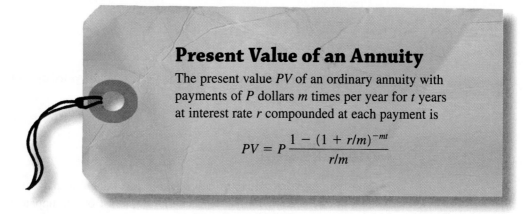

Present Value of an Annuity

The present value PV of an ordinary annuity with payments of P dollars m times per year for t years at interest rate r compounded at each payment is

$$PV = P\,\frac{1 - (1 + r/m)^{-mt}}{r/m}$$

Although we will not use it, the archaic symbol $a_{\overline{n}|i}$ (pronounced "*a* sub *n* angle *i*") sometimes still occurs in business textbooks and tables to denote the value of $(1 - (1 + i)^{-n})/i$.   In this notation, the above formula takes the form $PV = Pa_{\overline{n}|i}$ with $n = mt$ and $i = r/m$.

→ EXAMPLE 1   Finding the Present Value of an Annuity

What is the present value of a 6% car loan for 4 years with monthly payments of $200?

Solution

$$PV = 200 \, \frac{1 - (1 + 0.06/12)^{-12 \cdot 4}}{0.06/12} \approx 8516.06$$

$$PV = P \, \frac{1 - (1 + r/m)^{-mt}}{r/m}$$
with $P = 200$, $r = 0.06$, $m = 12$, and $t = 4$

The present value is $8516.06.

In general, how can we interpret the present value of an annuity? It is the sum that must be deposited in a bank *now* (at the stated interest rate) to grow to exactly the amount of the annuity at the end of its term. For the loan in Example 1, in 4 years the present value will grow to $8516.06(1 + 0.06/12)^{12 \cdot 4} = \$10{,}819.57$, which is exactly the sum of the annuity that we found on page 76. Therefore, the present value of $8516.06 is exactly the price for which the sequence of car payments should be sold, and the lender will not care which she receives.

> ### → Practice Problem 1
>
> What is the present value of a 20-year retirement annuity paying $850 per month if the current long-term interest rate is 7.53%?
>
> ➤ Solution on page 85

Be careful! The formulas for the *present value* and the *accumulated* or *future value* of an annuity (derived on page 76) look similar but are not the same and should not be confused:

Present Value $\qquad P \, \dfrac{1 - (1 + r/m)^{-mt}}{r/m}$

Future (or Accumulated) Value $\qquad P \, \dfrac{(1 + r/m)^{mt} - 1}{r/m}$

As we saw on the previous page, the future value formula divided by $(1 + r/m)^{mt}$ gives the present value formula.

Amortization

Suppose you took out a loan to buy a house and wanted to pay off the debt in a fixed number of payments, with the bank charging interest on the unpaid balance. A debt is *amortized* (or "killed off") if it is repaid by a regular sequence of payments.

How do we calculate the correct payments to amortize a debt? The payments form an annuity with present value PV equal to the debt D. To find the payment P to amortize the debt, we solve for P in the "present value of an annuity" formula on the previous page (dividing by the fraction following the P) and then replacing PV by D.

Amortization Payment

The payment P to make m times per year for t years at interest rate r compounded at each payment to amortize a debt of D dollars is

$$P = D \frac{r/m}{1 - (1 + r/m)^{-mt}}$$

This formula may also be written $P = D/a_{\overline{n}i}$ with $n = mt$ and $i = r/m$.

→ EXAMPLE 2 Finding the Payment to Amortize a Debt

What monthly payment will amortize a $150,000 home mortgage at 4.3% in 30 years?

Solution

$$P = 150,000 \frac{0.043/12}{1 - (1 + 0.043/12)^{-12 \cdot 30}} \approx 742.31$$

$P = D \dfrac{r/m}{1 - (1 + r/m)^{-mt}}$
with $D = 150,000$,
$r = 0.043$, $m = 12$,
and $t = 30$

The required payment is $742.31 each month.

Notice that the borrower will pay a total of $742.31 \cdot 12 \cdot 30 = \$267,231.60$ during the 30 years of the loan to pay off a $150,000 debt. The difference, $117,231.60, is the interest paid to the bank.

→ Practice Problem 2

What monthly payment will amortize a $150,000 home mortgage at 4.3% in 25 years? What is the total amount the borrower will pay?

> Solution on page 85

Unpaid Balance

How much does the borrower still owe partway through an agreed amortization schedule? Suppose you have made mortgage payments for 10 years on a 30-year loan and now you must move. How much should you pay to settle your debt? Your remaining payments form another annuity and are worth precisely the present value of this new annuity.

4693902d/Shutterstock

➡ EXAMPLE 3 Finding the Unpaid Balance

What is the unpaid balance after 10 years of monthly payments on a 30-year mortgage of $150,000 at 4.3%?

Solution

First we must calculate the monthly payments. But for this problem, we know from Example 2 that the monthly payment to amortize the original mortgage is $742.31. The remaining payments thus form an annuity with $P = \$742.31$, $r = 0.043$, $m = 12$, and $t = 20$ years. The amount still owed on the mortgage is the present value of this annuity, which we can calculate by the formula on page 81.

$$PV = 742.31 \frac{1 - (1 + 0.043/12)^{-12 \cdot 20}}{0.043/12}$$

$$\approx 119{,}360.74$$

$$PV = P \frac{1 - (1 + r/m)^{-mt}}{r/m}$$
with $P = 742.31$, $r = 0.043$, $m = 12$, and $t = 20$

The unpaid balance after 10 years of payments on this $150,000 mortgage is $119,360.74.

Notice that after the first 10 of 30 years, much less than one third (in fact, only about 20%) of the mortgage has been paid. The early payments of any loan pay mostly interest and only a little of the principal.

As in this example, if you already know the payments, you use the same payments to calculate the unpaid balance. If you do *not* know the payments, you would first have to calculate them using the payment formula on the previous page.

➡ Practice Problem 3

How much is still owed on a 25-year mortgage of $150,000 at 4.3% after 13 years of monthly payments? [*Hint*: Use the result of Practice Problem 2 on page 83.]

➤ Solution on next page

Equity

One often hears the term *equity* used with homes, such as a "home equity loan." *Equity* is defined as the *value of the home minus any unpaid mortgage balance* and is a measure of the value that you (rather than the bank) have in your house. If the house with the mortgage in Example 3 now has a market value of $300,000, then the homeowner's equity after 10 years of payments is this value minus the unpaid balance: $300,000 − 119,361 = $180,639 (ignoring cents). This equity may be used to take out a new loan for a larger house or for other purposes.

→ Solutions to Practice Problems

1. $PV = 850 \dfrac{1 - (1 + 0.0753/12)^{-12 \cdot 20}}{0.0753/12} \approx 105{,}272.468$

The present value is \$105,272.47.

2. $P = 150{,}000 \dfrac{0.043/12}{1 - (1 + 0.043/12)^{-12 \cdot 25}} \approx 816.81$

The required payment is \$816.81 each month. The borrower will pay a total of $\$816.81 \cdot 12 \cdot 25 = \$245{,}043$.

3. The remaining payments form a 12-year annuity at 4.3% with monthly payments of \$816.81 (from Practice Problem 2). The present value of this annuity is

$$PV = 816.81 \dfrac{1 - (1 + 0.043/12)^{-12 \cdot 12}}{0.043/12} \approx 91{,}759.11$$

so the amount still owed is \$91,759.11. Notice that after more than half the term, less than half (in fact, only about 39%) of the \$150,000 loan is paid off.

→ 2.4 Exercises

Calculate the present value of each annuity.

1. \$15,000 annually at 7% for 10 years.

2. \$1400 monthly at 6.9% for 30 years.

Determine the payment to amortize each debt.

3. Monthly payments on \$100,000 at 5% for 25 years.

4. Quarterly payments on \$14,500 at 3.9% for 6 years.

Find the unpaid balance on each debt. You may already know the payments from Exercises 3 and 4.

5. After 6 years of monthly payments on \$100,000 at 5% for 25 years.

6. After 2 years 3 months of quarterly payments on \$14,500 at 3.9% for 6 years.

7. *Contest Prizes* The super prize in a contest is \$10 million. This prize will be paid out in equal yearly payments over the next 25 years. If the prize money is guaranteed by AAA bonds yielding 3% and is placed into an escrow account when the contest is announced 1 year before the first payment, how much do the contest sponsors have to deposit in the escrow account?

8. *MLB Contracts* When Alex Rodriguez signed a 10-year, \$275 million contract with the New York Yankees, it included a \$10 million signing bonus.

If the \$275 million was paid out in equal quarterly payments for the 10 years and the current long-term interest rate was 4%, how much was the contract worth when it was signed?

9. *Life Insurance* Just before his first attempt at bungee jumping, John decides to buy a life insurance policy. His annual income at age 30 is \$35,000, so he figures he should get enough insurance to provide his wife and new baby with that amount each year for the next 35 years. If the long-term interest rate is 6.7%, what is the present value of John's future annual earnings? Rounding up to the next \$50,000, how much life insurance should he buy?

10. *Credit Cards* A MasterCard statement shows a balance of \$560 at 13.9% compounded monthly. What monthly payment will pay off this debt in 1 year 8 months?

ATTENTION
NEED MORE PRACTICE? FIND MORE HERE:
CENGAGEBRAIN.COM

Chapter 3

Systems of Equations and Matrices

3.1   Systems of Two Linear Equations in Two Variables

Neither of the statements "Bob and Sue together have \$100" and "Bob has \$20 less than Sue" tells us how much either has, but taken together, they force the conclusion that Bob has \$40 and Sue has \$60. Various methods of solving such simple problems have been used since antiquity, but many become unworkable as the problems become more complicated. This chapter provides a method that solves such problems and has the pleasing property that the method for complicated problems is an easy extension of the method for the simplest.

Systems of Equations

We begin with the simplest form of these problems.

Systems of Two Linear Equations in Two Variables

A system of two linear equations in two variables is any problem expressible in the form

$$\begin{cases} ax + by = h \\ cx + dy = k \end{cases}$$

where x and y are the variables and the constants $a, b, c, d,$ h, k are such that at least one of the coefficients a, b, c, d is not zero.

Such equations are called *linear* because $ax + by = h$ (with at least one of a and b not equal to zero) is the same as the *general linear equation* (see page 11) whose graph is a line.

→ EXAMPLE 1　Two Equations in Two Variables

Express the statements "Bob and Sue together have \$100" and "Bob has \$20 less than Sue" as a system of two equations in two variables.

Solution

Let x represent the amount of money Bob has and y represent the amount Sue has. The statement that together they have \$100 may be written "$x + y = 100$" and the statement that Bob has

$20 less than Sue as "$x = y - 20$." Rearranging this second equation by subtracting y from both sides, we obtain the system of equations

$$\begin{cases} x + y = 100 \\ x - y = -20 \end{cases}$$

$ax + by = h$ with
$a = 1, \quad b = 1, \quad h = 100$

$cx + dy = k$ with
$c = 1, \quad d = -1, \quad k = -20$

There are many other ways to express the two statements as a system of equations. For example, we could write the same equations but in the opposite order:

$$\begin{cases} x - y = -20 \\ x + y = 100 \end{cases}$$

Switching the order of the equations

Or instead we could multiply one of the equations by 2:

$$\begin{cases} 2x + 2y = 200 \\ x - y = -20 \end{cases}$$

$x + y = 100$
multiplied by 2

We could even add (or subtract) the equations and use the result to replace one of the original equations:

$$\begin{cases} x + y = 100 \\ 2x \quad\quad = 80 \end{cases}$$

$\begin{aligned} x + y &= 100 \\ \underline{x - y} &= -20 \quad \text{Adding} \\ 2x &= 80 \end{aligned}$

It is an easy matter to check that each of these three systems has the same solution, $x = 40$ and $y = 60$. Notice that the second equation in the above system immediately gives $x = 40$, which is "half" of the solution. In fact, if we could write the system in the following way, then we could immediately read off the *entire* solution:

$$\begin{cases} 1x + 0y = 40 \\ 0x + 1y = 60 \end{cases}$$

⟵ gives $x = 40$
⟵ gives $y = 60$

Our goal will be to simplify a system of equations until its solution is obvious (or until it is clear that there is no solution).

➤ **Practice Problem 1**

Express the statements "a jar of pennies and nickels contains 80 coins" and "the coins in the jar are worth $1.60" as a system of two equations in two variables.

➤ **Solution on page 93**

A *solution* of a system of equations in two variables is a pair of values for the variables that satisfy all the equations (such as $x = 40$, $y = 60$ for Example 1). The *solution set* is the collection of all solutions. *Solving* the system of equations means finding this solution set. In this section we will solve systems of linear equations by two different methods: *graphing* and *elimination*. Each has its own advantages and disadvantages.

Graphical Representations of Equations

An equation of the form $ax + by = h$ (with at least one of a and b not zero) is the equation of a line written in *general form*. We begin by graphing linear equations.

→ **EXAMPLE 2** Graphing a Linear Equation in Two Variables

Sketch the graph of each equation.

a. $2x + 3y = 12$ **b.** $2x - 3y = 0$ **c.** $3y = 12$ **d.** $2x = 12$

Solution

a. To graph $2x + 3y = 12$ we find the intercepts by setting each of the variables in turn equal to zero:

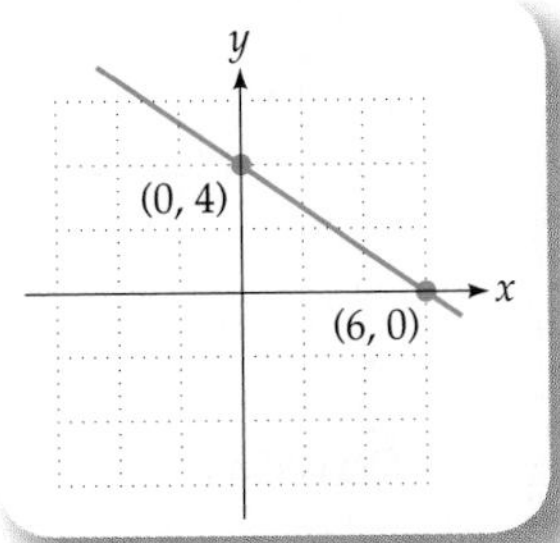

$$3y = 12$$
$$y = 4 \qquad \text{for the point } (0, 4)$$
$$2x = 12$$
$$x = 6 \qquad \text{for the point } (6, 0)$$

$2x + 3y = 12$ with $x = 0$
Since we began with $x = 0$
The original $2x + 3y = 12$
Since we began with $y = 0$

The line through $(0, 4)$ and $(6, 0)$ is shown on the left.

b. To graph $2x - 3y = 0$ we begin by setting $x = 0$:

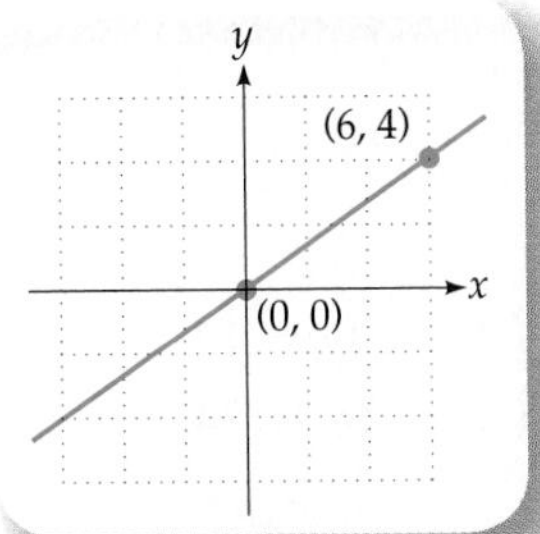

$$-3y = 0$$

$$y = 0 \qquad \text{for the point } (0, 0)$$

$2x - 3y = 0$ with $x = 0$
Now divide by -3

Since we began with $x = 0$

For another point we choose any other value of x.

$$12 - 3y = 0$$

$2x - 3y = 0$ with $x = 6$
Now subtract 12 and then divide by -3

$$y = 4 \qquad \text{for the point } (6, 4)$$

Since we began with $x = 6$

The line through $(0, 0)$ and $(6, 4)$ is shown on the left.

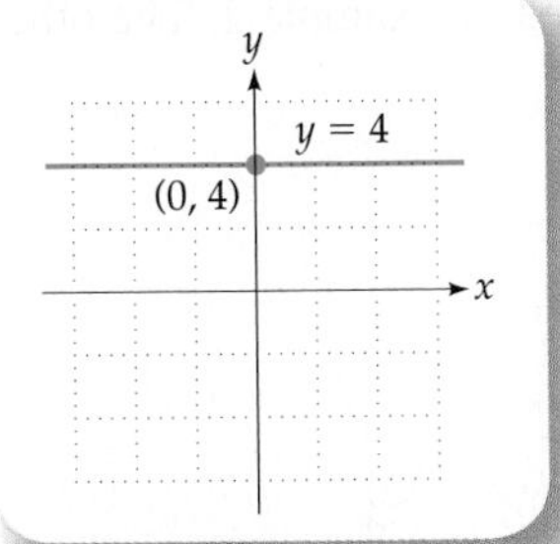

c. To graph $3y = 12$:

$$y = 4 \qquad\qquad\qquad\qquad \text{Dividing by 3}$$

$y = 4$ is a horizontal line. Its graph is shown on the left.

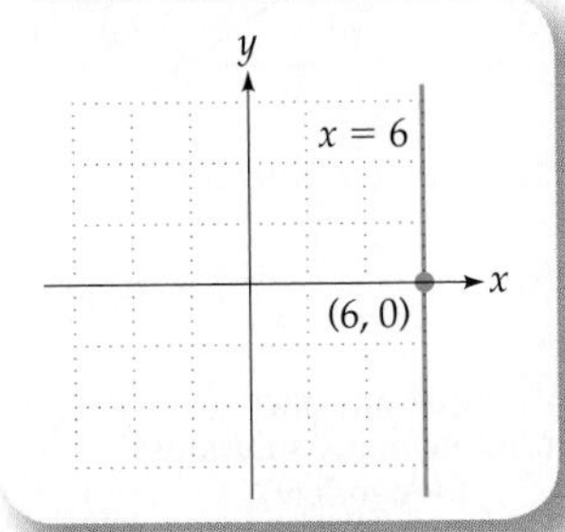

d. To graph $2x = 12$:

$$x = 6 \qquad\qquad\qquad\qquad \text{Dividing by 2}$$

$x = 6$ is a vertical line. Its graph is shown on the left.

The graph of a *pair* of equations

$$\begin{cases} ax + by = h \\ cx + dy = k \end{cases}$$

may take three different forms: two lines intersecting at just one point (a *unique* solution), two lines that don't intersect (*no* solutions), or two lines that are the *same* (*infinitely many* solutions), as shown below. For lines that don't intersect, we say that the equations are *inconsistent*, and for two lines that are the *same*, we call the equations *dependent*.

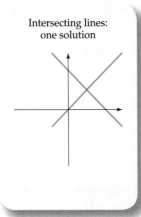

Intersecting lines:
one solution

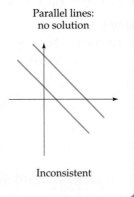

Parallel lines:
no solution

Inconsistent

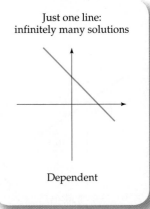

Just one line:
infinitely many solutions

Dependent

The type and solution (if any) of a system of equations can be determined from the graph if it is drawn with sufficient accuracy.

→ **EXAMPLE 3**   **Solving by Graphing**

Solve each system of equations by graphing the lines.

a. $\begin{cases} 2x + 3y = 12 \\ 2x - 3y = 0 \end{cases}$ **b.** $\begin{cases} 2x + 3y = 12 \\ 4x + 6y = 12 \end{cases}$ **c.** $\begin{cases} 2x + 3y = 12 \\ 4x + 6y = 24 \end{cases}$

Solution
The lines $2x + 3y = 12$ and $2x - 3y = 0$ were graphed in Example 2. The other lines may be graphed similarly.

a.

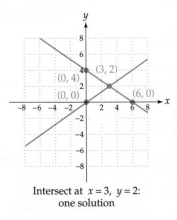

Intersect at $x = 3$, $y = 2$:
one solution

b.

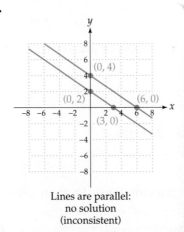

Lines are parallel:
no solution
(inconsistent)

c.

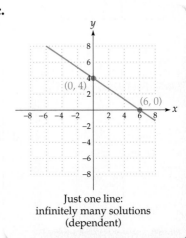

Just one line:
infinitely many solutions
(dependent)

In part (a) the unique solution $x = 3$, $y = 2$ is easily checked by substituting these values into the two equations:

$$2 \cdot 3 + 3 \cdot 2 = 12$$

$2x + 3y = 12$ with $x = 3$ and $y = 2$ (It checks: $12 = 12$)

$$2 \cdot 3 - 3 \cdot 2 = 0$$

$2x - 3y = 0$ with $x = 3$ and $y = 2$ (It checks: $0 = 0$)

In part (b) the lines do not intersect, so there is *no solution* (the equations are *inconsistent*).

In part (c) the lines are the *same* (the equations are *dependent*), so *every* point on the line $2x + 3y = 12$ is a solution.

For situations with infinitely many solutions, as in part (c), we may express *all* of these solutions at once as follows. We begin by solving either of the equations for one of the variables:

$$x = \frac{12 - 3y}{2} = 6 - \frac{3}{2}y$$

Solving $2x + 3y = 12$ for x

and writing the solution as

$$\begin{cases} x = 6 - \frac{3}{2}t \\ y = t \end{cases}$$

Replacing y by t

t may take any value

where t is any number (called a *parameter*). For example, taking $t = 2$ gives $x = 3$, $y = 2$, and the point $(3, 2)$ is a solution because it satisfies both equations. All other solutions can be found by choosing other values for the parameter t.

➡ Practice Problem 2

Find the solution corresponding to the parameter value $t = -2$ and verify that the resulting values satisfy the equations in Example 3c.

➤ Solution on page 93

Equivalent Systems of Equations

Two systems of equations are *equivalent* if they have the same solution. On page 88 we saw that we obtained equivalent systems by applying any of the following operations: switching the order of the equations, multiplying one equation by a (nonzero) constant, or adding or subtracting two equations. By combining the last two operations, we may even add a *multiple* of an equation to another. For example:

1. Switch the order of the equations.

$$\begin{cases} x + y = 100 \\ x - y = -20 \end{cases} \xrightarrow{\text{Switch equations}} \begin{cases} x - y = -20 \\ x + y = 100 \end{cases}$$

2. Multiply or divide one of the equations by a nonzero number.

$$\begin{cases} x + y = 100 \\ x - y = -20 \end{cases} \xrightarrow{\substack{\text{Multiply second} \\ \text{equation by } -1}} \begin{cases} x + y = 100 \\ -x + y = 20 \end{cases}$$

3. Add (or subtract)
a multiple of one
equation to (or from)
the other.

$$\begin{cases} x + y = 100 \\ x - y = -20 \end{cases}$$

Add twice the first
to the second

$$\begin{cases} x + y = 100 \\ 3x + y = 180 \end{cases}$$

These systems are all equivalent to one another because they all have exactly the same solutions.

Elimination Method

Our second way of solving equations, the *elimination* method, attempts to remove one variable from each and write the equations as an equivalent system of the form

$$\begin{cases} 1x + 0y = p \\ 0x + 1y = q \end{cases}$$

so that the solution can be read off as $x = p$, $y = q$. If this cannot be done, the method will identify the system as inconsistent or dependent.

In these problems we may carry out subtractions in either order, *top line minus bottom* or *bottom line minus top*:

$$\begin{array}{r} 6 \\ \underline{15} \\ -9 \end{array}$$ ← top minus bottom
$(6 - 15 = -9)$

or

$$\begin{array}{r} 6 \\ \underline{15} \\ 9 \end{array}$$ ← bottom minus top
$(15 - 6 = 9)$

➡ **EXAMPLE 4   Solving by the Elimination Method**

Solve $\begin{cases} x - 3y = 3 \\ 2x + 5y = 28 \end{cases}$ by the elimination method.

Solution

First we eliminate the x from the second equation by subtracting an appropriate multiple of the first (namely, *twice* the first so as to match the $2x$ in the second equation). Beginning with the original system:

$$\begin{cases} x - 3y = 3 \\ 2x + 5y = 28 \end{cases}$$

$$\times 2 \longrightarrow \qquad 2x - 6y = 6$$
$$\longrightarrow \qquad \underline{2x + 5y = 28}$$
$$11y = 22 \quad \text{← bottom minus top}$$

$$\begin{cases} x - 3y = 3 \\ 11y = 22 \end{cases}$$ Unchanged

$$\begin{cases} x - 3y = 3 \\ y = 2 \end{cases}$$ Unchanged
Dividing $11y = 22$ by 11

Next we eliminate the y from the first equation by adding an appropriate multiple of the second (namely, *three times* the second to match the $3y$ in the first).

$$\begin{cases} x - 3y = 3 \\ y = 2 \end{cases} \quad\begin{array}{c} \rightarrow \\ \times 3 \rightarrow \end{array}\quad \begin{array}{l} x - 3y = 3 \\ \underline{3y = 6} \\ x = 9 \end{array} \quad\leftarrow \text{top plus bottom}$$

$$\begin{cases} x = 9 \\ y = 2 \end{cases} \quad\xleftarrow{} \quad \text{Unchanged} \qquad \text{Same as} \quad \begin{cases} 1x + 0y = 9 \\ 0x + 1y = 2 \end{cases}$$

The original system has a unique solution $x = 9, \quad y = 2.$

An inconsistent system of equations always leads to the impossible equation that zero equals a nonzero number. For instance, solving Example 3b by the elimination method could be done as follows:

$$\begin{cases} 2x + 3y = 12 \\ 4x + 6y = 12 \end{cases} \xrightarrow[\text{equation by 2}]{\text{Divide second}} \begin{cases} 2x + 3y = 12 \\ 2x + 3y = 6 \end{cases} \xrightarrow[\text{from second}]{\text{Subtract first}} \begin{cases} 2x + 3y = 12 \\ 0x + 0y = -6 \end{cases}$$

Since the last equation says $0 = -6,$ which is contradictory, there is no solution and the equations are inconsistent.

A dependent system of equations always results in one equation becoming $0x + 0y = 0.$ For instance, solving Example 3c by the elimination method could be done as follows:

$$\begin{cases} 2x + 3y = 12 \\ 4x + 6y = 24 \end{cases} \xrightarrow[\text{equation by 2}]{\text{Divide second}} \begin{cases} 2x + 3y = 12 \\ 2x + 3y = 12 \end{cases} \xrightarrow[\text{from second}]{\text{Subtract first}} \begin{cases} 2x + 3y = 12 \\ 0x + 0y = 0 \end{cases}$$

Since the last equation says $0 = 0,$ which is uninformative but not contradictory, the solutions are all the points on the line $2x + 3y = 12$ and the system is dependent.

→ **Practice Problem 3**

Solve $\begin{cases} x + y = 100 \\ x - y = -20 \end{cases}$ by the elimination method.

➤ **Solution on next page**

→ **Solutions to Practice Problems**

1. Let x be the number of pennies in the jar and y be the number of nickels. Then the first statement may be expressed as $x + y = 80$ and the second as $x + 5y = 160$ (in cents, since each penny is worth 1¢ and each nickel is worth 5¢). The system of equations is

$$\begin{cases} x + y = 80 \\ x + 5y = 160 \end{cases}$$

2. $\begin{cases} x = 6 - \frac{3}{2}(-2) = 9 \\ y = -2 \end{cases}$ Substituting these numbers into Example 3c:

$$\begin{cases} 2(9) + 3(-2) = 18 - 6 = 12 \quad\checkmark \\ 4(9) + 6(-2) = 36 - 12 = 24 \quad\checkmark \end{cases}$$

It checks!

$$3. \quad \begin{cases} x + y = 100 \\ x - y = -20 \end{cases} \xrightarrow[\text{to first}]{\text{Add second}} \begin{cases} 2x + 0y = 80 \\ x - y = -20 \end{cases} \xrightarrow[\text{by 2}]{\text{Divide first}}$$

$$\begin{cases} 1x + 0y = 40 \\ x - y = -20 \end{cases} \xrightarrow[\text{from second}]{\text{Subtract first}} \begin{cases} 1x + 0y = 40 \\ 0x - y = -60 \end{cases} \xrightarrow[\text{by } -1]{\text{Multiply second}}$$

$$\begin{cases} 1x + 0y = 40 \\ 0x + 1y = 60 \end{cases}$$

The solution is $x = 40$, $y = 60$. There are many other possible sequences of equivalent systems that solve this problem, and all reach the same conclusion.

→ 3.1 Exercises

Solve each system by graphing. If the solution is not unique, identify the system as "inconsistent" or "dependent."

1. $\begin{cases} x + y = 6 \\ x - y = 2 \end{cases}$
2. $\begin{cases} x + y = 10 \\ -x - y = 10 \end{cases}$

Solve each system by the elimination method. If the solution is not unique, identify the system as "inconsistent" or "dependent."

3. $\begin{cases} x + y = 11 \\ 2x + 3y = 30 \end{cases}$
4. $\begin{cases} 3x + y = 15 \\ x + 2y = 10 \end{cases}$
5. $\begin{cases} x + 2y = 14 \\ 3x + 4y = 36 \end{cases}$
6. $\begin{cases} 2x + 5y = 60 \\ 2x + 3y = 48 \end{cases}$
7. $\begin{cases} 3x + 4y = -24 \\ 6x + 8y = 24 \end{cases}$

Formulate each situation as a system of two linear equations in two variables. Be sure to state clearly the meaning of your x- and y-variables. Solve the system by the elimination method. Be sure to state your final answer in terms of the original question.

8. **Apartment Ownership** A lawyer has found 60 investors for a limited partnership to purchase an inner-city apartment building, with each contributing either $5000 or $10,000. If the partnership raised $430,000, then how many investors contributed $5000 and how many contributed $10,000?

9. **Coins in a Jar** A jar contains 60 nickels and dimes worth $4.30. How many of each kind of coin are in the jar?

10. **Ice Hockey Concession Receipts** The concession stand at an ice hockey rink had receipts of $7200 from selling a total of 3000 sodas and hot dogs. If each soda sold for $2 and each hot dog sold for $3, how many of each were sold?

ATTENTION
NEED MORE PRACTICE? FIND MORE HERE:
CENGAGEBRAIN.COM

3.2 Matrices and Linear Equations in Two Variables

In this section we will use matrix notation to streamline the elimination method used in the previous section to solve systems of two linear equations in two variables. In the next section we shall extend this method to the solutions of systems of many linear equations in many variables.

Matrices

A *matrix* is a rectangular array of numbers called *elements*. This rectangular array has *rows* (with the first at the top, the second below the first, and so on) and *columns* (with the first on the left, the second to the right of the first, and so on). The *dimension* of a matrix with m rows and n columns is written $m \times n$ (rows $\times$ columns). Thus, a 5×2 matrix is "tall and thin," and a 2×5 matrix is "short and wide." A *square matrix* has the same number of rows as columns. A *row matrix* has just one row, and a *column matrix* has just one column.

$$(1 \quad 2 \quad 3)$$

$$\begin{pmatrix} 1 \\ 2 \\ 3 \end{pmatrix}$$

$$\begin{pmatrix} 1 & 2 & 3 \\ 4 & 5 & 6 \\ 7 & 8 & 9 \end{pmatrix}$$

Row matrix
(dimension 1×3)

Column matrix
(dimension 3×1)

Square matrix
(dimension 3×3)

We name matrices with capital letters (A, B, C, ...). Then the elements are named by the corresponding lowercase letter together with subscripts indicating the row and then the column. For instance, if

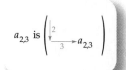

$a_{2,3}$ is

$$A = \begin{pmatrix} 1 & 2 & 3 & 4 \\ 5 & 6 & 7 & 8 \\ 9 & 10 & 11 & 12 \end{pmatrix}$$

then the dimension of A is 3×4 and $a_{2,3} = 7$ because 7 is the element in the second row and third column. This *double subscript* notation is sometimes used without the comma so that a_{23} means $a_{2,3}$ The elements on the *main diagonal* are those with the same row number as column number. For the matrix A above, the main diagonal consists of the elements $a_{1,1} = 1$, $a_{2,2} = 6$, and $a_{3,3} = 11$.

main
diagonal

Augmented Matrices from Systems of Equations

An *augmented matrix* is a matrix created from two "smaller" matrices having the same number of rows by placing them beside each other and joining them into one "larger" matrix. The system of equations

$$\begin{cases} ax + by = h \\ cx + dy = k \end{cases}$$

has a *coefficient matrix* $\begin{pmatrix} a & b \\ c & d \end{pmatrix}$ and a *constant term matrix* $\begin{pmatrix} h \\ k \end{pmatrix}$. Taken together, these

form the augmented matrix $\left(\begin{array}{cc|c} a & b & h \\ c & d & k \end{array} \right)$, separating the original matrices by a vertical

line. The augmented matrix represents the system of equations with the vertical bar representing the equals signs. The first column gives the coefficients of the x-variable, the second gives those of the y-variable, and the last column (after the bar) contains the constant terms after the equal signs.

Augmented Matrix of a System of Equations

The augmented matrix $\begin{pmatrix} a & b & | & h \\ c & d & | & k \end{pmatrix}$ represents the system of equations $\begin{cases} ax + by = h \\ cx + dy = k \end{cases}$

→ EXAMPLE 1 Augmented Matrices and Systems of Equations

a. Find the augmented matrix representing the system $\begin{cases} 2x + 3y = 24 \\ 4x + 5y = 60 \end{cases}$

b. Find the system represented by the augmented matrix $\begin{pmatrix} 6 & 8 & | & 84 \\ 4 & 5 & | & 60 \end{pmatrix}$.

Solution

a. The system $\begin{cases} 2x + 3y = 24 \\ 4x + 5y = 60 \end{cases}$ is represented by $\begin{pmatrix} 2 & 3 & | & 24 \\ 4 & 5 & | & 60 \end{pmatrix}$.

b. The augmented matrix $\begin{pmatrix} 6 & 8 & | & 84 \\ 4 & 5 & | & 60 \end{pmatrix}$ represents $\begin{cases} 6x + 8y = 84 \\ 4x + 5y = 60 \end{cases}$

→ Practice Problem 1

a. Find the augmented matrix representing the system $\begin{cases} 2x - y = 14 \\ x + 3y = 21 \end{cases}$

b. Find the system represented by the augmented matrix $\begin{pmatrix} 3 & 2 & | & 35 \\ 1 & 3 & | & 21 \end{pmatrix}$.

> **Solutions on page 102**

Fedarovich Dzmitry/Shutterstock

Row Operations

We will solve a system of equations by operating on its augmented matrix, using the following *matrix row operations* to make the solution obvious. These row operations correspond to the steps we used on pages 91–92 to solve systems of equations, and they result in a succession of *equivalent* matrices, each representing a system that has exactly the same solution as the original system.

Matrix Row Operations

1. Switch any two rows.
2. Multiply or divide one of the rows by a nonzero number.
3. Replace a row by its sum or difference with a multiple of another row.

We may apply row operations to matrices of any size. We write short formulas for row operations using R for "row" and either a double arrow $\leftrightarrow$ for "is switched with" or an arrow $\rightarrow$ for "becomes." When we write the new matrix, the formula for each new row ends with an arrow $\rightarrow$ pointing to the row that was changed.

➡ EXAMPLE 2   Performing Matrix Row Operations

Carry out the indicated row operation on the given matrix.

a. $R_1 \leftrightarrow R_2$ on $\begin{pmatrix} 2 & 3 & | & 25 \\ 3 & 4 & | & 36 \end{pmatrix}$

b. $3R_2 \rightarrow R_2$ on $\begin{pmatrix} 2 & 3 & | & 30 \\ 1 & 1 & | & 13 \end{pmatrix}$

c. $R_1 - R_2 \rightarrow R_1$ on $\begin{pmatrix} 2 & 3 & | & 30 \\ 2 & -2 & | & 20 \end{pmatrix}$

Solution

a. $R_1 \leftrightarrow R_2$ says that row 1 and row 2 are switched:

$$\begin{array}{c} R_2 \rightarrow \\ R_1 \rightarrow \end{array} \begin{pmatrix} 3 & 4 & | & 36 \\ 2 & 3 & | & 25 \end{pmatrix} \qquad \text{From } \begin{pmatrix} 2 & 3 & | & 25 \\ 3 & 4 & | & 36 \end{pmatrix}$$

b. $3R_2 \rightarrow R_2$ says that 3 times row 2 becomes row 2:

$$\begin{array}{c} \\ 3R_2 \rightarrow \end{array} \begin{pmatrix} 2 & 3 & | & 30 \\ 3 & 3 & | & 39 \end{pmatrix} \qquad \text{From } \begin{pmatrix} 2 & 3 & | & 30 \\ 1 & 1 & | & 13 \end{pmatrix}$$

c. $R_1 - R_2 \rightarrow R_1$ says that row 1 minus row 2 becomes row 1:

$$\begin{array}{c} R_1 - R_2 \rightarrow \\ \\ \end{array} \begin{pmatrix} 0 & 5 & | & 10 \\ 2 & -2 & | & 20 \end{pmatrix} \qquad \text{From } \begin{pmatrix} 2 & 3 & | & 30 \\ 2 & -2 & | & 20 \end{pmatrix}$$

➡ Practice Problem 2

a. Carry out $R_1 \leftrightarrow R_2$ on $\begin{pmatrix} 2 & 1 & | & 14 \\ 1 & -3 & | & 21 \end{pmatrix}$.

b. Carry out $R_1 + R_2 \rightarrow R_1$ on $\begin{pmatrix} 6 & 3 & | & 42 \\ 1 & -3 & | & 21 \end{pmatrix}$.

Continues

c. Carry out $\frac{1}{7}R_1 \rightarrow R_1$ on $\begin{pmatrix} 7 & 0 & | & 63 \\ 1 & -3 & | & 21 \end{pmatrix}$.

d. Is $0R_1 \rightarrow R_1$ a valid row operation?

e. Is $5R_1 \rightarrow R_2$ a valid row operation?

➤ Solutions on page 102

Solving Equations by Row Reduction

Two matrices are *equivalent* if one can be transformed into the other by a sequence of row operations. Since an augmented matrix represents a system of equations and the row operations correspond to the steps used to solve the system, we can solve the system by *row-reducing* the augmented matrix to an equivalent matrix that displays the solution. There are three possibilities. If we can obtain the form

$$\begin{pmatrix} 1 & 0 & | & p \\ 0 & 1 & | & q \end{pmatrix} \qquad \begin{cases} 1x + 0y = p \\ 0x + 1y = q \end{cases} \text{ or } \begin{cases} x = p \\ y = q \end{cases}$$

then the unique solution is $x = p$, $y = q$. If, however, we obtain a row of zeros ending with a nonzero number:

$$0 \quad 0 \mid m \qquad\qquad\qquad\qquad 0x + 0y = m, \quad m \neq 0$$

then the system is *inconsistent* and has *no solution* because the equation $0x + 0y = m$ makes the impossible claim that zero equals a nonzero number. On the other hand, if there is no such "inconsistent" row but there is a row consisting entirely of zeros:

$$0 \quad 0 \mid 0 \qquad\qquad\qquad\qquad\qquad 0x + 0y = 0$$

then the system is *dependent* and there are *infinitely many solutions* because the equation $0x + 0y = 0$ is *always* true and represents no restriction at all. These important observations will be extended in the next section.

➡ **EXAMPLE 3** Solving Equations by Row Reduction

Solve $\begin{cases} x + 3y = 15 \\ 2x - 5y = 8 \end{cases}$ by row-reducing an augmented matrix.

Solution
The augmented matrix for this system is

$$\begin{pmatrix} 1 & 3 & | & 15 \\ 2 & -5 & | & 8 \end{pmatrix}$$

We hope to change the first column from $\frac{1}{2}$ to $\frac{1}{0}$ (which is the same as removing the x-variable from the second equation) and then the second column to $\frac{0}{1}$ (removing the y-variable from the first equation). Therefore, we first want a 0 in the bottom-left corner. Subtracting twice the 1 in the top row from the 2 in the second row will achieve this 0, suggesting the row operation $R_2 - 2R_1 \to R_2$. We write "twice row 1" in small numbers above row 1 to make the subtraction of "row 2 minus twice row 1" easier:

$$\text{(Want a 0 here)} \quad \begin{matrix} 2 & 6 & 30 \end{matrix} \quad \begin{pmatrix} 1 & 3 & | & 15 \\ 2 & -5 & | & 8 \end{pmatrix} \quad R_2 - 2R_1 \to R_2 \quad \text{gives}$$

$$R_2 - 2R_1 \to \begin{pmatrix} 1 & 3 & | & 15 \\ 0 & -11 & | & -22 \end{pmatrix}$$

The bottom row has a common factor of -11, which we divide out, obtaining the simpler matrix below. We then want a 0 where the 3 is. Subtracting three times the 1 in the second row from the 3 will achieve this, suggesting the row operation $R_1 - 3R_2 \to R_1$. To help with the subtraction we write "three times row 2" in small numbers under row 2:

$$\text{(Want a 0 here)} \qquad \tfrac{1}{-11}R_2 \to \begin{pmatrix} 1 & 3 & | & 15 \\ 0 & 1 & | & 2 \end{pmatrix} \qquad R_1 - 3R_2 \to R_1 \quad \text{gives}$$

$$\begin{matrix} 0 & 3 & 6 \end{matrix}$$

$$R_1 - 3R_2 \to \begin{pmatrix} 1 & 0 & | & 9 \\ 0 & 1 & | & 2 \end{pmatrix} \qquad \text{Equivalent to} \qquad \begin{cases} 1x + 0y = 9 \\ 0x + 1y = 2 \end{cases}$$

The unique solution is $x = 9$, $y = 2$.

There are many other sequences of row operations to reduce this augmented matrix, and all lead to the same solution.

➡ Practice Problem 3

Solve $\begin{cases} 2x + y = 14 \\ x - 3y = 21 \end{cases}$ by row-reducing an augmented matrix.

➤ Solution on page 102

➡ EXAMPLE 4 Solving Equations by Row Reduction

Solve $\begin{cases} 6x - 3y = 30 \\ -8x + 4y = -40 \end{cases}$ by row-reducing an augmented matrix.

Solution
The augmented matrix is

$$\begin{pmatrix} 6 & -3 & | & 30 \\ -8 & 4 & | & -40 \end{pmatrix}$$

When the numbers in a row have an obvious common factor, removing that factor can sometimes simplify the reduction.

$$\begin{array}{l} \frac{1}{3}R_1 \rightarrow \\ \frac{1}{4}R_2 \rightarrow \end{array} \left(\begin{array}{cc|c} 2 & -1 & 10 \\ -2 & 1 & -10 \end{array} \right)$$ Want a 0 here

Adding the first row to the second, we get a zero at the bottom of the first column.

$$R_2 + R_1 \rightarrow \left(\begin{array}{cc|c} 2 & -1 & 10 \\ 0 & 0 & 0 \end{array} \right)$$ Want a 1 here

To finish, we divide the first row by 2 to make the row begin with a 1 on the left.

$$\frac{1}{2}R_1 \rightarrow \left(\begin{array}{cc|c} 1 & -\frac{1}{2} & 5 \\ 0 & 0 & 0 \end{array} \right)$$ ← a zero row

The zero row means that the system is *dependent*, so there are *infinitely many solutions*. The first row of this final matrix says that

$$x - \tfrac{1}{2}y = 5$$

or, solving for x,

$$x = 5 + \tfrac{1}{2}y$$

We may let y be *any* number t and then determine x from this equation. That is, the solutions may be parameterized as $x = 5 + \tfrac{1}{2}t$, $y = t$, where t is *any* number. The following table lists some of these solutions for various values of the parameter t.

t	$x = 5 + \tfrac{1}{2}t$	$y = t$
-20	-5	-20
-10	0	-10
0	5	0
10	10	10
20	15	20

Parameterized solution

Evaluating x and y at t

There are many other sequences of row operations to reduce this augmented matrix, and all reach the same conclusion.

➡ Practice Problem 4

Verify that $x = -5$, $y = -20$ and $x = 15$, $y = 20$ from the preceding table solve $\begin{cases} 6x - 3y = 30 \\ -8x + 4y = -40 \end{cases}$

➤ Solution on page 102

A worker in a plastics factory breaks apart sheets of component A and strips of component B and then snaps one of each together to make a finished item. If the worker can break off 20 As or 30 Bs per minute and snap together 10 pairs of As and Bs per minute, how should the worker's 440-minute workday be divided so as to complete as many items as possible with no unused pieces left over?

Solution

Let x be the number of minutes spent breaking apart sheets of component A and y be the number of minutes breaking apart strips of component B. The remainder of the worker's time, $440 - (x + y)$, will then be spent snapping As and Bs together. Because the worker needs as many As as Bs,

$$20x = 30y$$

Rate × time gives number finished

As many finished items as As will be completed:

$$20x = 10[440 - (x + y)]$$

Simplifies to
$30x + 10y = 4400$

Rewriting these as a system of equations, the problem becomes

$$\begin{cases} 20x - 30y = 0 \\ 30x + 10y = 4400 \end{cases}$$

The augmented matrix is

$$\begin{pmatrix} 20 & -30 & | & 0 \\ 30 & 10 & | & 4400 \end{pmatrix}$$

We begin the row reduction by removing from both rows the common factor of 10.

$$\begin{matrix} \frac{1}{10}R_1 \to \\ \frac{1}{10}R_2 \to \end{matrix} \begin{pmatrix} 2 & -3 & | & 0 \\ 3 & 1 & | & 440 \end{pmatrix} \quad \text{Want a 1 here}$$

To get a 1 in the upper left-hand corner:

$$R_2 - R_1 \to \begin{pmatrix} 1 & 4 & | & 440 \\ 3 & 1 & | & 440 \end{pmatrix} \quad \text{Want a 0 here}$$

To get a 0 in the lower left-hand corner:

$$3R_1 - R_2 \to \begin{pmatrix} 1 & 4 & | & 440 \\ 0 & 11 & | & 880 \end{pmatrix}$$

Removing the common factor in row 2:

$$\frac{1}{11}R_2 \to \begin{pmatrix} 1 & 4 & | & 440 \\ 0 & 1 & | & 80 \end{pmatrix} \quad \text{Want a 0 here}$$

To get a 0 above the 1 in the second row:

$$R_1 - 4R_2 \to \begin{pmatrix} 1 & 0 & | & 120 \\ 0 & 1 & | & 80 \end{pmatrix} \quad \begin{matrix} x = 120 \\ y = 80 \end{matrix}$$

The solution is $x = 120$, $y = 80$. In terms of the original question, the worker should break apart sheets of component A for 120 minutes, break apart strips of component B for 80 minutes, and snap As and Bs together for the remaining 240 minutes.

→ Solutions to Practice Problems

1. a. $\begin{pmatrix} 2 & -1 & | & 14 \\ 1 & 3 & | & 21 \end{pmatrix}$ **b.** $\begin{cases} 3x + 2y = 35 \\ x + 3y = 21 \end{cases}$

2. a. $\begin{matrix} R_2 \to \\ R_1 \to \end{matrix} \begin{pmatrix} 1 & -3 & | & 21 \\ 2 & 1 & | & 14 \end{pmatrix}$ **b.** $\begin{matrix} R_1 + R_2 \to \\ \\ \end{matrix} \begin{pmatrix} 7 & 0 & | & 63 \\ 1 & -3 & | & 21 \end{pmatrix}$

c. $\begin{matrix} \frac{1}{7} R_1 \to \\ \\ \end{matrix} \begin{pmatrix} 1 & 0 & | & 9 \\ 1 & -3 & | & 21 \end{pmatrix}$

d. No; you can multiply only by a *nonzero* number.

e. No; multiplying row 1 by 5 must still give row 1 (*not* row 2).

3. Starting with the augmented matrix $\begin{pmatrix} 2 & 1 & | & 14 \\ 1 & -3 & | & 21 \end{pmatrix}$, one possible sequence of row operations is

$$R_1 - R_2 \to \begin{pmatrix} 1 & 4 & | & -7 \\ 1 & -3 & | & 21 \end{pmatrix}$$

$$R_1 - R_2 \to \begin{pmatrix} 1 & 4 & | & -7 \\ 0 & 7 & | & -28 \end{pmatrix}$$

$$\tfrac{1}{7} R_2 \to \begin{pmatrix} 1 & 4 & | & -7 \\ 0 & 1 & | & -4 \end{pmatrix}$$

$$R_1 - 4R_2 \to \begin{pmatrix} 1 & 0 & | & 9 \\ 0 & 1 & | & -4 \end{pmatrix}$$

The unique solution is $x = 9$, $y = -4$. There are many other sequences of row operations to reduce this augmented matrix, and all reach the same conclusion.

4. For $x = -5$, $y = -20$, the equations become

$$\begin{cases} 6(-5) - 3(-20) = -30 + 60 = 30 \ ✔ \\ -8(-5) + 4(-20) = 40 - 80 = -40 \ ✔ \end{cases}$$

For $x = 15$, $y = 20$, the equations become

$$\begin{cases} 6(15) - 3(20) = 90 - 60 = 30 \ ✔ \\ -8(15) + 4(20) = -120 + 80 = -40 \ ✔ \end{cases}$$

Both check!

→ 3.2 Exercises

Find the dimension of each matrix and the values of the specified elements.

1. $\begin{pmatrix} 1 & 2 \\ 2 & 5 \\ 3 & 6 \end{pmatrix}$; $a_{1,1}$, $a_{3,2}$

2. $\begin{pmatrix} 1 & -1 & 2 & -2 \\ -3 & 3 & -4 & 4 \\ 5 & -5 & 6 & -6 \end{pmatrix}$; $a_{2,2}$, $a_{3,4}$

Find the augmented matrix representing the system of equations.

3. $\begin{cases} x + 2y = 2 \\ 3x + 4y = 12 \end{cases}$

4. $\begin{cases} 3x - 2y = 24 \\ x = 6 \end{cases}$

Carry out the row operation on the matrix.

5. $R_1 \leftrightarrow R_2$ on $\begin{pmatrix} 3 & 4 & | & 24 \\ 5 & 6 & | & 30 \end{pmatrix}$

6. $R_1 - R_2 \rightarrow R_1$ on $\begin{pmatrix} 8 & 7 & | & 56 \\ 6 & 5 & | & 60 \end{pmatrix}$

7. $\frac{1}{8}R_2 \rightarrow R_2$ on $\begin{pmatrix} 1 & -2 & | & -42 \\ 0 & 8 & | & 120 \end{pmatrix}$

Interpret each augmented matrix as the solution of a system of equations. State the solution or identify the system as "inconsistent" or "dependent."

8. $\begin{pmatrix} 1 & 0 & | & 7 \\ 0 & 1 & | & -3 \end{pmatrix}$

9. $\begin{pmatrix} 1 & 1 & | & 0 \\ 0 & 0 & | & 1 \end{pmatrix}$

10. $\begin{pmatrix} 1 & 2 & | & 3 \\ 0 & 0 & | & 0 \end{pmatrix}$

Solve each system by row-reducing the corresponding augmented matrix. State the solution or identify the system as "inconsistent" or "dependent."

11. $\begin{cases} 2x + y = 4 \\ x + y = 3 \end{cases}$

12. $\begin{cases} -x + 2y = 4 \\ x - 2y = 6 \end{cases}$

13. $\begin{cases} 2x - 6y = 18 \\ -3x + 9y = -27 \end{cases}$

Express each situation as a system of two equations in two variables. Be sure to state clearly the meaning of your x- and y-variables. Solve the system by row-reducing the corresponding augmented matrix. State your final answer in terms of the original question.

14. **Commodity Futures** A corn and soybean commodities speculator invested $15,000 yesterday with twice as much in soybean futures as in corn futures. How much did she invest in each?

15. **Political Advertising** For the final days before the election, the campaign manager has a total of $36,000 to spend on TV and radio campaign advertisements. Each TV ad costs $3000 and is seen by 10,000 voters, while each radio ad costs $500 and is heard by 2000 voters. Ignoring repeated exposures to the same voter, how many TV and radio ads will contact 130,000 voters using the allocated funds?

3.3 Systems of Linear Equations and the Gauss–Jordan Method

In this section we use augmented matrices to solve larger systems of linear equations. We simply enlarge the augmented matrix to allow for more equations (rows) and more variables (columns) and then apply row operations to find an equivalent matrix that displays the solution.

Names for Many Variables

To deal with many variables, we now distinguish them by subscripts instead of different letters: x_1 ("x sub one"), x_2 ("x sub two"), and so on for as many as we need. For example,

$$\begin{cases} 3x_1 + 2x_2 + x_3 = 39 \\ 2x_1 + 3x_2 + x_3 = 34 \\ x_1 + 2x_2 + 3x_3 = 26 \end{cases}$$

We form the augmented matrix exactly as before, so for this system we have

$$\begin{array}{ccc} x_1 & x_2 & x_3 \end{array}$$
$$\left(\begin{array}{ccc|c} 3 & 2 & 1 & 39 \\ 2 & 3 & 1 & 34 \\ 1 & 2 & 3 & 26 \end{array} \right)$$

← Variables corresponding to columns (last column represents constant terms)

← Each row represents an equation and the bar represents the equals sign

Row-Reduced Form

We continue to use row operations to "solve" the augmented matrix by finding an equivalent matrix that displays the solution. With our examples of 2×3 matrices from the previous section as a guide, we make the following definition for matrices of any dimension. In this definition a *zero row* is a row containing only zeros and a *nonzero row* is a row with at least one nonzero element.

Kalim/Shutterstock

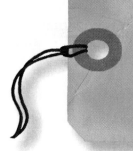

Row-Reduced Form

A matrix is in *row-reduced form* if it satisfies the following four conditions.

1. All zero rows are grouped below the nonzero rows.
2. The first nonzero element in each nonzero row is a 1.
   We call these particular 1s "leftmost 1s."
3. If a column contains a leftmost 1, all other entries in the column are 0s.
4. Each leftmost 1 appears to the *right* of the leftmost 1 in the row above it.

The following matrices are in row-reduced form:

$$\begin{pmatrix} 0 & 1 & 2 & 0 & 1 \\ 0 & 0 & 0 & 1 & 4 \end{pmatrix}, \quad \begin{pmatrix} 1 & 2 & 3 & 0 \\ 0 & 0 & 0 & 1 \\ 0 & 0 & 0 & 0 \end{pmatrix}, \quad \text{and} \quad \begin{pmatrix} 1 & 0 & 0 & 0 & 1 \\ 0 & 1 & 0 & 0 & 2 \\ 0 & 0 & 1 & 0 & 3 \\ 0 & 0 & 0 & 1 & 4 \end{pmatrix}$$

Leftmost 1s   Zero row

Notice that the 1s in the upper right corners of the first and last matrices above are not *leftmost* 1s and do not need to have 0s below them. The following matrix is *not* in row-reduced form because although conditions (1), (2), and (3) are satisfied, condition (4) fails for the third row.

$$\begin{pmatrix} 1 & 0 & 0 & 2 \\ 0 & 0 & 1 & 4 \\ 0 & 1 & 0 & 6 \end{pmatrix}$$

Not row-reduced:

The 1 in the third row is not to the right of the 1 in the second row

---

→ Practice Problem 1

Find a row operation that will correct this defect and result in a matrix in row-reduced form.

> **Solution on page 112**

In many augmented matrices the row-reduced form will be particularly simple: to the left of the bar will be a square matrix with 1s down the main diagonal and 0s elsewhere, as shown below. In such cases the system has a *unique solution* and we may immediately read off the values of the variables from the numbers to the right of the bar (top to bottom):

$$\begin{pmatrix} 1 & 0 & 0 & 5 \\ 0 & 1 & 0 & -2 \\ 0 & 0 & 1 & 3 \end{pmatrix} \quad \text{gives} \quad \begin{cases} x_1 = 5 \\ x_2 = -2 \\ x_3 = 3 \end{cases} \quad \text{From} \quad \begin{matrix} 1x_1 + 0x_2 + 0x_3 = 5 \\ 0x_1 + 1x_2 + 0x_3 = -2 \\ 0x_1 + 0x_2 + 1x_3 = 3 \end{matrix}$$

This is the pattern of 1s and 0s we seek because it leads to a unique solution. If another pattern occurs, we will see how to interpret it as indicating either no solution or infinitely many solutions. We will row-reduce augmented matrices by following a procedure called the Gauss–Jordan method. Named after mathematician Carl Friedrich Gauss (1777–1855) and engineer and surveyor Wilhelm Jordan (1842–1899), the Gauss–Jordan method operates on one column at a time, first obtaining a 1 and then 0s above and below it.

Gauss–Jordan Method

To row-reduce a matrix:

For instance, you might get:

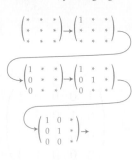

1. If any rows have leading 0s, then switch the rows so that the rows with the *fewest* leading 0s are at the top, down to the rows with the *most* leading 0s at the bottom.
2. In the first row, find the leftmost nonzero entry and divide the row through by that number. This step gives a *leftmost 1* in that row.
3. Add or subtract multiples of the first row to each other row to obtain 0s in the rest of the column above and below the leftmost 1 found in step 2.
4. Repeat steps 1, 2, and 3 but replacing "first row" by "second row" and then by "third row" and so on. Stop when you reach the bottom row or a row consisting entirely of zeros; when that happens, the matrix is *row-reduced*.

Just as in the last section, there are three possibilities: a unique solution, no solution, or infinitely many solutions that can be parameterized.

→ EXAMPLE 1   Solving a System Using the Gauss–Jordan Method

Solve by the Gauss–Jordan method:
$$\begin{cases} 2x_1 + 4x_2 + 2x_3 = 2 \\ 3x_1 + 7x_2 + 3x_3 = 0 \\ x_1 + 2x_2 + 4x_3 = 7 \end{cases}$$

Solution

Beginning with the augmented matrix, we carry out steps 1 through 4 of the Gauss–Jordan method.

$$\left(\begin{array}{ccc|c} 2 & 4 & 2 & 2 \\ 3 & 7 & 3 & 0 \\ 1 & 2 & 4 & 7 \end{array}\right)$$
Want a 1 here

Step 1 concerns leading 0s, and there are none. Step 2 says to divide row 1 by the first nonzero element, which is 2, to get a leftmost 1. The following row operation achieves this.

$$\tfrac{1}{2}R_1 \rightarrow \left(\begin{array}{ccc|c} 1 & 2 & 1 & 1 \\ 3 & 7 & 3 & 0 \\ 1 & 2 & 4 & 7 \end{array}\right)$$
Want 0s here

Step 3 says to subtract multiples of row 1 from the other rows to get 0s below the leftmost 1. The following two row operations do this.

$$\begin{array}{c} R_2 - 3R_1 \rightarrow \\ R_3 - R_1 \rightarrow \end{array} \left(\begin{array}{ccc|c} 1 & 2 & 1 & 1 \\ 0 & 1 & 0 & -3 \\ 0 & 0 & 3 & 6 \end{array} \right)$$ Want a 0 here

Row 2 already has a leftmost 1, so to achieve the 0 above it, in the next step we subtract twice row 2.

$$R_1 - 2R_2 \rightarrow \left(\begin{array}{ccc|c} 1 & 0 & 1 & 7 \\ 0 & 1 & 0 & -3 \\ 0 & 0 & 3 & 6 \end{array} \right)$$ Want a 1 here

Step 2 then says to divide row 3 by its first nonzero element, 3, to obtain a leftmost 1 as follows.

$$\begin{array}{c} \\ \\ \tfrac{1}{3}R_3 \rightarrow \end{array} \left(\begin{array}{ccc|c} 1 & 0 & 1 & 7 \\ 0 & 1 & 0 & -3 \\ 0 & 0 & 1 & 2 \end{array} \right)$$ Want a 0 here

The next step subtracts row 3 from row 1, achieving the final zero and completing the row reduction of the augmented matrix.

$$R_1 - R_3 \rightarrow \left(\begin{array}{ccc|c} 1 & 0 & 0 & 5 \\ 0 & 1 & 0 & -3 \\ 0 & 0 & 1 & 2 \end{array} \right)$$

The solution is $\begin{cases} x_1 = 5 \\ x_2 = -3 \\ x_3 = 2 \end{cases}$

Reading the numbers from the last column

The system in Example 1 had a unique solution. If there is an "inconsistent" row, then the system will have *no solutions*, just as in the previous section.

No Solutions

If a row-reduced matrix has a row of 0s ending in a 1 (such as $\begin{array}{ccc|c} 0 & 0 & 0 & 1 \end{array}$), then the system of equations from which it came is *inconsistent* and has *no solutions*.

➡ EXAMPLE 2   A System with No Solution

Solve by the Gauss–Jordan method: $\begin{cases} 2x_1 - 4x_2 = 2 \\ -3x_1 + 6x_2 = 4 \end{cases}$

Solution

The augmented matrix is

$$\left(\begin{array}{cc|c} 2 & -4 & 2 \\ -3 & 6 & 4 \end{array} \right)$$ Want a 1 here

There are no leading 0s, so we go to step 2, dividing row 1 by the first nonzero element, 2, as follows.

$$\tfrac{1}{2}R_1 \to \begin{pmatrix} 1 & -2 & | & 1 \\ -3 & 6 & | & 4 \end{pmatrix}$$

Want a 0 here

Step 3 says to add or subtract a multiple of row 1 to row 2 to get a zero, so in the next matrix we add three times row 1.

$$R_2 + 3R_1 \to \begin{pmatrix} 1 & -2 & | & 1 \\ 0 & 0 & | & 7 \end{pmatrix}$$

Want a 1 here

Step 2 says that to get a leading 1 in row 2, divide the row by 7, leading to the following matrix.

$$\tfrac{1}{7}R_2 \to \begin{pmatrix} 1 & -2 & | & 1 \\ 0 & 0 & | & 1 \end{pmatrix}$$

An "inconsistent" row

Row 2 is of the form $\quad 0 \ 0 \ | \ 1$ saying that $\quad 0 = 1,\quad$ which means that the system has *no solutions.*

The system is inconsistent and has no solutions.

Once we encountered the line $\quad 0 \ 0 \ | \ 7,\quad$ it was clear that by dividing by 7 we would obtain the "inconsistent" row $\quad 0 \ 0 \ | \ 1.\quad$ Therefore, in practice, whenever we encounter a row of 0s ending with any nonzero number after the bar we may stop at that point and declare that the system is inconsistent and has no solutions.

> ### ➡ Practice Problem 2
>
> Solve by the Gauss–Jordan method: $\quad \begin{cases} 3x_1 - 6x_2 = 12 \\ -5x_1 + 10x_2 = -14 \end{cases}$
>
> ➤ Solution on page 112

The rule for infinitely many solutions is a little more complicated than in the case of two equations. It depends on the number of leftmost 1s compared to the number of variables.

Infinitely Many Solutions

For a row-reduced augmented matrix with no "inconsistent" rows, if the number of leading 1s is *less than* the number of variables, then the original system is *dependent* and has *infinitely many solutions.*

If there are infinitely many solutions, then we may *parameterize* them as we did on pages 99–100. First we classify each variable as *determined* or *free* by looking at the column in the row-reduced augmented matrix corresponding to that variable:

The variable is *determined* if its column has a leftmost 1.

The variable is *free* if its column does *not* have a leftmost 1.

For example, given the following row-reduced matrix (with two columns to the left of the bar corresponding to two variables x_1 and x_2), the first variable x_1 is *determined* and the second variable x_2 is *free*:

x_1 is *determined* (its column has a leftmost 1)

x_2 is *free* (its column does not have a leftmost 1)

$$\begin{pmatrix} 1 & 2 & | & 3 \\ 0 & 0 & | & 0 \end{pmatrix}$$

$$x_1 + 2x_2 = 3$$
$$\text{so} \quad x_1 = 3 - 2x_2$$

That is, x_2 is *free to take any value t*, while the value of x_1 is *determined* by the equation $x_1 = 3 - 2x_2$, giving the parameterized solution (replacing x_2 by t):

$$\begin{cases} x_1 = 3 - 2t \\ x_2 = t \end{cases}$$

From $x_1 = 3 - 2x_2$
with x_2 replaced by t

Free variables are sometimes called *independent*, and determined variables are then called *dependent*.

➡ **EXAMPLE 3** **A System with Infinitely Many Solutions**

Solve by the Gauss–Jordan method:
$$\begin{cases} 2x_1 + 6x_2 + 10x_3 = 8 \\ 3x_1 + 9x_2 + 15x_3 = 12 \\ 2x_1 + 5x_2 + 8x_3 = 7 \end{cases}$$

Solution
The augmented matrix is

$$\begin{pmatrix} 2 & 6 & 10 & | & 8 \\ 3 & 9 & 15 & | & 12 \\ 2 & 5 & 8 & | & 7 \end{pmatrix}$$

Want a 1 here

Since there are no leading 0s, we go to step 2, which says to divide the first row by 2, as follows.

$$\tfrac{1}{2}R_1 \to \begin{pmatrix} 1 & 3 & 5 & | & 4 \\ 3 & 9 & 15 & | & 12 \\ 2 & 5 & 8 & | & 7 \end{pmatrix}$$

Want 0s here

To "zero-out" the rest of the column, in the next matrix we subtract appropriate multiples of the first row.

$$\begin{matrix} \\ R_2 - 3R_1 \to \\ R_3 - 2R_1 \to \end{matrix} \begin{pmatrix} 1 & 3 & 5 & | & 4 \\ 0 & 0 & 0 & | & 0 \\ 0 & -1 & -2 & | & -1 \end{pmatrix}$$

Has more leading 0s than

Since row 2 has more leading 0s that row 3, step 1 says to switch rows 2 and 3, as we now do.

$$\begin{matrix} \\ R_3 \to \\ R_2 \to \end{matrix} \begin{pmatrix} 1 & 3 & 5 & | & 4 \\ 0 & -1 & -2 & | & -1 \\ 0 & 0 & 0 & | & 0 \end{pmatrix}$$

Want a 1 here

Now divide by -1 (which is equivalent to multiplying by -1) to get a leftmost 1 in the second row, as follows.

$$-1R_2 \to \begin{pmatrix} 1 & 3 & 5 & | & 4 \\ 0 & 1 & 2 & | & 1 \\ 0 & 0 & 0 & | & 0 \end{pmatrix}$$

Want a 0 here

Using the 1 to "zero-out" the rest of the column gives the matrix on the next page.

$$R_1 - 3R_2 \rightarrow \begin{pmatrix} 1 & 0 & -1 & | & 1 \\ 0 & 1 & 2 & | & 1 \\ 0 & 0 & 0 & | & 0 \end{pmatrix}$$

Row-reduced!

The corresponding equations are $\begin{cases} x_1 \quad\;\; -x_3 = 1 \\ \quad\; x_2 + 2x_3 = 1 \end{cases}$

Since the variables x_1 and x_2 are *determined* (their columns *do* have leftmost 1s) and the variable x_3 is *free* (column 3 does not have a leading 1), the free variable may take *any* value, $x_3 = t$, and we solve the other equations for the *determined* variables x_1 and x_2. The solution is

$$\begin{cases} x_1 = 1 + t \\ x_2 = 1 - 2t \\ x_3 = t \end{cases}$$

From solving $\quad x_1 - x_3 = 1 \quad$ for x_1
From solving $\quad x_2 + 2x_3 = 1 \quad$ for x_2
and then replacing x_3 by t

We obtain all solutions by taking different values for t. For example,

$$t = 3 \quad \text{gives} \quad \begin{cases} x_1 = 4 \\ x_2 = -5 \\ x_3 = 3 \end{cases} \quad \text{and} \quad t = -2 \quad \text{gives} \quad \begin{cases} x_1 = -1 \\ x_2 = 5 \\ x_3 = -2 \end{cases}$$

It is easily checked that these are solutions to the original equations.

→ **Practice Problem 3**

Find the solution of a system of equations with augmented matrix equivalent to the row-reduced matrix

$$\begin{pmatrix} 1 & 2 & 0 & 0 & | & 0 \\ 0 & 0 & 1 & 0 & | & 3 \\ 0 & 0 & 0 & 1 & | & 4 \\ 0 & 0 & 0 & 0 & | & 0 \end{pmatrix}$$

➤ Solution on page 112

→ **EXAMPLE 4 Managing Production**

A pottery shop manufactures hand-thrown plates, cups, and vases. Each plate requires 4 ounces of clay, 6 minutes of shaping, and 5 minutes of painting; each cup requires 4 ounces of clay, 5 minutes of shaping, and 3 minutes of painting; and each vase requires 3 ounces of clay, 4 minutes of shaping, and 4 minutes of painting. This week the shop has 165 pounds of clay, 59 hours of skilled labor for shaping, and 46 hours of skilled labor for painting. If the shop manager wishes to use all these resources fully, how many of each product should the shop produce?

Laurent Renault/Shutterstock

Let x_1, x_2, and x_3 be the numbers of plates, cups, and vases produced. The number of ounces of clay required is then $4x_1 + 4x_2 + 3x_3$, and this must match the 165 pounds available:

	Plates	Cups	Vases
Clay	4	4	3
Shaping	6	5	4
Painting	5	3	4

$$4x_1 + 4x_2 + 3x_3 = 2640$$

Use ounces on both sides of the equation

Similarly, for the time in minutes required for shaping and painting,

$$6x_1 + 5x_2 + 4x_3 = 3540$$
$$5x_1 + 3x_2 + 4x_3 = 2760$$

Use minutes on both sides of the equation

Therefore, the augmented matrix is

$$\begin{pmatrix} 4 & 4 & 3 & | & 2640 \\ 6 & 5 & 4 & | & 3540 \\ 5 & 3 & 4 & | & 2760 \end{pmatrix}$$

We now solve this problem by row-reducing the augmented matrix using the Gauss–Jordan method.

$$\tfrac{1}{4}R_1 \to \begin{pmatrix} 1 & 1 & \tfrac{3}{4} & | & 660 \\ 6 & 5 & 4 & | & 3540 \\ 5 & 3 & 4 & | & 2760 \end{pmatrix}$$

Get a 1 in the first column of the first row

$$\begin{array}{c} \\ R_2 - 6R_1 \to \\ R_3 - 5R_1 \to \end{array} \begin{pmatrix} 1 & 1 & \tfrac{3}{4} & | & 660 \\ 0 & -1 & -\tfrac{1}{2} & | & -420 \\ 0 & -2 & \tfrac{1}{4} & | & -540 \end{pmatrix}$$

"Zero-out" the rest of the first column

$$-R_2 \to \begin{pmatrix} 1 & 1 & \tfrac{3}{4} & | & 660 \\ 0 & 1 & \tfrac{1}{2} & | & 420 \\ 0 & -2 & \tfrac{1}{4} & | & -540 \end{pmatrix}$$

Get a 1 in the second column of the second row

$$\begin{array}{c} R_1 - R_2 \to \\ \\ R_3 + 2R_2 \to \end{array} \begin{pmatrix} 1 & 0 & \tfrac{1}{4} & | & 240 \\ 0 & 1 & \tfrac{1}{2} & | & 420 \\ 0 & 0 & \tfrac{5}{4} & | & 300 \end{pmatrix}$$

"Zero-out" the rest of the second column

$$\tfrac{4}{5}R_3 \to \begin{pmatrix} 1 & 0 & \tfrac{1}{4} & | & 240 \\ 0 & 1 & \tfrac{1}{2} & | & 420 \\ 0 & 0 & 1 & | & 240 \end{pmatrix}$$

Get a 1 in the third column of the third row

$$\begin{array}{c} R_1 - \tfrac{1}{4}R_3 \to \\ R_2 - \tfrac{1}{2}R_3 \to \\ \\ \end{array} \begin{pmatrix} 1 & 0 & 0 & | & 180 \\ 0 & 1 & 0 & | & 300 \\ 0 & 0 & 1 & | & 240 \end{pmatrix}$$

"Zero-out" the rest of the third column

The system has a unique solution: $x_1 = 180$, $x_2 = 300$, $x_3 = 240$. You should check that these values satisfy the original equations and that the clay and skilled labor resources are fully used. In terms of the original question, the shop should produce 180 plates, 300 cups, and 240 vases this week.

The Three Possibilities in Pictures

We can show the three possibilities (unique solution, no solutions, or infinitely many solutions) as 3-dimensional graphs. Linear equations in three variables represent planes in space, and three planes can intersect at a single point as on the left below (unique solution), not at all as in the middle (no solutions), or at infinitely many points as on the right (infinitely many solutions).

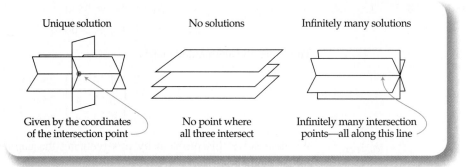

Unique solution   No solutions   Infinitely many solutions

Given by the coordinates of the intersection point   No point where all three intersect   Infinitely many intersection points—all along this line

→ Solutions to Practice Problems

1. $R_2 \leftrightarrow R_3$

2.
$$\begin{pmatrix} 3 & -6 & | & 12 \\ -5 & 10 & | & -14 \end{pmatrix}$$

$$\tfrac{1}{3}R_1 \to \begin{pmatrix} 1 & -2 & | & 4 \\ -5 & 10 & | & -14 \end{pmatrix}$$

$$R_2 + 5R_1 \to \begin{pmatrix} 1 & -2 & | & 4 \\ 0 & 0 & | & 6 \end{pmatrix}$$

Inconsistent, so *no* solution.

3. x_1, x_3, and x_4 are determined, but x_2 is free. The solution is
$$\begin{cases} x_1 = -2t \\ x_2 = t \\ x_3 = 3 \\ x_4 = 4 \end{cases}$$

→ 3.3 Exercises

Find the augmented matrix representing the system of equations.

1. $\begin{cases} x_1 + x_2 + x_3 = 4 \\ x_1 + 2x_2 + x_3 = 3 \\ x_1 + 2x_2 + 2x_3 = 5 \end{cases}$

2. $\begin{cases} 2x_1 + x_2 + 5x_3 + 4x_4 + 5x_5 = 2 \\ x_1 + x_2 + 3x_3 + 3x_4 + 3x_5 = -1 \end{cases}$

Interpret each row-reduced matrix as the solution of a system of equations.

3. $\begin{pmatrix} 1 & 0 & 0 & | & 4 \\ 0 & 1 & 0 & | & 5 \\ 0 & 0 & 1 & | & -4 \end{pmatrix}$

4. $\begin{pmatrix} 1 & 0 & 1 & | & 0 \\ 0 & 1 & 0 & | & 0 \\ 0 & 0 & 0 & | & 1 \end{pmatrix}$

5. $\begin{pmatrix} 1 & 0 & -1 & | & -5 \\ 0 & 1 & 1 & | & 5 \\ 0 & 0 & 0 & | & 0 \end{pmatrix}$

Use an appropriate row operation or sequence of row operations to find the equivalent row-reduced matrix.

6. $\begin{pmatrix} 0 & 1 & 0 & | & 2 \\ 1 & 0 & 0 & | & 1 \\ 0 & 0 & 1 & | & 3 \end{pmatrix}$ 7. $\begin{pmatrix} 1 & 0 & 1 & | & 4 \\ 0 & 1 & 0 & | & 2 \\ 0 & 0 & 1 & | & 3 \end{pmatrix}$

Solve each system of equations by the Gauss–Jordan method. If the solution is not unique, identify the system as "dependent" or "inconsistent."

8. $\begin{cases} x_1 + x_2 + x_3 = 2 \\ x_1 + 2x_2 + 2x_3 = 3 \\ x_1 + 3x_2 + 2x_3 = 1 \end{cases}$

9. $\begin{cases} 2x_1 + 3x_2 + x_3 = 4 \\ 3x_1 + 5x_2 + 2x_3 = 12 \\ x_1 + 2x_2 + x_3 = 3 \end{cases}$

10. $\begin{cases} 4x_1 + 3x_2 + 2x_3 = 24 \\ x_1 + x_2 + 3x_3 = 7 \\ 5x_1 + 4x_2 + 5x_3 = 31 \end{cases}$

Formulate each situation as a system of linear equations. Be sure to state clearly the meaning of each variable. Solve using the Gauss–Jordan method. State your final answer in terms of the original question.

11. *Plant Fertilizer* A large commercial farm needs 35,000 pounds of potash, 68,000 pounds of nitrogen, and 25,000 pounds of phosphoric acid. Three brands of fertilizer, GrowRite, MiracleMix, and GreatGreen, are available and contain the amounts of potash, nitrogen, and phosphoric acid per truckload listed in the table. How many truckloads of each brand should be used to provide the required potash, nitrogen, and phosphoric acid?

(Pounds per truckload)	GrowRite	MiracleMix	GreatGreen
Potash	400	600	700
Nitrogen	500	1000	1600
Phosphoric acid	300	400	500

12. *Apparel Production*

a. A "limited edition" ladies' fashion shop has a 240-yard supply of a silk fabric suitable for scarves, dresses, blouses, and skirts. Each scarf requires 1 yard of material, 3 minutes of cutting, and 5 minutes of sewing; each dress requires 3 yards of material, 14 minutes of cutting, and 40 minutes of sewing; each blouse requires 1.5 yards of material, 9 minutes of cutting, and 30 minutes of sewing; and each skirt requires 2 yards of material, 8 minutes of cutting, and 20 minutes of sewing. If the shop has 17 hours of skilled pattern-cutter labor and 45 hours of skilled seamstress labor available, how many of each item can be made?

b. If 20 blouses and 10 skirts are made, how many scarves and dresses can be made?

3.4 Matrix Arithmetic

Our focus thus far has been on the augmented matrix representation of a system of equations. This representation is "unnatural" in the sense that it somehow "contains" but does not show the equal signs from the equations. We shall now define the arithmetic of matrices so that we may write an entire system of equations as a single *matrix equation.* In the next section the solution of such a matrix equation will deepen our understanding of what it means to solve a system of equations.

Equality of Matrices

Two matrices are *equal* if they have the same dimension and if elements in corresponding locations are equal; that is, $A = B$ if $a_{i,j} = b_{i,j}$ for every row i and column j. For instance, the following matrices are equal:

$$\begin{pmatrix} 1 & 2 \\ 3 & 4 \end{pmatrix} = \begin{pmatrix} 1 & 1+1 \\ 4-1 & 2^2 \end{pmatrix}$$

Equal: Same dimension and corresponding values are equal

The following matrices are *not* equal:

$$\begin{pmatrix} 1 & 2 \\ 3 & 4 \end{pmatrix} \neq \begin{pmatrix} 1 \\ 2 \\ 3 \\ 4 \end{pmatrix}$$

Not equal because their dimensions are different

$$\begin{pmatrix} 1 & 2 \\ 3 & 4 \end{pmatrix} \neq \begin{pmatrix} 1 & 2 \\ 4 & 3 \end{pmatrix}$$

Not equal because some corresponding values differ

Transpose of a Matrix

The *transpose* of a matrix A is the matrix A^t formed by turning each row into a column (or, equivalently, each column into a row). For example,

$$\begin{pmatrix} 1 & 2 & 3 \\ 4 & 5 & 6 \end{pmatrix}^t = \begin{pmatrix} 1 & 4 \\ 2 & 5 \\ 3 & 6 \end{pmatrix}$$

Transpose: First row becomes first column
Second row becomes second column

Identity Matrix

An *identity matrix* is a square matrix with 1s on the main diagonal and 0s elsewhere. We write $I = I_n$ for the $n \times n$ identity matrix. The identity matrices $I_1, I_2, I_3,$ and I_4 are written as

$$I_1 = (1) \quad I_2 = \begin{pmatrix} 1 & 0 \\ 0 & 1 \end{pmatrix} \quad I_3 = \begin{pmatrix} 1 & 0 & 0 \\ 0 & 1 & 0 \\ 0 & 0 & 1 \end{pmatrix} \quad I_4 = \begin{pmatrix} 1 & 0 & 0 & 0 \\ 0 & 1 & 0 & 0 \\ 0 & 0 & 1 & 0 \\ 0 & 0 & 0 & 1 \end{pmatrix}$$

Scalar Multiplication

Scalar multiplication of a matrix simply means multiplying each element of a matrix by the same number. ("Scalar" is just another word for "number" and is used to distinguish a number from a matrix.) In particular, the *negative* of a matrix is $-A = -1 \cdot A$ and is found by multiplying each element by –1. For example, if

$$A = \begin{pmatrix} 1 & 2 & 3 \\ 4 & 5 & 6 \end{pmatrix}$$

then

$$3A = \begin{pmatrix} 3 & 6 & 9 \\ 12 & 15 & 18 \end{pmatrix} \quad \text{and} \quad -A = \begin{pmatrix} -1 & -2 & -3 \\ -4 & -5 & -6 \end{pmatrix}$$

Scalar Product

The product of a scalar (a number) times a matrix, kA, is the matrix A with each element multiplied by k.

Matrix Addition and Subtraction

For two matrices with the same dimensions, *matrix addition* means adding elements in corresponding locations.

$$\begin{pmatrix} 1 & 4 \\ 2 & 5 \\ 3 & 6 \end{pmatrix} + \begin{pmatrix} 11 & 12 \\ 10 & 9 \\ 7 & 8 \end{pmatrix} = \begin{pmatrix} 12 & 16 \\ 12 & 14 \\ 10 & 14 \end{pmatrix}$$

Matrices with *different* dimensions cannot be added.

$$\begin{pmatrix} 1 & 4 \\ 2 & 5 \\ 3 & 6 \end{pmatrix} + \begin{pmatrix} 7 & 8 & 9 \\ 10 & 11 & 12 \end{pmatrix} \qquad \text{Is not possible}$$

Similarly, two matrices with the same dimensions can be *subtracted* by subtracting corresponding elements (or, equivalently, adding the negative of the second matrix):

$$\begin{pmatrix} 11 & 12 \\ 10 & 9 \\ 7 & 8 \end{pmatrix} - \begin{pmatrix} 1 & 4 \\ 2 & 5 \\ 3 & 6 \end{pmatrix} = \begin{pmatrix} 10 & 8 \\ 8 & 4 \\ 4 & 2 \end{pmatrix}$$

Addition and Subtraction of Matrices

For two matrices A and B of the same dimensions, their sum $A + B$ and difference $A - B$ are found by adding or subtracting corresponding elements. Matrices of different dimensions *cannot* be added or subtracted.

A *zero matrix* has all elements equal to zero. We write 0 for the zero matrix of whatever size is appropriate for the situation. If two matrices are equal, then their difference is a zero matrix: that is, if $A = B$, then $A - B = 0$ means that 0 is the zero matrix of the same dimension as A and B.

$$\begin{pmatrix} 1 & 2 \\ 3 & 4 \\ 5 & 6 \end{pmatrix} - \begin{pmatrix} 1 & 2 \\ 3 & 4 \\ 5 & 6 \end{pmatrix} = 0 \quad \text{means} \quad \begin{pmatrix} 1 & 2 \\ 3 & 4 \\ 5 & 6 \end{pmatrix} - \begin{pmatrix} 1 & 2 \\ 3 & 4 \\ 5 & 6 \end{pmatrix} = \begin{pmatrix} 0 & 0 \\ 0 & 0 \\ 0 & 0 \end{pmatrix}$$

→ **EXAMPLE 1** Using Matrix Arithmetic

A real estate company buying cars for its sales staff wants to equip some with navigation systems, DVD players, and satellite radios. AutoTech Inc. sells these products for, respectively, $500, $150, and $400 each, while SmartCars Inc. sells them for $600, $100, and $300. The costs to AutoTech to obtain these products are $350, $70, and $280 each, while to SmartCars the costs are $460, $60, and $140.

a. Represent the selling prices and costs as matrices.
b. Using the matrices from (a), find the profit matrix.
c. Using the matrix from (b), find the salesperson's commission on each item at each company if the commission is 30% of the profit.

Solution

a. Let R be the revenue matrix of selling prices and let C be the cost matrix. We have three items from two companies, so we can choose the matrices to be either 2×3 or 3×2. We choose the matrices to be 2×3 with the rows corresponding to the different companies and the columns to the different products:

$$
\begin{array}{ccc}
\text{Navigation} & \text{DVD} & \text{Satellite} \\
\text{system} & \text{player} & \text{radio} \\
\downarrow & \downarrow & \downarrow
\end{array}
$$

$$
\begin{array}{l}
\text{AutoTech} \rightarrow \\
\text{SmartCars} \rightarrow
\end{array}
\left(
\begin{array}{ccc}
\underline{\quad} & \underline{\quad} & \underline{\quad} \\
\underline{\quad} & \underline{\quad} & \underline{\quad}
\end{array}
\right)
$$

From the given information, we obtain the revenue and cost matrices

$$
R = \begin{pmatrix} 500 & 150 & 400 \\ 600 & 100 & 300 \end{pmatrix} \quad \text{and} \quad C = \begin{pmatrix} 350 & 70 & 280 \\ 460 & 60 & 140 \end{pmatrix}
$$

b. Profit is revenue minus cost, so we find $P = R - C$ by matrix subtraction:

$$
\overbrace{\begin{pmatrix} 500 & 150 & 400 \\ 600 & 100 & 300 \end{pmatrix}}^{R} - \overbrace{\begin{pmatrix} 350 & 70 & 280 \\ 460 & 60 & 140 \end{pmatrix}}^{C} = \overbrace{\begin{pmatrix} 150 & 80 & 120 \\ 140 & 40 & 160 \end{pmatrix}}^{P}
$$

c. The commission matrix is 30% of the profit matrix P, which we find by scalar multiplication:

$$
0.30P = 0.30 \cdot \begin{pmatrix} 150 & 80 & 120 \\ 140 & 40 & 160 \end{pmatrix} = \begin{pmatrix} 45 & 24 & 36 \\ 42 & 12 & 48 \end{pmatrix}
$$

Matrix Multiplication

To evaluate an expression like:
$$5x_1 + 6x_2 + 7x_3$$

when x_1, x_2, and x_3 are equal to:
$$2, \quad 3, \text{ and } 4$$

we simply multiply and add:
$$5 \cdot 2 + 6 \cdot 3 + 7 \cdot 4 = 56$$

That is, evaluating linear expressions amounts to multiplying pairs of numbers and adding. With this kind of evaluation in mind, we define matrix multiplication as *multiplying in order the numbers from a row by the numbers from a column and adding the results.* For example, the matrix product of a row matrix by a column matrix is a 1×1 matrix:

$$(5 \quad 6 \quad 7) \cdot \begin{pmatrix} 2 \\ 3 \\ 4 \end{pmatrix} = (5 \cdot 2 + 6 \cdot 3 + 7 \cdot 4) = (56)$$

Of course, there must be exactly the same number of elements in the row as in the column. In general,

$$(a_1 \quad a_2 \cdots a_n) \cdot \begin{pmatrix} b_1 \\ b_2 \\ \vdots \\ b_n \end{pmatrix} = (a_1 b_1 + a_2 b_2 + \cdots + a_n b_n)$$

> **→ Practice Problem 1**
>
> Find $(1 \quad 2) \cdot \begin{pmatrix} 3 \\ 4 \end{pmatrix}$.
>
> ➤ Solution on page 120

If there are several rows on the left and several columns on the right, then we multiply *each row* times *each column*, with each answer placed in the product matrix at the row and column position from which it came. So, for example, when we multiply the *first row* of a matrix by the *second column* of another, we place the result in the *first row* and *second column* of the answer. Notice that we always take *rows* (from the left) multiplied by *columns* (on the right).

First row times second column goes here, in first row, second column

$$\begin{pmatrix} 3 & 2 & 1 \\ 2 & 0 & -2 \end{pmatrix}\begin{pmatrix} 2 & 7 \\ 3 & 6 \\ 4 & 5 \end{pmatrix} = \begin{pmatrix} (3\ 2\ 1)\begin{pmatrix} 2 \\ 3 \\ 4 \end{pmatrix} & (3\ 2\ 1)\begin{pmatrix} 7 \\ 6 \\ 5 \end{pmatrix} \\ (2\ 0\ -2)\begin{pmatrix} 2 \\ 3 \\ 4 \end{pmatrix} & (2\ 0\ -2)\begin{pmatrix} 7 \\ 6 \\ 5 \end{pmatrix} \end{pmatrix} = \begin{pmatrix} 16 & 38 \\ -4 & 4 \end{pmatrix}$$

Go from here

Try to multiply each row times column "in your head"

directly to here

In general, to multiply two matrices, the row length of the first must match the column length of the second, with the other numbers giving the dimension of the product:

$$A \quad \cdot \quad B$$

$$m \times p \quad p \times n$$

must match

$$m \times n$$

dimension of $A \cdot B$

Outside numbers give dimension
Inside numbers get "absorbed"

Matrix Multiplication

If A is an $m \times p$ matrix and B is a $p \times n$ matrix, the matrix product $A \cdot B$ is the $m \times n$ matrix whose element in row i and column j is the product of row i from A times column j from B. If the row length of A does not equal the column length of B, then the product $A \cdot B$ is not defined.

Even when both products $A \cdot B$ and $B \cdot A$ are defined, the different orders may give very different results. For example,

$$(5 \quad 6 \quad 7)\begin{pmatrix} 2 \\ 3 \\ 4 \end{pmatrix} = (5 \cdot 2 + 6 \cdot 3 + 7 \cdot 4) = (56)$$

Multiplying in one order gives a 1×1 matrix

$$\begin{pmatrix} 2 \\ 3 \\ 4 \end{pmatrix}(5 \quad 6 \quad 7) = \begin{pmatrix} 2 \cdot 5 & 2 \cdot 6 & 2 \cdot 7 \\ 3 \cdot 5 & 3 \cdot 6 & 3 \cdot 7 \\ 4 \cdot 5 & 4 \cdot 6 & 4 \cdot 7 \end{pmatrix} = \begin{pmatrix} 10 & 12 & 14 \\ 15 & 18 & 21 \\ 20 & 24 & 28 \end{pmatrix}$$

Multiplying in the other order gives a 3×3 matrix

That is:

Matrix multiplication is not commutative.

In general, $A \cdot B \neq B \cdot A$.

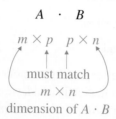 **EXAMPLE 2** Matrix Multiplication

Use the selling price ("revenue") matrix from Example 1 (page 116) and matrix multiplication to find the total selling price for an order of four navigation systems, one DVD player, and two satellite radios from each of the companies.

Solution

The revenue matrix R from Example 1 gives the selling prices in rows, one row for each company. If we write the four navigation systems, one DVD player, and two satellite radios as a *column*, then we can use matrix multiplication to find the total price at each company.

$$\begin{pmatrix} 500 & 150 & 400 \\ 600 & 100 & 300 \end{pmatrix} \begin{pmatrix} 4 \\ 1 \\ 2 \end{pmatrix} = \begin{pmatrix} 500 \cdot 4 + 150 \cdot 1 + 400 \cdot 2 \\ 600 \cdot 4 + 100 \cdot 1 + 300 \cdot 2 \end{pmatrix} = \begin{pmatrix} 2950 \\ 3100 \end{pmatrix}$$

R Order   Prices

(from Example 1)

The total selling prices are \$2950 at AutoTech and \$3100 at SmartCars.

➔ Practice Problem 2

Modify Example 2 to find the total selling price for an order of three navigation systems, two DVD players, and five satellite radios at each of the companies.

➤ Solution on next page

Identity Matrices

Multiplication by an identity matrix (of the proper size) does not change the matrix:

$$\begin{pmatrix} 1 & 0 & 0 \\ 0 & 1 & 0 \\ 0 & 0 & 1 \end{pmatrix} \begin{pmatrix} 2 & 3 \\ 4 & 7 \\ 6 & 5 \end{pmatrix} = \begin{pmatrix} 2 & 3 \\ 4 & 7 \\ 6 & 5 \end{pmatrix}$$

Same matrix on the right as on the left

The first row $(1 \ 0 \ 0)$ picks out just the first value, the second $(0 \ 1 \ 0)$ picks out the second value, and so on. (You should carefully verify this multiplication to see how the identity works.) To multiply by the identity matrix on the other side, we need to use an identity matrix of a different size:

$$\begin{pmatrix} 2 & 3 \\ 4 & 7 \\ 6 & 5 \end{pmatrix} \begin{pmatrix} 1 & 0 \\ 0 & 1 \end{pmatrix} = \begin{pmatrix} 2 & 3 \\ 4 & 7 \\ 6 & 5 \end{pmatrix}$$

Again, the matrix is duplicated

Writing I for an identity matrix of the appropriate size, we have for any matrix A:

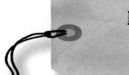

Multiplication by I

$$I \cdot A = A \cdot I = A$$

That is, the matrix I plays the role in matrix arithmetic that the number 1 plays in ordinary arithmetic: multiplying by it gives back exactly what you started with.

Matrix Multiplication and Systems of Equations

Because we defined matrix multiplication as evaluation of linear expressions (page 117), we may use it to write an entire system of equations as a single matrix equation. Consider the system of equations

$$\begin{cases} 2x_1 + 3x_2 + 3x_3 = 15 \\ 3x_1 + 4x_2 + 4x_3 = 22 \\ 3x_1 + 4x_2 + 5x_3 = 30 \end{cases}$$

The "coefficient" matrix A and the "constant term" matrix B are

$$A = \begin{pmatrix} 2 & 3 & 3 \\ 3 & 4 & 4 \\ 3 & 4 & 5 \end{pmatrix} \quad \text{and} \quad B = \begin{pmatrix} 15 \\ 22 \\ 30 \end{pmatrix}$$

Let X be the column matrix of the variables:

$$X = \begin{pmatrix} x_1 \\ x_2 \\ x_3 \end{pmatrix}$$

Then the matrix product AX is precisely the left sides of the original equations (as you should check):

$$\underbrace{\begin{pmatrix} 2 & 3 & 3 \\ 3 & 4 & 4 \\ 3 & 4 & 5 \end{pmatrix}}_{A} \underbrace{\begin{pmatrix} x_1 \\ x_2 \\ x_3 \end{pmatrix}}_{X} = \begin{pmatrix} 2x_1 + 3x_2 + 3x_3 \\ 3x_1 + 4x_2 + 4x_3 \\ 3x_1 + 4x_2 + 5x_3 \end{pmatrix}$$

The system of equations is the same as the matrix equation $AX = B$:

$$\begin{cases} 2x_1 + 3x_2 + 3x_3 = 15 \\ 3x_1 + 4x_2 + 4x_3 = 22 \\ 3x_1 + 4x_2 + 5x_3 = 30 \end{cases} \quad \text{is the same as} \quad \underbrace{\begin{pmatrix} 2 & 3 & 3 \\ 3 & 4 & 4 \\ 3 & 4 & 5 \end{pmatrix}}_{A} \underbrace{\begin{pmatrix} x_1 \\ x_2 \\ x_3 \end{pmatrix}}_{X} = \underbrace{\begin{pmatrix} 15 \\ 22 \\ 30 \end{pmatrix}}_{B}$$

We can therefore represent the entire system of equations as just *one matrix equation* $AX = B$, as demonstrated above. This, at the very least, saves writing out the variables in each equation, a saving that is even greater for larger systems.

➡ **Solutions to Practice Problems**

1. $(1 \quad 2) \cdot \begin{pmatrix} 3 \\ 4 \end{pmatrix} = (1 \cdot 3 + 2 \cdot 4) = (11)$

2. $\begin{pmatrix} 500 & 150 & 400 \\ 600 & 100 & 300 \end{pmatrix} \begin{pmatrix} 3 \\ 2 \\ 5 \end{pmatrix} = \begin{pmatrix} 500 \cdot 3 + 150 \cdot 2 + 400 \cdot 5 \\ 600 \cdot 3 + 100 \cdot 2 + 300 \cdot 5 \end{pmatrix} = \begin{pmatrix} 3800 \\ 3500 \end{pmatrix}$

The prices are \$3800 at AutoTech and \$3500 at SmartCars.

Use the given matrices to find each expression.

$$A = \begin{pmatrix} 1 & 2 & 3 \\ 4 & 5 & 6 \\ 7 & 8 & 9 \end{pmatrix} \qquad B = \begin{pmatrix} 9 & 8 & 7 \\ 6 & 5 & 4 \\ 3 & 2 & 1 \end{pmatrix}$$

$$C = \begin{pmatrix} 1 & 6 & 8 \\ 4 & 2 & 7 \\ 9 & 5 & 3 \end{pmatrix}$$

1. A^t

2. $3C$

3. $-B$

4. $A + C$

5. $C - (A + I)$

Find each matrix product.

6. $\begin{pmatrix} 1 \\ 2 \end{pmatrix} \cdot (3 \quad 4)$

7. $\begin{pmatrix} 1 & 2 & 1 \\ 2 & 1 & 2 \end{pmatrix} \begin{pmatrix} 2 \\ -3 \\ 2 \end{pmatrix}$

8. $\begin{pmatrix} 1 & 3 & 1 \\ 2 & 1 & 2 \end{pmatrix} \begin{pmatrix} 1 & 2 \\ 1 & 1 \\ 2 & 1 \end{pmatrix}$

Rewrite each system of linear equations as a matrix equation $AX = B$.

9. $\begin{cases} x_1 + 5x_2 + 4x_3 = 6 \\ x_1 + x_2 + x_3 = 4 \\ 2x_1 + 3x_2 + 3x_3 = 9 \end{cases}$

10. $\begin{cases} 5x_1 + 2x_2 - 4x_3 + x_4 + 5x_5 = 7 \\ 3x_1 + x_2 - 3x_3 + x_4 + 3x_5 = 5 \end{cases}$

Formulate each situation in matrix form. Be sure to indicate the meaning of your rows and columns. Find the requested quantities using the appropriate matrix arithmetic.

11. *Automobile Sales* A car dealer sells sedans, station wagons, vans, and pickup trucks at sales lots in Oakdale and Roanoke. The "dealer markup" is the difference between the sticker price and the dealer invoice price. The dealer invoice prices at both locations are the same: $15,000 per sedan, $19,000 per wagon, $23,000 per van, and $25,000 per pickup. The sticker prices at the Oakdale lot are $18,900 per sedan, $22,900 per wagon, $26,900 per van, and $29,900 per pickup, while at the Roanoke lot the sticker prices are $19,900 per sedan, $21,900 per wagon, $27,900 per van, and $28,900 per pickup. Represent these prices as a dealer invoice matrix and a sticker price matrix. Use these matrices to find the dealer markup matrix for these vehicles at these sales lots.

12. *Overseas Manufacturing* A sports apparel company manufactures shorts, tee shirts, and caps in Costa Rica and Honduras for importation and sale in the United States. In Costa Rica the labor costs per item are 75¢ per pair of shorts, 25¢ per tee shirt, and 45¢ per cap, while the costs of the necessary materials are $1.60 per pair of shorts, 95¢ per tee shirt, and $1.15 per cap. In Honduras the labor costs per item are 80¢ per pair of shorts, 20¢ per tee shirt, and 55¢ per cap, while the costs of the necessary materials are $1.50 per pair of shorts, 80¢ per tee shirt, and $1.10 per cap. Represent these costs as a labor cost matrix and a materials cost matrix. Use these matrices to find the total cost matrix for these products in these countries.

3.5   Inverse Matrices and Systems of Linear Equations

We can write a system of linear equations as a single matrix equation $AX = B$. Just as we solved the simple equation $2x = 10$ by multiplying each side by $\frac{1}{2}$ (the inverse of 2) to obtain $1x = 5$, we will solve the matrix equation $AX = B$ by mutiplying by an "inverse" matrix.

Inverse Matrices

The inverse of the number 2 is $\frac{1}{2}$ because $\frac{1}{2}$ multiplied by 2 (in either order) is 1. Similarly, the *inverse* of a matrix is defined as another matrix whose product with the first (in either order) gives I, the identity matrix. The inverse of the matrix A is denoted A^{-1} just as the inverse of 2 is $2^{-1} = \frac{1}{2}$.

> ### Inverse Matrix
>
> A square matrix A has an *inverse matrix*, denoted A^{-1}, if and only if
>
> $$A \cdot A^{-1} = I \quad \text{and} \quad A^{-1} \cdot A = I$$
>
> If the inverse A^{-1} exists, then A is *invertible*, and otherwise A is *singular*. The inverse A^{-1} will exist if and only if A is equivalent to the identity matrix I.

We read A^{-1} as "A inverse." The I in the equations stands for the identity matrix of the *same size* as A. Since the product in either order must give the identity matrix, if one matrix is the inverse of another, then the second is also the inverse of the first. Notice that only square matrices can have inverses.

→ EXAMPLE 1   Checking Inverse Matrices

Verify that the inverse of $A = \begin{pmatrix} \frac{1}{2} & 3 \\ 1 & 5 \end{pmatrix}$ is $A^{-1} = \begin{pmatrix} -10 & 6 \\ 2 & -1 \end{pmatrix}$.

Solution

We must show that $A^{-1} \cdot A = I$ and $A \cdot A^{-1} = I$:

$$A^{-1} \cdot A = \begin{pmatrix} -10 & 6 \\ 2 & -1 \end{pmatrix}\begin{pmatrix} \frac{1}{2} & 3 \\ 1 & 5 \end{pmatrix} = \begin{pmatrix} 1 & 0 \\ 0 & 1 \end{pmatrix} = I$$

and

$$A \cdot A^{-1} = \begin{pmatrix} \frac{1}{2} & 3 \\ 1 & 5 \end{pmatrix} \begin{pmatrix} -10 & 6 \\ 2 & -1 \end{pmatrix} = \begin{pmatrix} 1 & 0 \\ 0 & 1 \end{pmatrix} = I \qquad \text{Multiplication in either order gives } I$$

as required. (You should check the arithmetic in both multiplications.)

To verify that two matrices are inverses of each other, it is enough to multiply them in either order to get I.

Finding Inverse Matrices

Besides *checking* inverses, how do we actually *find* the inverse of a given matrix? We know that we can solve the system $AX = B$ for X by row-reducing the augmented matrix $(A \mid B)$ to $(I \mid X)$ with the solution X found to the right of the bar. If we had *two* such systems, $AX = B_1$ and $AX = B_2$, differing only in their right-hand sides, we could solve both at once by putting both columns on the right and row-reducing $(A \mid B_1 B_2)$ to obtain $(I \mid X_1 X_2)$, with the solutions X_1 and X_2 found to the right of the bar. If we can put *two* columns to the right of the bar, we can put an entire *matrix* to the right of the bar, and that matrix could even be the identity matrix I. Then, row-reducing $(A \mid I)$ to $(I \mid C)$ must give the C that solves $AC = I$. But since A and C multiply to I, the matrix C to the right of the bar must be the *inverse matrix* of A. Thus, we can calculate A^{-1} as follows:

Calculating an Inverse Matrix

If the square matrix A is invertible, then row-reducing the augmented matrix $(A \mid I)$ gives $(I \mid A^{-1})$ so that A^{-1} appears to the right of the bar. If row-reducing $(A \mid I)$ concludes with a matrix *other* than I on the left of the bar, then A is *singular* and does not have an inverse.

→ **EXAMPLE 2** Calculating an Inverse Matrix

Find the inverse of $A = \begin{pmatrix} 1 & 0 & 2 \\ 1 & 1 & 1 \\ 1 & 1 & 2 \end{pmatrix}$.

Solution
A is 3×3, so we augment it by I_3 and row-reduce $(A \mid I)$.

$$\begin{pmatrix} 1 & 0 & 2 & | & 1 & 0 & 0 \\ 1 & 1 & 1 & | & 0 & 1 & 0 \\ 1 & 1 & 2 & | & 0 & 0 & 1 \end{pmatrix} \qquad (A \mid I)$$

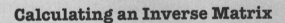

$\underbrace{\qquad}_{A} \quad \underbrace{\qquad}_{I}$

There are many different sequences of row operations to reduce this augmented matrix, and all reach the same conclusion. One way is as follows:

$$R_3 - R_2 \rightarrow \left(\begin{array}{ccc|ccc} 1 & 0 & 2 & 1 & 0 & 0 \\ 1 & 1 & 1 & 0 & 1 & 0 \\ 0 & 0 & 1 & 0 & -1 & 1 \end{array}\right)$$

$$\begin{array}{c} R_1 - 2R_3 \rightarrow \\ R_2 - R_3 \rightarrow \end{array} \left(\begin{array}{ccc|ccc} 1 & 0 & 0 & 1 & 2 & -2 \\ 1 & 1 & 0 & 0 & 2 & -1 \\ 0 & 0 & 1 & 0 & -1 & 1 \end{array}\right)$$

$$R_2 - R_1 \rightarrow \underbrace{\left(\begin{array}{ccc} 1 & 0 & 0 \\ 0 & 1 & 0 \\ 0 & 0 & 1 \end{array}\right.}_{I} \left| \underbrace{\begin{array}{ccc} 1 & 2 & -2 \\ -1 & 0 & 1 \\ 0 & -1 & 1 \end{array}\right)}_{A^{-1}}$$

Since I is on the left, A^{-1} is on the right

Therefore, $\quad A^{-1} = \begin{pmatrix} 1 & 2 & -2 \\ -1 & 0 & 1 \\ 0 & -1 & 1 \end{pmatrix} \quad$ is the inverse of $\quad A = \begin{pmatrix} 1 & 0 & 2 \\ 1 & 1 & 1 \\ 1 & 1 & 2 \end{pmatrix}.$

→ **Practice Problem 1**

For the matrices A and A^{-1} in Example 2, verify that A^{-1} is the inverse of A by showing that $\quad A^{-1}A = I.$

➤ **Solution on page 128**

Solving $AX = B$ Using A^{-1}

Just as the equation $\quad 2x = 10 \quad$ is solved simply by multiplying both sides by 2^{-1} or $\frac{1}{2}$, we can solve $\quad AX = B \quad$ by multiplying both sides (on the left) by A^{-1}.

$$\begin{aligned} AX &= B & &\text{Original equation} \\ A^{-1} \cdot A \cdot X &= A^{-1} \cdot B & &\text{Left-multiplying by } A^{-1} \\ I \cdot X &= A^{-1} \cdot B & &\text{Since } \quad A^{-1}A = I \\ X &= A^{-1}B & &\text{Since } \quad IX = X \end{aligned}$$

Notice that because matrix multiplication is *not* commutative, we must multiply both sides of $\quad AX = B \quad$ on the *left* by A^{-1}. Besides, the product $\quad BA^{-1} \quad$ is not defined because B is a column matrix and A^{-1} is a square matrix.

Solving $AX = B$ Using A^{-1}

If A is invertible, then the solution of $\quad AX = B \quad$ is

$$X = A^{-1}B$$

If A is singular, then we must use the Gauss–Jordan method (from Section 3.3) and we will then find that there are either no solutions or infinitely many solutions.

→ **EXAMPLE 3**   Solving a System Using an Inverse Matrix

Use an inverse matrix to solve $\begin{cases} x_1 + 2x_3 = 22 \\ x_1 + x_2 + x_3 = 11 \\ x_1 + x_2 + 2x_3 = 20 \end{cases}$

Solution

Writing this system as $AX = B$,   we have

$$\underbrace{\begin{pmatrix} 1 & 0 & 2 \\ 1 & 1 & 1 \\ 1 & 1 & 2 \end{pmatrix}}_{A} \underbrace{\begin{pmatrix} x_1 \\ x_2 \\ x_3 \end{pmatrix}}_{X} = \underbrace{\begin{pmatrix} 22 \\ 11 \\ 20 \end{pmatrix}}_{B} \qquad AX = B$$

Ordinarily, we would now find the inverse of matrix A by row-reducing $(A\,|\,I)$,   but that is exactly the matrix that we just row-reduced in Example 2. So, we will simply use the A^{-1} found on the previous page and write the solution as $X = A^{-1}B$.

$$\underbrace{\begin{pmatrix} x_1 \\ x_2 \\ x_3 \end{pmatrix}}_{X} = \underbrace{\begin{pmatrix} 1 & 2 & -2 \\ -1 & 0 & 1 \\ 0 & -1 & 1 \end{pmatrix}}_{A^{-1}} \underbrace{\begin{pmatrix} 22 \\ 11 \\ 20 \end{pmatrix}}_{B} = \begin{pmatrix} 4 \\ -2 \\ 9 \end{pmatrix} \qquad \text{Using } A^{-1} \text{ from Example 2}$$

The solution is $x_1 = 4$, $x_2 = -2$, $x_3 = 9$.

We may check this solution by substituting into the original equations:

$$4 + 2 \cdot 9 = 22$$
$$4 - 2 + 9 = 11$$
$$4 - 2 + 2 \cdot 9 = 20 \qquad \text{It checks!}$$

→ **Practice Problem 2**

Solve $\begin{cases} x_1 + 2x_3 = -5 \\ x_1 + x_2 + x_3 = 10 \\ x_1 + x_2 + 2x_3 = 0 \end{cases}$

[*Hint*: Write this system as $AX = B$ and notice that A is the same as in Example 3 but B is different, so the solution can be found by matrix multiplication using the inverse found in Example 2.]

➤ Solution on page 128

Solving $AX = B$ by finding the inverse and then using $X = A^{-1}B$ is only slightly more difficult than row-reducing the augmented matrix for the system of equations. If you need to solve $AX = B$ with the same A but with several different matrices B, however, using the inverse A^{-1} means that you only do the row reduction *once*; then the solution for any new matrix B is given by a simple matrix multiplication. Such problems, where A remains the same but B changes, occur frequently in applications.

Solving $AX = B$ for Many Different Bs

If A is invertible, then the solutions of the matrix equations

$$AX = B_1, \quad AX = B_2, \ldots, \quad AX = B_n$$

may all be found by calculating A^{-1} once and then finding the solutions as the products

$$X = A^{-1}B_1, \quad X = A^{-1}B_2, \ldots, \quad X = A^{-1}B_n$$

➡ EXAMPLE 4   Jewelry Production

Elnur/Shutterstock

An employee-owned jewelry company fabricates enameled gold rings, pendants, and bracelets. Each ring requires 3 grams of gold, 1 gram of enameling compound, and 2 hours of labor; each pendant requires 6 grams of gold, 2 grams of enameling compound, and 3 hours of labor; and each bracelet requires 8 grams of gold, 3 grams of enameling compound, and 2 hours of labor. Each of the five employee–owners works 160 hours each month, and the company has contracts guaranteeing the delivery of the grams of gold and enameling compound shown in the table on the first day of the months shown. How many rings, pendants, and bracelets should the company fabricate each month to use all the available materials and time?

	March	April	May	June
Gold	1720	2620	2460	2220
Enamel	600	960	900	800

Solution
Let x_1, x_2, and x_3 be the numbers of rings, pendants, and bracelets produced in one month. Then the required grams of gold, grams of enameling compound, and hours of labor are

Summarizing

	Rings	Pendants	Bracelets
Gold	3	6	8
Enamel	1	2	3
Labor	2	3	2

Gold	$3x_1 + 6x_2 + 8x_3$
Enamel	$x_1 + 2x_2 + 3x_3$
Labor	$2x_1 + 3x_2 + 2x_3$

For the month of March, these quantities must match the amounts available:

$$\begin{cases} 3x_1 + 6x_2 + 8x_3 = 1720 \\ x_1 + 2x_2 + 3x_3 = 600 \\ 2x_1 + 3x_2 + 2x_3 = 800 \end{cases}$$

From the table and 5 workers at 160 hours each

Elnur/Shutterstock

The systems of equations for the other months follow in a similar manner, and we have four problems to solve:

March		April	
$\begin{cases} 3x_1 + 6x_2 + 8x_3 = 1720 \\ x_1 + 2x_2 + 3x_3 = 600 \\ 2x_1 + 3x_2 + 2x_3 = 800 \end{cases}$		$\begin{cases} 3x_1 + 6x_2 + 8x_3 = 2620 \\ x_1 + 2x_2 + 3x_3 = 960 \\ 2x_1 + 3x_2 + 2x_3 = 800 \end{cases}$	
May		June	
$\begin{cases} 3x_1 + 6x_2 + 8x_3 = 2460 \\ x_1 + 2x_2 + 3x_3 = 900 \\ 2x_1 + 3x_2 + 2x_3 = 800 \end{cases}$		$\begin{cases} 3x_1 + 6x_2 + 8x_3 = 2220 \\ x_1 + 2x_2 + 3x_3 = 800 \\ 2x_1 + 3x_2 + 2x_3 = 800 \end{cases}$	

The coefficient matrices for these four problems are the same, so if we find the inverse matrix for the March problem by row-reducing $(A \,|\, I)$, then we will only have to carry out the row operations to reduce A once. Then the four problems can be solved with just four matrix multiplications.

We first find A^{-1} by row-reducing $(A \,|\, I)$. There are many different sequences of row operations to reduce this augmented matrix, and all reach the same conclusion. One possible way is as follows:

$$\left(\begin{array}{ccc|ccc} 3 & 6 & 8 & 1 & 0 & 0 \\ 1 & 2 & 3 & 0 & 1 & 0 \\ 2 & 3 & 2 & 0 & 0 & 1 \end{array} \right) \qquad (A \,|\, I)$$

$$\begin{array}{c} 3R_2 - R_1 \to \\ \\ 2R_2 - R_3 \to \end{array} \left(\begin{array}{ccc|ccc} 0 & 0 & 1 & -1 & 3 & 0 \\ 1 & 2 & 3 & 0 & 1 & 0 \\ 0 & 1 & 4 & 0 & 2 & -1 \end{array} \right)$$

Using 1 in the first column to zero-out the rest of that column

$$\begin{array}{c} \\ R_2 - 2R_3 \to \\ \\ \end{array} \left(\begin{array}{ccc|ccc} 0 & 0 & 1 & -1 & 3 & 0 \\ 1 & 0 & -5 & 0 & -3 & 2 \\ 0 & 1 & 4 & 0 & 2 & -1 \end{array} \right)$$

Using 1 in the second column to zero-out the rest of that column

$$\begin{array}{c} \\ R_2 + 5R_1 \to \\ R_3 - 4R_1 \to \end{array} \left(\begin{array}{ccc|ccc} 0 & 0 & 1 & -1 & 3 & 0 \\ 1 & 0 & 0 & -5 & 12 & 2 \\ 0 & 1 & 0 & 4 & -10 & -1 \end{array} \right)$$

Using 1 in third column to zero-out the rest of that column

$$\begin{array}{c} R_2 \to \\ R_3 \to \\ R_1 \to \end{array} \left(\begin{array}{ccc|ccc} 1 & 0 & 0 & -5 & 12 & 2 \\ 0 & 1 & 0 & 4 & -10 & -1 \\ 0 & 0 & 1 & -1 & 3 & 0 \end{array} \right)$$

Switching rows to achieve the correct order

Since the left side is I, the right side is the inverse matrix

$$A^{-1} = \left(\begin{array}{ccc} -5 & 12 & 2 \\ 4 & -10 & -1 \\ -1 & 3 & 0 \end{array} \right)$$

The solution for each month is simply the product of this inverse matrix with the column matrix of the constant terms for that month.

Month	$X = A^{-1}B$				Solution
March	$\begin{pmatrix} x_1 \\ x_2 \\ x_3 \end{pmatrix} = \begin{pmatrix} -5 & 12 & 2 \\ 4 & -10 & -1 \\ -1 & 3 & 0 \end{pmatrix} \begin{pmatrix} 1720 \\ 600 \\ 800 \end{pmatrix} = \begin{pmatrix} 200 \\ 80 \\ 80 \end{pmatrix}$				200 rings 80 pendants 80 bracelets
April	$\begin{pmatrix} x_1 \\ x_2 \\ x_3 \end{pmatrix} = \begin{pmatrix} -5 & 12 & 2 \\ 4 & -10 & -1 \\ -1 & 3 & 0 \end{pmatrix} \begin{pmatrix} 2620 \\ 960 \\ 800 \end{pmatrix} = \begin{pmatrix} 20 \\ 80 \\ 260 \end{pmatrix}$				20 rings 80 pendants 260 bracelets
May	$\begin{pmatrix} x_1 \\ x_2 \\ x_3 \end{pmatrix} = \begin{pmatrix} -5 & 12 & 2 \\ 4 & -10 & -1 \\ -1 & 3 & 0 \end{pmatrix} \begin{pmatrix} 2460 \\ 900 \\ 800 \end{pmatrix} = \begin{pmatrix} 100 \\ 40 \\ 240 \end{pmatrix}$				100 rings 40 pendants 240 bracelets
June	$\begin{pmatrix} x_1 \\ x_2 \\ x_3 \end{pmatrix} = \begin{pmatrix} -5 & 12 & 2 \\ 4 & -10 & -1 \\ -1 & 3 & 0 \end{pmatrix} \begin{pmatrix} 2220 \\ 800 \\ 800 \end{pmatrix} = \begin{pmatrix} 100 \\ 80 \\ 180 \end{pmatrix}$				100 rings 80 pendants 180 bracelets

→ Solutions to Practice Problems

1. $A^{-1}A = \begin{pmatrix} 1 & 2 & -2 \\ -1 & 0 & 1 \\ 0 & -1 & 1 \end{pmatrix} \begin{pmatrix} 1 & 0 & 2 \\ 1 & 1 & 1 \\ 1 & 1 & 2 \end{pmatrix}$

$= \begin{pmatrix} 1+2-2 & 0+2-2 & 2+2-4 \\ -1+0+1 & 0+0+1 & -2+0+2 \\ 0-1+1 & 0-1+1 & 0-1+2 \end{pmatrix} = \begin{pmatrix} 1 & 0 & 0 \\ 0 & 1 & 0 \\ 0 & 0 & 1 \end{pmatrix} = I$

2. Using $X = A^{-1}B$: $\begin{pmatrix} x_1 \\ x_2 \\ x_3 \end{pmatrix} = \begin{pmatrix} 1 & 2 & -2 \\ -1 & 0 & 1 \\ 0 & -1 & 1 \end{pmatrix} \begin{pmatrix} -5 \\ 10 \\ 0 \end{pmatrix} = \begin{pmatrix} 15 \\ 5 \\ -10 \end{pmatrix}$ so $\begin{cases} x_1 = 15 \\ x_2 = 5 \\ x_3 = -10 \end{cases}$

→ 3.5 Exercises

For each matrix A, row-reduce $(A\,|\,I)$ to find the inverse matrix A^{-1} or to identify A as a singular matrix.

1. $\begin{pmatrix} 1 & 3 \\ 0 & 1 \end{pmatrix}$

2. $\begin{pmatrix} 1 & 1 & 0 \\ 3 & 0 & 2 \\ 1 & 0 & 1 \end{pmatrix}$

3. $\begin{pmatrix} 1 & 1 & 0 \\ 0 & 1 & 1 \\ 1 & 2 & 1 \end{pmatrix}$

Rewrite each system of equations as a matrix equation $AX = B$ and use the inverse of A to find the solution. Be sure to check your solution in the original system of equations.

4. $\begin{cases} 11x_1 + 2x_2 = 9 \\ 6x_1 + x_2 = 5 \end{cases}$

5. $\begin{cases} x_1 + x_2 = 2 \\ 3x_1 + 2x_3 = 5 \\ x_1 + x_3 = 2 \end{cases}$

6. $\begin{cases} 2x_1 + 3x_2 + x_3 = 6 \\ x_1 + 2x_2 + x_3 = 4 \\ 2x_1 + 3x_2 + 2x_3 = 7 \end{cases}$

7. $\begin{cases} 3x_1 - 4x_2 + 2x_3 = 12 \\ -x_1 + 2x_2 - x_3 = -4 \\ 4x_1 - 3x_2 + 2x_3 = 15 \end{cases}$

Formulate each situation as a collection of systems of linear equations. Be sure to state clearly the meaning of each variable. Solve each collection using an inverse matrix. State your final answers in terms of the original questions.

8. *Speculative Partnerships* A broker offers limited partnerships in the amounts of $1000, $5000, and $10,000 in three organizations speculating in gold mines, oil wells, and modern art. The capitalization values and number of members in each are given in the table, together with the "altered value" of the capitalization should all the $1000 memberships be replaced by the same number of $5000 memberships. How many memberships of each amount are in each partnership?

	Gold Mines	Oil Wells	Modern Art
Value	$2.0 million	$3.9 million	$2.5 million
Number of Members	500	700	600
"Altered Value"	$3.0 million	$4.5 million	$3.8 million

9. *Infant Nutrition* A pediatric dietician at an inner-city foundling hospital needs to supplement each bottle of baby formula given to three infants in her ward with the units of vitamin A, vitamin D, calcium, and iron given in the first table. If four diet supplements are available with the nutrient content per drop given in the second table, how many drops of each supplement per bottle of formula should each infant receive?

Units Needed	Vitamin A	Vitamin D	Calcium	Iron
Billy	48	26	19	49
Susie	46	26	27	65
Jimmy	47	26	19	49

Units per Drop	Vitamin A	Vitamin D	Calcium	Iron
Supplement 1	5	3	0	1
Supplement 2	0	1	3	6
Supplement 3	4	2	3	7
Supplement 4	4	2	2	5

10. *Financial Planning* A retirement planning counselor recommends investing in a stock fund yielding 18%, a money market fund returning 6%, and a bond fund paying 8%, with twice as much in the bond and money market funds together as in the stock fund. Mr. and Mrs. Jordan have $300,000 to invest and need an annual return of $31,000; Mr. and Mrs. French have $234,900 to invest and need an annual return of $25,600; and Mr. and Mrs. Daimen have $270,000 to invest and need an annual return of $28,500. How much should each elderly couple place in each investment to receive their desired retirement income?

3.6 Introduction to Modeling: Leontief Models

We conclude this chapter with an important application of matrices, *input-output analysis*, which is used to model economies and large companies.

Input-output analysis was invented to study the flow of goods and services among the different sectors of an economy. It was developed by the Russian-born American economist Wassily Leontief (1906–1999), winner of the 1973 Nobel Memorial Prize for Economics. We will study "open" models, in which an economy or large corporation produces more goods than are needed for production, so the extra output is available to outsiders. The economy of any country can be modeled as an open system, as can any large diversified corporation that uses some of its products for its own production and sells the remainder to consumers.

A Model Economy

Consider a simplified economy that consists of three sectors: manufacturing (M), agriculture (A), and transportation (T). Each sector uses some of its own production (manufacturing builds the machinery it needs for its own production) and some from other sectors (manufacturing uses transportation to move parts among its factories, and transportation uses agriculture for biofuels). An *input-output diagram* shows the precise relationships between the sectors. Here's an example:

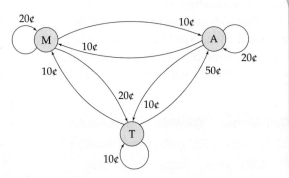

Input-output diagram from an economy with sectors for manufacturing (M), agriculture (A), and transportation (T).

The number on an arrow indicates *the amount of output from the "tail" sector needed to produce $1 of output of the "head" sector.* For example, the 20¢, 10¢, and 10¢ on the arrows pointing to M mean that producing $1 from the manufacturing sector requires 20¢ of output from the manufacturing sector itself, 10¢ of agriculture output, and 10¢ of transportation output.

Suppose that x_1, x_2, and x_3 represent the *total* values of the goods and services produced, respectively, by the manufacturing, agriculture, and transportation sectors, and let y_1, y_2, and y_3 represent the *excess* production from each of these sectors, to be sold to consumers. The value produced by the manufacturing sector must be the same as the value of the manufacturing sector used by the manufacturing, agriculture, and transportation sectors together with the amount sold to consumers. Using the numbers from the input-output diagram, we have

$$x_1 = 0.20x_1 + 0.10x_2 + 0.20x_3 + y_1$$

Total value of manufacturing products	Value of manufacturing products used in manufacturing	Value of manufacturing products used in agriculture	Value of manufacturing products used in transportation	Value of manufacturing products sold to consumers

Similarly for the agriculture and transportation sectors,

$$x_2 \quad = \quad 0.10x_1 \quad + \quad 0.20x_2 \quad + \quad 0.10x_3 \quad + \quad y_2$$

Total value of agricultural products	Value of agricultural products used in manufacturing	Value of agricultural products used in agriculture	Value of agricultural products used in transportation	Value of agricultural products sold to consumers

$$x_3 \quad = \quad 0.10x_1 \quad + \quad 0.50x_2 \quad + \quad 0.10x_3 \quad + \quad y_3$$

Total value of transportation services	Value of transportation services used in manufacturing	Value of transportation services used in agriculture	Value of transportation services used in transportation	Value of transportation services sold to consumers

This system of equations can be written in matrix form as

$$\begin{pmatrix} x_1 \\ x_2 \\ x_3 \end{pmatrix} = \begin{pmatrix} 0.2 & 0.1 & 0.2 \\ 0.1 & 0.2 & 0.1 \\ 0.1 & 0.5 & 0.1 \end{pmatrix} \begin{pmatrix} x_1 \\ x_2 \\ x_3 \end{pmatrix} + \begin{pmatrix} y_1 \\ y_2 \\ y_3 \end{pmatrix}$$

or more compactly as

$$X = AX + Y$$

with X, A, and Y representing the above matrices. The column matrix X represents the *total production*, the square matrix A of coefficients is called the *technology matrix*, and the column matrix Y represents the *excess production*.

We may solve this matrix equation for Y (the amounts available to consumers) just as we would solve any equation. First we subtract AX from each side.

$$X - AX = Y$$

We may write this as

$$IX - AX = Y$$

(since $IX = X$ for the identity matrix I). Factoring out the X on the left side:

$$(I - A)X = Y$$

Switching sides gives a formula for *the excess production Y resulting from total production X*:

$$Y = (I - A)X$$

To solve instead for the *total* production X we just multiply each side of this equation by $(I - A)^{-1}$, the inverse of the matrix $(I - A)$, making sure to multiply on the *left* of each side so that $(I - A)^{-1}$ and $(I - A)$ will multiply to the identity matrix I.

$$(I - A)^{-1}Y = \underbrace{(I - A)^{-1}(I - A)}_{I}X = IX = X$$

Writing this equation with X on the left gives the formula for *the total production X needed to produce a given excess production Y*

$$X = (I - A)^{-1}Y$$

We summarize these two important formulas as follows:

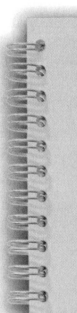

Finding Total and Excess Production

From the input-output diagram, construct the technology matrix A as follows:

- Choose an *order* for the sectors (1st, 2nd, 3rd, …). This order will be used for both the rows *and* columns of the matrix.
- Locate the first sector, and enter into the first *column* of the matrix the numbers that are on the arrows pointing *to* that sector (in order of their "tails").
- Repeat the above step with "first" replaced by "second," and then by "third," and so on for all sectors.

The *excess* production Y resulting from *total* production X is

$$Y = (I - A)X$$

The *total production X* needed to produce *excess production Y* is

$$X = (I - A)^{-1}Y$$

Let's see how these formulas are used in practice.

→ EXAMPLE 1   Finding Excess Production from Total Production

Suppose that in the economy described above, the manufacturing sector produced a total of $180 million, the agriculture sector produced $160 million, and the transportation sector produced $200 million. How much of these goods and services will be available to consumers?

Solution

We want to find the excess production $Y = (I - A)X$ resulting from the *total* production (in millions of dollars)

$$X = \begin{pmatrix} 180 \\ 160 \\ 200 \end{pmatrix} \quad \begin{matrix} \text{Manufacturing} \\ \text{Agriculture} \\ \text{Transportation} \end{matrix}$$

We divide the process into several steps.

Step 1: Write the matrix A.

From the input-output diagram we find the matrix A:

$$\begin{matrix} \text{Manufacturing} & \text{Agriculture} & \text{Transportation} \end{matrix}$$

$$\begin{pmatrix} 0.2 & 0.1 & 0.2 \\ 0.1 & 0.2 & 0.1 \\ 0.1 & 0.5 & 0.1 \end{pmatrix} \quad \begin{matrix} \text{Manufacturing} \\ \text{Agriculture} \\ \text{Transportation} \end{matrix}$$

Step 2: Calculate $I - A$ **by matrix subtraction.**

$$I - A = \begin{pmatrix} 1 & 0 & 0 \\ 0 & 1 & 0 \\ 0 & 0 & 1 \end{pmatrix} - \begin{pmatrix} 0.2 & 0.1 & 0.2 \\ 0.1 & 0.2 & 0.1 \\ 0.1 & 0.5 & 0.1 \end{pmatrix} = \begin{pmatrix} 0.8 & -0.1 & -0.2 \\ -0.1 & 0.8 & -0.1 \\ -0.1 & -0.5 & 0.9 \end{pmatrix}$$

Step 3: Calculate $(I - A)X$ **by matrix multiplication.**

$$\begin{pmatrix} 0.8 & -0.1 & -0.2 \\ -0.1 & 0.8 & -0.1 \\ -0.1 & -0.5 & 0.9 \end{pmatrix} \begin{pmatrix} 180 \\ 160 \\ 200 \end{pmatrix} = \begin{pmatrix} 88 \\ 90 \\ 82 \end{pmatrix} \qquad \begin{matrix} \text{Manufacturing} \\ \text{Agriculture} \\ \text{Transportation} \end{matrix}$$

Step 4: State the answer in terms of the original problem.

The goods and services available to consumers will be: $88 million of manufactured goods, $90 million of agricultural products, and $82 million of transportation services.

➡ EXAMPLE 2   Finding Total Production from Excess Production

Again for the economy described above, suppose that government economists decide that consumers have different needs from the economy, in particular, $120 million of manufactured goods, $125 million of agricultural goods, and $105 million of transportation services. What must be the total production to leave these amounts for consumers?

Solution

We now want to find the *total* production $X = (I - A)^{-1}Y$ needed for *excess* production

$$Y = \begin{pmatrix} 120 \\ 125 \\ 105 \end{pmatrix} \qquad \begin{matrix} \text{Manufacturing} \\ \text{Agriculture} \\ \text{Transportation} \end{matrix}$$

Steps 1 and 2:

are the same as those in Example 1, which we have already done.

Step 3:   Find the inverse matrix $(I - A)^{-1}$.

We calculate the inverse of the matrix $I - A$ found in step 2 by using the techniques of previous section: augmenting by the identity and row-reducing. Omitting the details, the result is

$$(I - A)^{-1} = \begin{pmatrix} 1.34 & 0.38 & 0.34 \\ 0.20 & 1.40 & 0.20 \\ 0.26 & 0.82 & 1.26 \end{pmatrix}$$

Step 4:   Find $(I - A)^{-1}Y$ **by matrix multiplication.**

$$\begin{pmatrix} 1.34 & 0.38 & 0.34 \\ 0.20 & 1.40 & 0.20 \\ 0.26 & 0.82 & 1.26 \end{pmatrix} \begin{pmatrix} 120 \\ 125 \\ 105 \end{pmatrix} = \begin{pmatrix} 244 \\ 220 \\ 266 \end{pmatrix}$$

Step 5:   State the answer in terms of the original problem.
The total amounts of production needed are: $244 million of manufactured goods, $220 million of agricultural products, and $266 million of transportation services.

The matrix A was originally called the "interindustry matrix of technical coefficients" and Leontief's original model of the American economy used a matrix with 42 rows and columns. The procedure, however, is the same for any size model. For a diversified company that makes several products, some of which are used by other divisions of the company for their own production with the rest sold to consumers, input-output analysis enables the company to determine the *excess* production that will result from a given *total* production, and also the *total* production needed for a given *excess* production.

→ 3.6 Exercises

Let A denote agriculture, C denote construction, E denote electronics, and L denote light industry. Find the technology matrix for each economy diagram.

1.

2.

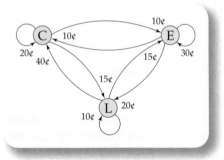

Find the excess production *Y* of each economy with technology matrix *A* and economic activity level *X*.

3. $A = \begin{pmatrix} 0.20 & 0.30 \\ 0.35 & 0.25 \end{pmatrix}$ and $X = \begin{pmatrix} 130 \\ 110 \end{pmatrix}$

4. $A = \begin{pmatrix} 0.05 & 0.15 & 0.20 \\ 0.15 & 0.05 & 0.15 \\ 0.10 & 0.10 & 0.05 \end{pmatrix}$ and $X = \begin{pmatrix} 150 \\ 170 \\ 140 \end{pmatrix}$

Find, for each economy with technology matrix *A*, the economic activity level *X* necessary to generate excess production *Y*.

5. $A = \begin{pmatrix} 0.20 & 0.30 \\ 0.30 & 0.20 \end{pmatrix}$ and $Y = \begin{pmatrix} 84 \\ 51 \end{pmatrix}$

6. $A = \begin{pmatrix} 0.10 & 0.20 & 0 \\ 0 & 0.15 & 0.20 \\ 0.30 & 0.10 & 0.20 \end{pmatrix}$ and $Y = \begin{pmatrix} 60 \\ 31 \\ 50 \end{pmatrix}$

Represent each situation as a Leontief model by constructing an economy diagram and the corresponding technology matrix. Find the required excess production or level of economic activity and state your final answer in terms of the original question.

7. *Industrial Production* An international conglomerate uses some of its own production to provide materials for its own use. Company records show that the technology matrix for three of its divisions is

From\To	Electronics	Energy	Plastics
Electronics	0.60	0.10	0.20
Energy	0.10	0.20	0.10
Plastics	0.10	0.20	0.10

What total production levels are necessary if the corporate directors want the excess production levels to be $100 million from each division?

8. **Developing Productivity** In an effort to raise the standard of living, the new government of a developing country wants to increase its excess production by stimulating the heavy industry sector of the national economy. An analysis of the relationships between its heavy industry, light industry, and railroad sectors found that the current production levels for these sectors are $100 million from heavy industry, $150 million from light industry, and $100 million from the railroads, with the technology matrix given in the table. Find the current excess production from these sectors of the economy. If the light industry and railroad productions remain the same, how much does this excess production increase with each $10 million increase in heavy industry production? What is the greatest heavy industry production level that this part of the national economy can tolerate?

From\To	Heavy Industry	Light Industry	Railroads
Heavy industry	0.20	0.20	0.30
Light industry	0.50	0.20	0.20
Railroads	0.20	0.10	0.20

Linear Programming

4.1   Linear Inequalities

This chapter describes a method for solving a large and important class of problems called *linear programming problems*. A typical linear programming problem, such as a decision of how to divide investment funds between low- and high-risk ventures, asks for the maximum value of a linear function subject to a collection of linear inequalities. Such problems often arise in modern business, sometimes involving many variables, and are usually solved by a procedure known as the *simplex method*. Although the simplex method is a numerical algorithm, it is based on the geometry of linear inequalities such as those we studied on page 2. We will first consider linear programming problems in *two* variables since they can be solved graphically. In this section we will graph inequalities in two variables, restricting ourselves to inequalities using ≥ and ≤ signs since these are the ones that occur in practice.

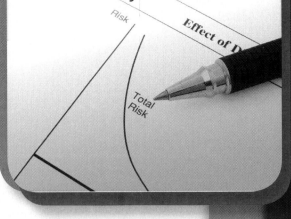

Inequalities in Two Variables

To graph an inequality in two variables means to graph all the points that satisfy the inequality. The *boundary* of the inequality is the corresponding *equality*, that is, an *equation*. For example, the boundary of the inequality $3x - 7y \leq 42$ is the equation $3x - 7y = 42$. The graph of the inequality is then one side or the other of the boundary, the correct side being determined by a *test point*. The correct side of the boundary is called the *feasible region* for the inequality.

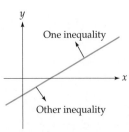

→ **EXAMPLE 1**   **Graphing a Linear Inequality in the Plane**

Graph the linear inequality $2x + 3y \geq 12$.

Solution

First we graph its boundary, $2x + 3y = 12$,   by plotting the intercepts. We find the y-intercept from $2x + 3y = 12$ with $x = 0$:

$$0 + 3y = 12 \qquad \text{so} \qquad y = 4 \qquad \text{For the point } (0, 4)$$

and then the x-intercept from $2x + 3y = 12$ with $y = 0$:

$$2x + 0 = 12 \qquad \text{so} \qquad x = 6 \qquad \text{For the point } (6, 0)$$

Using these two points, we draw the boundary line shown on the left.

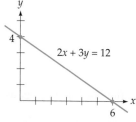

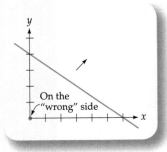

On the "wrong" side

To determine the correct side we use the test point $(0, 0)$. (We may use any point not on the line.) Substituting $x = 0$, $y = 0$ into the inequality $2x + 3y \geq 12$ gives $0 \geq 12$, which is *false* (0 is *not* greater than 12), meaning that the correct side of the line is the opposite side from the point $(0, 0)$, as indicated by the arrow in the graph.

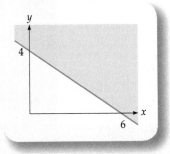

Finally, we shade the correct side of the boundary to show the feasible region, as in the third graph.

The general procedure for graphing an inequality is as follows:

To Graph a Linear Inequality

1. Draw the boundary line (possibly using the intercepts).
2. Use a test point not on the line (possibly the origin) to determine the correct side. If this point satisfies the inequality, the point lies on the correct side; if not, the *other* side is the correct side.
3. Shade the correct side to show the feasible region.

➡ Practice Problem 1

Graph the linear inequality $3x - 5y \leq 60$.

➤ Solution on page 143

Two or more inequalities joined by a curly brace { make up a *system* of inequalities, meaning that *each* of the inequalities must hold. To graph the feasible region for the system we graph each boundary, determine its correct side (which we mark with an arrow), and then shade the region that lies on the correct side of all the lines. The system is *feasible* if there is *at least one point* in the feasible region, and *infeasible* if there are *no* points in the feasible region.

➡ EXAMPLE 2 Graphing a System of Inequalities

a. Graph the system
$$\begin{cases} x + 2y \leq 12 \\ x \geq 0 \\ y \geq 0 \end{cases}$$

b. Is the system feasible?

c. What happens if the last inequality, $y \ge 0$, is replaced by $y \ge 7$?

Solution

a. We graph the boundary just as before, plotting the intercepts of $x + 2y = 12$: Substituting $x = 0$ gives $2y = 12$ or $y = 6$ for the point $(0, 6)$, and substituting $y = 0$ into $x + 2y = 12$ gives $x = 12$ for the point $(12, 0)$. These two points give the line in the first of the following three graphs.

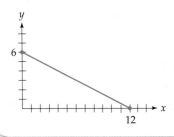 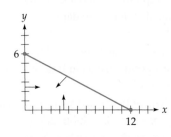 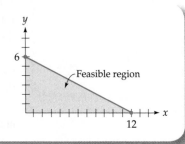

The point $(0, 0)$ *does* satisfy $x + 2y \le 12$ (since $0 + 0 \le 12$), so the origin is on the *correct* side, giving the arrow in the second graph. We add the arrows on the axes because $x \ge 0$ means from the y-axis to the *right*, and $y \ge 0$ means from the x-axis *upwards*. From the three arrows we get the shaded region in the third graph.

b. Since there *are* points in the feasible region, the system *is* feasible.

c. Replacing $y \ge 0$ by $y \ge 7$ gives the graph shown on the left. There are no points on the correct side of all the lines, so the system with $y \ge 7$ becomes *infeasible*.

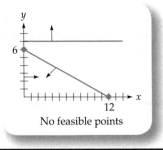

No feasible points

→ **Practice Problem 2**

Sketch the system $\begin{cases} x + 2y \le 20 \\ x \ge 0 \\ y \ge 0 \end{cases}$ Is this system feasible?

➤ **Solution on page 143**

Vertices of Feasible Regions

The "corner points" of the feasible region are called *vertices*. More precisely:

> **Vertex**
>
> A *vertex* of a system of linear inequalities is an intersection point of two (or more) of the boundaries that satisfies all the inequalities.

Each vertex may be found by solving for an intersection of two of the boundaries; if this point satisfies all the inequalities, then it is a vertex of the region.

→ **EXAMPLE 3** **Finding the Vertices of a Feasible Region**

Graph and find the vertices of the feasible region for $\begin{cases} x + 2y \le 12 \\ x - y \ge 0 \\ y \ge 0 \end{cases}$

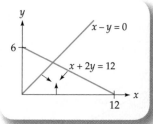

Solution

We graph the boundaries in the usual way, by plotting their intercepts. In Example 2 we found that the boundary $x + 2y = 12$ had intercepts $(0, 6)$ and $(12, 0)$. For the boundary $x - y = 0$ we use the intercept $(0, 0)$ and any other point satisfying $x - y = 0$, such as $(1, 1)$, leading to the graph on the left. The arrows are determined by test points just as in Examples 1 and 2. The inequality $y \ge 0$ means "upwards from the x-axis."

The only vertex that is not obvious from the graph is the intersection of $x + 2y = 12$ and $x - y = 0$. Subtracting:

$$\begin{array}{r} x + 2y = 12 \\ \underline{x - y = 0} \\ 3y = 12 \end{array}$$ ⟵ Top minus bottom

This last equation gives $y = 4$, which, when substituted into $x - y = 0$, gives $x - 4 = 0$ or $x = 4$, so the intersection point is $(4, 4)$. The feasible region is shown on the left, with vertices $(0, 0)$, $(4, 4)$, and $(12, 0)$.

Notice that the point $(0, 6)$ is *not* a vertex of the feasible region since it is on the *wrong* side of the line $x - y = 0$ (it does not satisfy $x - y \ge 0$). Vertices of the feasible region must be on the correct side of *all* the lines (that is, their coordinates must satisfy *all* the inequalities).

Bounded and Unbounded Regions

A region is *bounded* if it is enclosed on all sides by boundary lines. For example, the feasible region in Example 3 is bounded. However, if in Example 3 the direction of the inequality $x + 2y \le 12$ were reversed to become $x + 2y \ge 12$, then the feasible region would be on the *other* side of the boundary $x + 2y = 12$, giving the *unbounded* region shown on the right below.

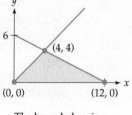

The bounded region
from Example 3.

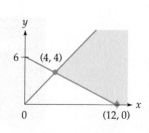

An unbounded region.

→ **Practice Problem·3**

Graph the feasible region $\begin{cases} x + 2y \geq 20 \\ x \geq 0 \\ y \geq 0 \end{cases}$ Is the region bounded?

[*Hint*: Modify your graph from Practice Problem 2.]   ➤ **Solution on page 143**

Applications

Since most physical quantities are nonnegative, applied problems usually include the two *nonnegativity inequalities* $x \geq 0$ and $y \geq 0$, which means that the feasible region will be in the first quadrant.

→ **EXAMPLE 4   Jewelry Production**

A jewelry company prepares and mounts semiprecious stones. There are 10 lapidaries (who cut and polish the stones) and 12 jewelers (who mount the stones in gold settings). Each employee works 7 hours each day. Each tray of agates requires 5 hours of cutting and polishing and 4 hours of mounting, and each tray of onyxes requires 2 hours of cutting and polishing and 3 hours of mounting. How many trays of each stone can be processed each day?

Formulate this situation as a system of linear inequalities, sketch the feasible region, and find the vertices.

Elnur/Shutterstock

Summarizing

| 10 *Lapidaries* | | |
| 12 *Jewelers* | × 7 hours | |

(hours)	**Agates**	**Onyxes**
Cut & **Polish**	5	2
Mount	4	3

Solution

Clearly, there are many different ways the company could operate each day: It could choose to do nothing and give the employees the day off, it could process just one kind of stone, or it could process some combination. However, it could not exceed the amount of time the lapidaries could work (10 workers at 7 hours each is 70 work-hours) or the amount of time the jewelers could work (12 workers at 7 hours each is 84 work-hours). Nor could it process a negative number of trays.

Let

$$x = \begin{pmatrix} \text{Number of trays} \\ \text{of agates} \end{pmatrix} \quad \text{and} \quad y = \begin{pmatrix} \text{Number of trays} \\ \text{of onyxes} \end{pmatrix}$$

For the lapidaries,

$$5x \quad + \quad 2y \quad \leq \quad 70 \qquad \begin{pmatrix} \text{Time} \\ \text{for} \\ \text{agates} \end{pmatrix} + \begin{pmatrix} \text{Time} \\ \text{for} \\ \text{onyxes} \end{pmatrix} \leq \begin{pmatrix} \text{Total time} \\ \text{for} \\ \text{lapidaries} \end{pmatrix}$$

x trays y trays   No   10 workers
@5 hrs   @2 hrs   more   @7 hrs
each   each   than   each

For the jewelers,

$$4x + 3y \leq 84$$

$\underbrace{4x}$ $\underbrace{3y}$ $\underbrace{}$ $\underbrace{84}$

x trays   y trays   No   12 workers
@4 hrs   @3 hrs   more   @7 hrs
each   each   than   each

$$\begin{pmatrix} \text{Time} \\ \text{for} \\ \text{agates} \end{pmatrix} + \begin{pmatrix} \text{Time} \\ \text{for} \\ \text{onyxes} \end{pmatrix} \leq \begin{pmatrix} \text{Total time} \\ \text{for} \\ \text{jewelers} \end{pmatrix}$$

Combining these constraints with the nonnegativity conditions $x \geq 0$ and $y \geq 0$, we can represent the problem by the system of linear inequalities

$$\begin{cases} 5x + 2y \leq 70 \\ 4x + 3y \leq 84 \\ x \geq 0 \\ y \geq 0 \end{cases}$$

We can now proceed as usual. The intercepts of the boundary line $5x + 2y = 70$ are $(14, 0)$ and $(0, 35)$. The intercepts of the boundary line $4x + 3y = 84$ are $(21, 0)$ and $(0, 28)$. The nonnegativity conditions $x \geq 0$ and $y \geq 0$ place the feasible region in the first quadrant. Since the test point $(0, 0)$ satisfies the constraints, this region has the origin for one of its vertices.

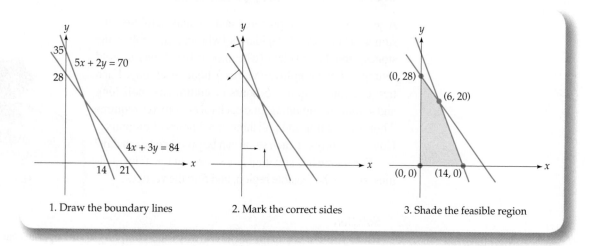

1. Draw the boundary lines   2. Mark the correct sides   3. Shade the feasible region

Three of the four vertices of this bounded region are already known from the x- and y-intercepts of the boundary lines. The fourth is the intersection of $5x + 2y = 70$ and $4x + 3y = 84$:

$$\begin{array}{lll} 5x + 2y = 70 & \times\, 3 \longrightarrow & 15x + 6y = 210 \\ 4x + 3y = 84 & \times\, 2 \longrightarrow & 8x + 6y = 168 \\ \hline & & 7x = 42 \end{array}$$

← Top minus bottom

The solution to this last equation is $x = 6$, which, substituted into either equation, gives $y = 20$, for the intersection point $(6, 20)$. The feasible region has vertices $(0, 0)$, $(14, 0)$, $(6, 20)$, and $(0, 28)$, as shown on the right above.

→ Solutions to Practice Problems

1. The boundary is the line $3x - 5y = 60$. From $x = \frac{60}{3} = 20$, the x-intercept is $(20, 0)$. From $y = \frac{60}{-5} = -12$, the y-intercept is $(0, -12)$. Because $3 \cdot 0 - 5 \cdot 0$ is ≤ 60, the origin $(0, 0)$ is on the correct side of the boundary line.

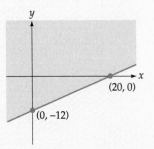

2. The boundary line is $x + 2y = 20$ with intercepts $(20, 0)$ and $(0, 10)$. The origin $(0, 0)$ is on the correct side of $x + 2y \leq 20$ (since $0 \leq 20$). The inequality $x \geq 0$ means "to the right of the y-axis," and $y \geq 0$ means "above the x-axis."

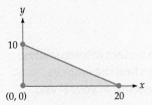

This system *is* feasible.

3. The "reversed" inequality $x + 2y \geq 20$ means the other side of the boundary line, so the region is:

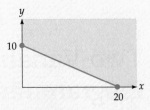

The region is *unbounded*.

Sketch each system of inequalities. List all vertices and identify the region as "bounded" or "unbounded."

1. $\begin{cases} x + 2y \le 40 \\ x \ge 0, \; y \ge 0 \end{cases}$

2. $\begin{cases} -2x + y \le 10 \\ x \le 10 \\ x \ge 0, \; y \ge 0 \end{cases}$

3. $\begin{cases} x + 2y \le 8 \\ x + y \le 6 \\ x \ge 0, \; y \ge 0 \end{cases}$

4. $\begin{cases} 5x + 2y \ge 20 \\ x \ge 0, \; y \ge 0 \end{cases}$

5. $\begin{cases} 4x + 3y \ge 24 \\ y \ge 4 \\ x \ge 0, \; y \ge 0 \end{cases}$

6. $\begin{cases} 3x + y \ge 12 \\ x + y \ge 8 \\ x \ge 0, \; y \ge 0 \end{cases}$

Formulate each situation as a system of inequalities, sketch the feasible region, and find the vertices. Be sure to state clearly the meaning of your x- and y-variables.

7. **Livestock Management** A rancher raises goats and llamas on his 400-acre ranch. Each goat needs 2 acres of land and requires $100 of veterinary care per year, while each llama needs 5 acres of land and requires $80 of veterinary care per year. If the rancher can afford no more than $13,200 for veterinary care this year, how many of each animal can he raise?

8. **Production Planning** A boat company manufactures aluminum dinghies and rowboats. The hours of metal work and painting needed for each are shown in the table, together with the hours of skilled labor available for each task. How many of each kind of boat can the company manufacture?

(hours)	Dinghy	Rowboat	Labor Available
Metal work	2	3	120
Painting	2	2	100

4.2 Two-Variable Linear Programming Problems

In this section we will explain what a linear programming problem is and then use the geometry of feasible regions from the previous section to show that the solution of a linear programming problem occurs at a vertex of the region.

Linear Programming Problems

We begin by formulating an example of a linear programming problem, which we will then solve in Example 2.

→ EXAMPLE 1   Farm Management

A farmer grows soybeans and corn on his 300-acre farm. To maintain soil fertility, the farmer rotates the crops and always plants at least as many acres of soybeans as acres of corn. If each acre of soybeans yields a profit of $100 and each acre of corn yields a profit of $200, how many acres of each crop should the farmer plant to obtain the greatest possible profit?

Formulate this situation as a linear programming problem by identifying the variables, the objective function, and the constraints.

Solution

Since the question asks "how many acres of each crop," we let

$$x = \begin{pmatrix} \text{Number of} \\ \text{acres of soybeans} \end{pmatrix} \quad \text{and} \quad y = \begin{pmatrix} \text{Number of} \\ \text{acres of corn} \end{pmatrix}$$

The objective is to maximize the farmer's profit, and this profit is $P = 100x + 200y$ because the profits per acre are $100 for soybeans and $200 for corn. The 300-acre size of the farm leads to the constraint $x + y \leq 300$, and the crop rotation requirement that $x \geq y$ can be written as $x - y \geq 0$. Since x and y cannot be negative, we also have the nonnegativity constraints $x \geq 0$ and $y \geq 0$. This maximum linear programming problem may be written as

$$\overset{\text{Objective function}}{}$$

Maximize $P = 100x + 200y$ Corn and soybean profits

Subject to $\begin{cases} x + y \leq 300 & \text{Size of farm} \\ x - y \geq 0 & \text{Crop rotation} \\ x \geq 0 \quad \text{and} \quad y \geq 0 & \text{Nonnegativity} \end{cases}$

$$\underset{\text{Constraints}}{}$$

We call $P = 100x + 200y$ the "objective function" because the objective of the entire procedure is to maximize it. The inequalities are called "constraints" because they constrain the profit by, for example, the limiting the crops to 300 acres.

In general, a *linear programming problem* consists of a linear *objective function* to be maximized or minimized subject to *constraints* in the form of a system of linear inequalities. Since an inequality with a $\geq$ can be changed into one with a $\leq$ by multiplying by -1 (for example, $x - y \geq 0$ can be written as $-x + y \leq 0$), every linear programming problem in two variables may be written in one of the following forms:

Linear Programming Problems

Maximize $P = Mx + Ny$ Minimize $C = Mx + Ny$

or

Subject to $\begin{cases} ax + by \leq c \\ \vdots \end{cases}$ Subject to $\begin{cases} ax + by \geq c \\ \vdots \end{cases}$

We call the objective function in the maximum problem P for profit and in the minimum problem C for cost, but problems with different goals are perfectly acceptable. Nonnegativity conditions should be included only when appropriate.

Fundamental Theorem of Linear Programming

A *solution* of a *maximum* linear programming problem is a point that satisfies the system of linear inequalities (a "feasible" point) and that gives the *largest* possible value of the objective function. Similarly, a solution of a *minimum* linear programming problem is a feasible point that gives the *smallest* possible value of the objective function. With either problem, if the constraints have no solutions (they are "infeasible"), then the problem has no solution.

How can we solve such a problem? Consider the objective function $P = 100x + 200y$ from Example 1. If we solve this equation for y, we obtain:

$$200y = -100x + P \qquad \text{Switching sides}$$
$$y = -\tfrac{1}{2}x + \tfrac{1}{200}P \qquad \text{Dividing through by 200}$$

The $y = mx + b$ form of this equation shows that it represents a line of slope $-\tfrac{1}{2}$. Different values of P will give parallel "objective function lines," all with slope $-\tfrac{1}{2}$. In Example 2 we will see that the feasible region for Example 1 is the triangular region in the following graph. On this graph are drawn several of these objective function lines for various values of P. Higher lines correspond to larger value of P, so *maximizing P* means choosing the uppermost line that still intersects the feasible region.

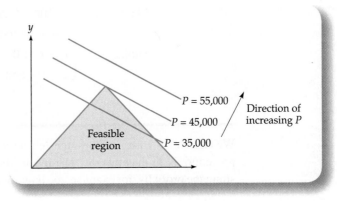

As the diagram shows, the uppermost line that intersects the region does so at a *vertex* of the region. (Any line with a higher value of *P*, such as $P = 55{,}000$, does not intersect the feasible region at all.) It should be clear that the situation will be the same for *any* linear objective function and *any* feasible region bounded by linear inequalities: *For a linear programming problem with a bounded region, the solution occurs at a vertex of the region.*

For an *unbounded* region the solution might not exist: The objective function lines might move in an unbounded direction as *P* increases, allowing higher and higher values of *P*, so there would never be a largest value. However, if increasing the value of *P* always means moving toward a boundary, then the solution *will* exist and again will occur at a vertex.

Similar reasoning applies to a *minimum* linear programming problem by finding the objective function line with the *minimum* value of *P* that intersects the feasible region. Combining these observations, we have established the following important result about the solution of a linear programming problem.

Fundamental Theorem of Linear Programming

If a linear programming problem has a solution, then it occurs at a *vertex* of the feasible region.

The Fundamental Theorem of Linear Programming gives us the following procedure to find a solution.

How to Solve a Linear Programming Problem

If the region is bounded, list the vertices and calculate the value of the objective function at each. The solution occurs at the vertex that gives the largest (or smallest) value.

If the region is unbounded, first check whether the objective function "improves" in an unbounded direction. If it does, there is *no solution*; if it does not, list the vertices and calculate the value of the objective function at each; the solution occurs at the vertex that gives the largest (or smallest) value.

➡ EXAMPLE 2   Solution of Example 1

Solve the linear programming problem from Example 1 on page 145:

$$\text{Maximize} \quad P = 100x + 200y \qquad \text{Objective function}$$

$$\text{Subject to} \quad \begin{cases} x + y \le 300 \\ x - y \ge 0 \\ x \ge 0 \quad \text{and} \quad y \ge 0 \end{cases} \qquad \text{Constraints}$$

Solution

We begin by sketching the region determined by the constraints. The nonnegativity conditions $x \ge 0$ and $y \ge 0$ place the region in the first quadrant. The boundary line $x + y = 300$ has intercepts $(300, 0)$ and $(0, 300)$, and the boundary line $x - y = 0$

passes through the origin $(0, 0)$ and the point $(1, 1)$. These points give the first of the following graphs. For test points, the origin satisfies the inequality $x + y \leq 300$, and the point $(300, 0)$ satisfies the inequality $x - y \geq 0$.

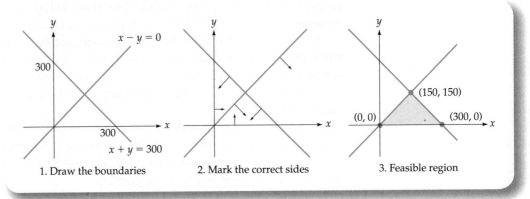

1. Draw the boundaries   2. Mark the correct sides   3. Feasible region

The region has three vertices, and two are already known from the x- and y-intercepts of the boundary lines. The third is the intersection of $x - y = 0$ and $x + y = 300$:

$$x + y = 300$$
$$\underline{x - y = 0}$$
$$2x = 300 \qquad \longleftarrow \text{ Adding}$$

The last equation gives $x = 150$, and from $x + y = 300$, we find $y = 150$, for the vertex $(150, 150)$. The vertices are $(0, 0)$, $(300, 0)$, and $(150, 150)$, as shown in the third graph.

Since the region is bounded, this problem *does* have a solution, and it occurs at a vertex. Evaluating the objective function at the vertices, we find the following values:

Vertex	Value of $P = 100x + 200y$	
$(0, 0)$	0	$0 = 100 \cdot 0 + 200 \cdot 0$
$(300, 0)$	30,000	$30{,}000 = 100 \cdot 300 + 200 \cdot 0$
$(150, 150)$	45,000 ← Largest	$45{,}000 = 100 \cdot 150 + 200 \cdot 150$

Since the largest value of the objective function occurs at the vertex $(150, 150)$, the solution of this problem is:

The maximum value of P is 45,000 at the vertex $(150, 150)$.

In terms of the original word problem in Example 1, the maximum profit is \$45,000 when the farmer plants 150 acres of corn and 150 acres of soybeans.

➔ Practice Problem 1

Solve the linear programming problem:

Maximize $\quad P = 5x + 2y$

Subject to $\quad \begin{cases} 3x + y \leq 60 \\ y \leq 36 \\ x \geq 0 \quad \text{and} \quad y \geq 0 \end{cases}$

➤ Solution on page 153

➡ EXAMPLE 3 A Minimum Problem on an Unbounded Region

Solve the linear programming problem:

$$\text{Minimize} \quad C = 3x + 5y$$

$$\text{Subject to} \quad \begin{cases} 2x + y \ge 12 \\ x + y \ge 8 \\ x \ge 0 \quad \text{and} \quad y \ge 0 \end{cases}$$

Solution

We begin by sketching the region determined by the constraints. The nonnegativity conditions $x \ge 0$ and $y \ge 0$ place the region in the first quadrant. The boundary line $2x + y = 12$ has intercepts $(6, 0)$ and $(0, 12)$, while the boundary line $x + y = 8$ has intercepts $(8, 0)$ and $(0, 8)$. Since the test point $(0, 0)$ does *not* satisfy $2x + y \ge 12$ and $x + y \ge 8$, the region is on the sides of these boundaries *away* from the origin.

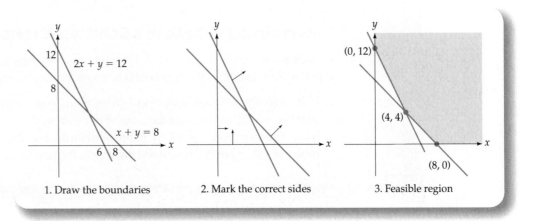

1. Draw the boundaries   2. Mark the correct sides   3. Feasible region

This unbounded region has three vertices, and two are already known from the *x*- and *y*-intercepts of the boundary lines. The third is the intersection of $2x + y = 12$ and $x + y = 8$:

$$2x + y = 12$$
$$\underline{x + y = 8}$$
$$x = 4 \qquad \longleftarrow \text{Top minus bottom}$$

Substituting $x = 4$ into $x + y = 8$ gives $y = 4$, for the intersection point $(4, 4)$. The vertices are $(0, 12)$, $(4, 4)$, and $(8, 0)$, as shown in the third graph above.

The region is unbounded, so the solution may not exist. The third graph above shows that the region is unbounded in the *positive x direction* and also in the *positive y direction*, which means that we could increase x as much as we choose and we could increase y as much as we choose. However, increasing either x or y will *increase* the objective function $C = 3x + 5y$ (since x and y are multiplied by *positive* numbers), which is the *opposite* of what we should do to minimize C. Therefore, the objective function does not "improve" in an unbounded direction, so the solution *does* exist. Evaluating the objective function at the vertices, we find the following values.

Vertex	$C = 3x + 5y$		
(8, 0)	24	← Smallest	$24 = 3 \cdot 8 + 5 \cdot 0$
(4, 4)	32		$32 = 3 \cdot 4 + 5 \cdot 4$
(0, 12)	60		$60 = 3 \cdot 0 + 5 \cdot 12$

Since the smallest value of the objective function occurs at the vertex $(8, 0)$, the solution of this problem is:

The minimum value of C is 24 at the vertex $(8, 0)$.

A linear programming problem with an unbounded feasible region may or may not have a solution. For example, in a maximum problem if the objective function can be made arbitrarily large by choosing x and y values far out in the unbounded direction, then it will not have a largest value, so the problem will have no solution. The general situation is described in the following box, where the phrase *the objective function "improves"* means that it *increases* for a *maximum* problem and that it *decreases* for a *minimum* problem.

Determining Whether a Solution Exists

If the feasible region is unbounded, the linear programming problem may not have a solution. First determine the direction(s) in which the feasible region is unbounded.

1. If the objective function *improves* when the x- and y-variables take values further out an unbounded direction, then the solution *does not exist*.
2. If the objective function does *not* improve when the x- and y-variables take values further out in any of the unbounded directions, then the solution *does* exist.

➡ Practice Problem 2

Solve the linear programming problem:

Maximize $\quad P = 5x + 11y$

Subject to $\quad \begin{cases} x + 3y \geq 60 \\ x \geq 0 \quad \text{and} \quad y \geq 0 \end{cases}$

➤ Solution on page 153

➡ EXAMPLE 4　A Manufacturing Problem

A fully automated plastics factory produces two toys, a racing car and a jet airplane, in three stages: molding, painting, and packaging. After allowing for routine maintenance, the equipment for each stage can operate no more than 150 hours per week. Each batch of racing cars requires 6 hours of molding, 2.5 hours of painting, and 5 hours of packaging, while each batch of jet airplanes requires 3 hours of molding, 7.5 hours of painting, and 5 hours of packaging. If the profit per batch of toys is $120 for cars and $100 for

airplanes, how many batches of each toy should be produced each week to obtain the greatest possible profit?

Summarizing

(hours)	Car	Plane
Molding	6	3
Painting	2.5	7.5
Packaging	5	5
Profit	$120	$100

Solution

Let

$$x = \begin{pmatrix} \text{Number of} \\ \text{batches of cars} \end{pmatrix} \quad \text{and} \quad y = \begin{pmatrix} \text{Number of} \\ \text{batches of jets} \end{pmatrix}$$

made during the week. The profit is $P = 120x + 100y$, and the problem is to maximize P subject to the constraints that the molding, painting, and packaging processes can each take no more than 150 hours:

$$6x + 3y \leq 150 \qquad \text{Molding time}$$
$$2.5x + 7.5y \leq 150 \qquad \text{Painting time}$$
$$5x + 5y \leq 150 \qquad \text{Packaging time}$$

and the nonnegativity conditions $x \geq 0$ and $y \geq 0$. Simplifying the constraints by removing common factors (dividing the molding constraint by 3, the painting constraint by 2.5, and the packaging constraint by 5), we can rewrite this problem as the linear programming problem

Maximize $\quad P = 120x + 100y \qquad$ Objective function

Subject to $\quad \begin{cases} 2x + y \leq 50 \\ x + 3y \leq 60 \\ x + y \leq 30 \\ x \geq 0 \quad \text{and} \quad y \geq 0 \end{cases} \qquad$ Constraints

Proceeding as usual, we graph the boundary lines by plotting their intercepts (the first graph), determine the correct sides by using the origin as a test point (the second graph), and finally shade the feasible region (the third graph).

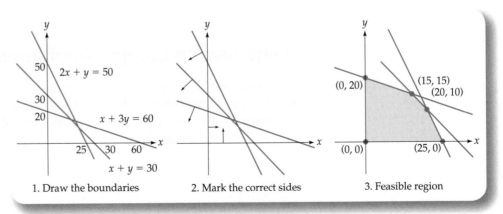

1. Draw the boundaries   2. Mark the correct sides   3. Feasible region

The third graph shows that the region has five vertices, three of which are already known from the x- and y-intercepts. The other two come from solving pairs of equations in the usual way, giving the following vertices:

$$2x + y = 50 \qquad\qquad x + 3y = 60$$
$$\underline{x + y = 30} \qquad\qquad \underline{x + y = 30}$$
$$x \quad\;\; = 20 \qquad\qquad\quad 2y = 30$$

Vertex: $(20, 10)$ Vertex: $(15, 15)$

The vertices are $(0, 0)$, $(25, 0)$, $(20, 10)$, $(15, 15)$, and $(0, 20)$ as shown in the preceding diagram. [The intersection of $2x + y = 50$ and $x + 3y = 60$ is the point $(18, 14)$, but this is *not* a vertex of the region because it does not also satisfy $x + y \leq 30$: $18 + 14$ is not ≤ 30.]

The region is bounded, so this problem *has* a solution, and it occurs at a vertex. Evaluating the objective function at the vertices, we find the following values:

Vertex	$P = 120x + 100y$	
$(0, 0)$	0	
$(25, 0)$	3000	
$(20, 10)$	3400	← —— Largest
$(15, 15)$	3300	
$(0, 20)$	2000	

Since the largest value of the objective function occurs at the vertex $(20, 10)$, the solution is:

The maximum value is 3400 at the vertex $(20, 10)$.

In terms of the original question, the maximum profit is $3400 when the factory produces 20 batches of racing car toys and 10 batches of jet airplane toys each week.

Be careful! Except in very simple problems, some boundary line intersections are *not* vertices of the feasible region because they violate at least one of the other constraints. Particularly when sketching regions by hand, you may want to find all the intersection points anyway and then verify which are feasible and which are not.

Extensions to Larger Problems

Our geometric method is limited to two variables and therefore to two products, which is unrealistic for a large company. In the next two sections we will develop a numerical method that applies to problems with *any* number of variables and that also eliminates the need to find and check *all* the vertices.

1.

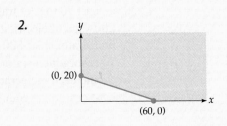

Vertex	$P = 5x + 2y$
(0, 0)	0
(20, 0)	100
(8, 36)	112
(0, 36)	72

The maximum value is 112 when $x = 8$ and $y = 36$.

2.

The region is unbounded in the positive x and positive y directions. Since the objective function $P = 5x + 11y$ increases ("improves") as x and y take values in those directions, *the solution does not exist.*

→ 4.2 EXERCISES

Use the region below to find each maximum or minimum value. If such a value does not exist, explain why not.

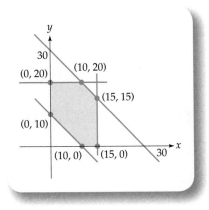

1. Maximum of $P = 2x + y$
2. Minimum of $C = 3x + 4y$

Use the region below to find each maximum or minimum value. If such a value does not exist, explain why not.

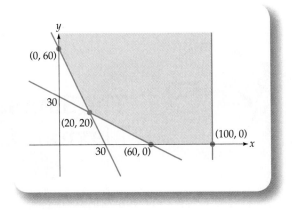

3. Minimum of $C = 3x + y$
4. Maximum of $P = 5x + 8y$

Solve each linear programming problem by sketching the region and labeling the vertices, deciding whether a solution exists, and then finding it if it does exist.

5. Maximize $P = 30x + 40y$

Subject to $\begin{cases} 2x + y \le 16 \\ x + y \le 10 \\ x \ge 0, \quad y \ge 0 \end{cases}$

6. Minimize $C = 15x + 45y$

Subject to $\begin{cases} 2x + 5y \ge 20 \\ x \ge 0, \quad y \ge 0 \end{cases}$

7. Minimize $C = 12x + 10y$

Subject to $\begin{cases} 4x + y \ge 40 \\ 2x + 3y \ge 60 \\ x \ge 0, \quad y \ge 0 \end{cases}$

8. Maximize $P = 5x + 3y$

Subject to $\begin{cases} 2x + y \le 90 \\ x + y \le 50 \\ x + 2y \le 90 \\ x \ge 0, \quad y \ge 0 \end{cases}$

Formulate each situation as a linear programming problem by identifying the variables, the objective function, and the constraints. Be sure to state clearly the meaning of each variable. Determine whether a solution exists, and if it does, find it. State your final answer in terms of the original question.

9. **Livestock Management** A rancher raises goats and llamas on his 400-acre ranch. Each goat needs 2 acres of land and requires $100 of veterinary care per year, and each llama needs 5 acres of land and requires $80 of veterinary care per year. The rancher can afford no more than $13,200 for veterinary care this year. If the expected profit is $60 for each goat and $90 for each llama, how many of each animal should he raise to obtain the greatest possible profit?

10. **Waste Management** The Marshall County trash incinerator in Norton burns 10 tons of trash per hour and cogenerates 6 kilowatts of electricity, while the Wiseburg incinerator burns 5 tons per hour and cogenerates 4 kilowatts. The county needs to burn at least 70 tons of trash and cogenerate at least 48 kilowatts of electricity every day. If the Norton incinerator costs $80 per hour to operate and the Wiseburg incinerator costs $50, how many hours should each incinerator operate each day with the least cost to the county?

4.3 The Simplex Method for Standard Maximum Problems

Using the geometric method from the previous section as our guide, we shall explain the *simplex method* for solving linear programming problems. This method finds a *path of vertices* leading from an initial vertex to a solution without finding all the intersection points of the boundary lines or even all the vertices of the feasible region. We begin by defining a *standard maximum problem*, the type of problem that we will solve in this section. Another type of problem will be solved in the next section.

Standard Maximum Problems

A standard maximum problem is a linear programming problem in which the objective function is to be maximized, the variables $x_1, x_2, \dots, x_n$ are all nonnegative, and the other constraints have $\leq$ inequalities with nonnegative numbers on the right-hand sides.

Standard Maximum Problem

A problem is a *standard maximum problem* if it can be written:

Maximize an objective function

$$P = c_1 x_1 + c_2 x_2 + \cdots + c_n x_n$$

Subject to inequalities of the form

$$a_1 x_1 + a_2 x_2 + \cdots + a_n x_n \leq b \quad \text{with} \quad b \geq 0$$

and

$$x_1 \geq 0, x_2 \geq 0, \dots, x_n \geq 0$$

Observe that if we substitute 0 for each variable, the inequalities are *all* satisfied, so for a standard maximum problem the origin *is* a vertex of the feasible region.

➡ EXAMPLE 1 Verifying Standard Maximum Form

Is the following a standard maximum problem?

$$\text{Maximize} \quad P = 9x_1 + 16x_2$$

$$\text{Subject to} \quad \begin{cases} 3x_1 + 4x_2 \leq 24 \\ x_1 + 2x_2 \leq 10 \\ x_1 \geq 0 \quad \text{and} \quad x_2 \geq 0 \end{cases}$$

Solution
The objective function is to be maximized, the variables are both nonnegative, and the other inequalities all have $\leq$ with nonnegative numbers on the right, so it *is* a standard maximum problem.

➡ Practice Problem 1

Is the following a standard maximum problem?

$$\text{Maximize} \quad P = 7x_1 + 5x_2$$

$$\text{Subject to} \quad \begin{cases} 2x_1 - 3x_2 \leq 24 \\ 5x_1 + 2x_2 \leq -4 \\ x_1 \geq 0 \quad \text{and} \quad x_2 \geq 0 \end{cases}$$

➤ Solution on page 166

Matrix Form of a Standard Maximum Problem

We may express a standard maximum problem in the matrix notation of the preceding chapter. Just as two matrices are *equal* if they have the same dimension and corresponding elements are equal, two matrices obey an *inequality* if they have the same dimension and corresponding elements obey that inequality. For example:

$$\begin{pmatrix} 1 & 2 \\ 3 & 4 \end{pmatrix} \le \begin{pmatrix} 1 & 2 \\ 6 & 8 \end{pmatrix}$$

Matrices are the same size, and corresponding elements obey $\le$

Using matrix inequalities, the problem written on the left below (from Example 1) may be expressed in the matrix form on the right:

Maximize $P = 9x_1 + 16x_2$

Subject to $\begin{cases} 3x_1 + 4x_2 \le 24 \\ x_1 + 2x_2 \le 10 \\ x_1 \ge 0 \\ x_2 \ge 0 \end{cases}$

Maximize $P = \begin{pmatrix} 9 & 16 \end{pmatrix} \begin{pmatrix} x_1 \\ x_2 \end{pmatrix}$

Subject to $\begin{cases} \begin{pmatrix} 3 & 4 \\ 1 & 2 \end{pmatrix} \begin{pmatrix} x_1 \\ x_2 \end{pmatrix} \le \begin{pmatrix} 24 \\ 10 \end{pmatrix} \\ \begin{pmatrix} x_1 \\ x_2 \end{pmatrix} \ge 0 \end{cases}$

In this way, any standard maximum problem may be expressed in the following matrix form, using $X = \begin{pmatrix} x_1 \\ \vdots \\ x_n \end{pmatrix}$ for a column matrix of variables:

4x5 FILM

Standard Maximum Problem in Matrix Form

Maximize $P = c^t X$

Subject to $\begin{cases} AX \le b \\ X \ge 0 \end{cases}$ where $b \ge 0$

220 EPC SSO

We write c^t for the new matrix of coefficients for historical reasons only—the mathematicians who developed these methods preferred *column* matrices and so represented row matrices as *transposed* column matrices.

The Initial Simplex Tableau

Equations are easier to solve than inequalities, so we first simplify the problem by changing the inequalities to equations by adding to each a *slack variable* that is nonnegative and represents the "amount not used." For example, in the inequality $4 \le 7$ the two sides differ by 3, so adding 3 to the left side gives an *equation* $4 + 3 = 7$.   Similarly, the inequality $x \le 7$ can be rewritten as the *equation* $x + s = 7$ by adding a slack variable $s \ge 0$ to the left side to make up the difference between the sides. We do this for each inequality, using a different slack variable for each inequality.

➡ EXAMPLE 2   Introducing Slack Variables

Write the following inequalities (from Example 1) as equations with slack variables.

$$\begin{cases} 3x_1 + 4x_2 \le 24 \\ x_1 + 2x_2 \le 10 \end{cases}$$

Solution

We introduce a slack variable $s_1 \ge 0$ into the first inequality and another slack variable $s_2 \ge 0$ into the second to obtain the *equations* shown in the middle below.

$$\begin{cases} 3x_1 + 4x_2 \le 24 \\ x_1 + 2x_2 \le 10 \end{cases} \quad \text{becomes} \quad \begin{cases} 3x_1 + 4x_2 + s_1 \quad\;\; = 24 \\ x_1 + 2x_2 + \quad\; s_2 = 10 \end{cases} \quad \text{or} \quad \begin{cases} 3x_1 + 4x_2 + 1s_1 + 0s_2 = 24 \\ x_1 + 2x_2 + 0s_1 + 1s_2 = 10 \end{cases}$$

The version on the right above has *all* the variables included, with the unnecessary ones multiplied by zeros.

1 Incorporate slack variables

2 Rewrite objective function

3 Build tableau

jules2000/Shutterstock (frame) and Gilmanshin/Shutterstock (flag)

Slack variables are sometimes called simply *slacks*.

We can write the objective function $P = 9x_1 + 16x_2$ (from Example 1) with all variables moved to the left of the equals sign, including the slacks multiplied by zeros:

$$P = 9x_1 + 16x_2 \quad \text{becomes} \quad P - 9x_1 - 16x_2 + 0s_1 + 0s_2 = 0$$

Since both the objective function and the constraints from Example 1 are now written as equations, with the variables in the same order, the entire problem can be summarized in a table, called the "simplex tableau" (using the French word for *table*, whose plural is *tableaux*). We state the general form and then give an example.

Initial Tableau

The initial tableau for the linear programming problem is

Maximize $P = c^t X$

Subject to $\begin{cases} AX \le b \\ X \ge 0 \end{cases}$ is

	X	S	
S	A	I	b
P	$-c^t$	0	0

where S stands for the slack variables and I stands for the identity matrix.

➡ EXAMPLE 3   Writing a Simplex Tableau

Write the initial simplex tableau for the following linear programming problem (from Example 1):

Maximize $P = 9x_1 + 16x_2$

Subject to $\begin{cases} 3x_1 + 4x_2 \le 24 \\ x_1 + 2x_2 \le 10 \\ x_1 \ge 0 \quad \text{and} \quad x_2 \ge 0 \end{cases}$

Solution

We write the constraints as

$$\begin{cases} 3x_1 + 4x_2 + 1s_1 + 0s_2 = 24 \\ x_1 + 2x_2 + 0s_1 + 1s_2 = 10 \end{cases}$$ From Example 2

and the objective function as

$$P - 9x_1 - 16x_2 + 0s_1 + 0s_2 = 0$$ From $P = 9x_1 + 16x_2$

For the initial simplex tableau we write the variables on the top, the constraint numbers in the middle, and the objective numbers on the bottom (all in the correct columns), and finally the slack variables on the left:

	x_1	x_2	s_1	s_2	
s_1	3	4	1	0	24
s_2	1	2	0	1	10
P	-9	-16	0	0	0

Column variables

Constraints

Objective function (signs changed)

Represents =

From the original problem:

Maximize $P = 9x_1 + 16x_2$

Subject to $\begin{cases} 3x_1 + 4x_2 \leq 24 \\ x_1 + 2x_2 \leq 10 \\ x_1 \geq 0, \quad x_2 \geq 0 \end{cases}$

➤ **Practice Problem 2**

Write the initial simplex tableau for the problem

Maximize $P = 3x_1 + 5x_2$

Subject to $\begin{cases} 2x_1 - 3x_2 \leq 24 \\ 5x_1 + 2x_2 \leq 20 \\ 3x_1 + 2x_2 < 12 \\ x_1 \geq 0 \quad \text{and} \quad x_2 \geq 0 \end{cases}$

➤ **Solution on page 166**

Basic and Nonbasic Variables

In an initial simplex tableau like the one below (from Example 3), the slack variables s_1 and s_2 have two special properties.

	x_1	x_2	s_1	s_2	
s_1	3	4	1	0	24
s_2	1	2	0	1	10
P	-9	-16	0	0	0

1. Their columns contain all zeros except for exactly one 1, which occurs above the bottom row.
2. Each of the rows above the bottom row has exactly one of these special 1s. These special variables determine a feasible point in the region without any further calculation: Just give the variable the value that is at the right-hand end of its row and set all other variables equal to 0. For the above tableau, this means:

	x_1	x_2	s_1	s_2	
s_1	3	4	1	0	24	← $s_1 = 24$
s_2	1	2	0	1	10	← $s_2 = 10$
P	−9	−16	0	0	0	and $x_1 = x_2 = 0$

Other variables set equal to 0

The general procedure for determining basic and nonbasic variables in a tableau is as follows:

Basic and Nonbasic Variables in a Simplex Tableau

Count the number of slack variables (or, equivalently, the number of constraints). This number, m, will be the number of basic variables throughout the solution.
1. For basic variables, choose any m variables whose columns have all zeros except for one 1 that must appear above the bottom row and such that each of the rows above the bottom row has exactly one of these special 1s; assign to each basic variable the value at the right-hand end of the row containing this 1.
2. All *other* variables are *nonbasic* variables; assign to each the value 0.
3. At these values, the objective function equals the number in the bottom right corner.

Notice that statements 1 and 2 agree with the values we assigned to the variables in the tableau above the box. Furthermore, the objective function $P = 9x_1 + 16x_2$ evaluated at the assigned $x_1 = 0$ and $x_2 = 0$ is $P = 0$, the number in the bottom right corner of the tableau, agreeing with statement 3 in the box. Because of their importance, we list the basic variables on the left side of the tableau, each in the row corresponding to its special 1. We interpret the tableau as specifying a basic feasible point (a vertex) of the feasible region along with the value of the objective function at that point.

The Pivot Element

The simplex method begins by taking the slack variables as the basic variables and then successively finds better basic variables to increase the value of the objective function until the solution is found. The method mimics the geometric procedure that we used on pages 147–148 to find vertices, adding and subtracting equations as we did on pages 98–99 to solve systems of equations, a process we now call *pivoting*. We will explain the process by referring to the problem in Example 1, but explaining how to carry out the steps on the simplex tableau since we use the tableau to represent the problem.

In the objective function from Example 1, $P = 9x_1 + 16x_2$, x_1 is multiplied by 9 and x_2 is multiplied by a larger number, 16, so increasing x_2 should do more to increase P than increasing x_1. In general, to increase *any* objective function, we should increase the variable with the *largest positive* coefficient. Since the bottom row of the tableau contains the *negatives* of these coefficients, this means choosing the column with the *smallest negative* number in the bottom row. That is the *pivot column*.

Pivot Column

The *pivot column* is the column with the smallest negative entry in the bottom row (omitting the right column). If there is a tie, choose the leftmost such column.

Be careful! The smallest negative number among 12, −8, and −5 is −8. (Students sometimes call this step choosing *the "most negative" number*).

Having chosen the column and therefore the variable to increase, how much can we increase it without violating a constraint? The boundary lines for the constraints in Example 1 are on the left below. Recall from our geometric method that we cannot increase x_2 beyond the *intercepts* of these boundaries, which are calculated on the right below.

Boundary line	x_2-intercept	
$3x_1 + 4x_2 = 24$	$x_2 = \frac{24}{4} = 6$	⟵ x_2 cannot exceed 6
$x_1 + 2x_2 = 10$	$x_2 = \frac{10}{2} = 5$	⟵ x_2 cannot exceed 5

To stay below *two* numbers means staying below the *smaller* of them, so x_2 must not exceed the *smaller* of the two numbers, each number being the right-hand side of the inequality divided by the number in the pivot column. We now have the rule for choosing the *pivot row* from the simplex tableau:

Pivot Row

For each row (except the bottom), divide the rightmost entry by the pivot column entry (omitting any row with a zero or negative pivot column entry). The row with the smallest ratio is the *pivot row*. If there is a tie, choose the uppermost such row.

We ignore rows with zeros in the pivot column (because the corresponding inequality would not include that variable) and with negative numbers (because such an inequality would not restrict the variable from increasing).

Pivot Element

The *pivot element* is the entry in the pivot column and the pivot row.

The Pivot Operation

Once we have found the pivot element, we want to increase the selected variable as much as possible (up to the boundary of the feasible region) and to increase the objective function correspondingly. This is accomplished by the pivot operation, which is just the tableau version of the elimination method that we used for solving equations on pages 92–93.

Pivot Operation

1. Divide every entry in the pivot row by the pivot element to obtain a *new* pivot row, which will then have a 1 where the pivot element was. Replace the variable at the left-

hand end of this row by the variable corresponding to the pivot column. The variables listed on the left will be the new *basic* variables.

2. Subtract multiples of the new pivot row from all other rows of the tableau to get zeros in the pivot column above and below the 1.

Specifically, if R_p is the new pivot row found from step 1 and R is any *other* row and if we denote the entry in that row and in the pivot column by p, then the row operation in step 2 for that row is $R - p \cdot R_p \to R$.

➡ **EXAMPLE 4** Finding the Pivot Element and Pivoting

Find the pivot element in the following simplex tableau (from Example 3) and carry out the pivot operation.

	x_1	x_2	s_1	s_2	
s_1	3	4	1	0	24
s_2	1	2	0	1	10
P	-9	-16	0	0	0

Solution

The smallest negative number in the bottom row is -16, so the pivot column is column 2 (as shown below). To find the pivot row, we divide the last entry of each row by the pivot column entry for that row (skipping the bottom row and any row with a zero or a negative pivot column entry). We choose the row with the smallest nonnegative ratio, as shown by the calculations to the right of the following tableau. The pivot row is then row 2.

	x_1	x_2	s_1	s_2	
s_1	3	4	1	0	24	$\frac{24}{4} = 6$
s_2	1	2	0	1	10	$\frac{10}{2} = 5$ ⟵ Smallest ratio ⟵ Pivot row
P	-9	-16	0	0	0

Smallest negative

Pivot column ⟶

Pivot element (we want a 1 here)

Having found the pivot element, we carry out the pivot operation. The first step is to divide the pivot row by the pivot element, so we divide the second row by 2, which changes the pivot element to a 1. We also update the basis on the left—the pivot column variable x_2 replaces the pivot row variable s_2, so the new basic variables are s_1 and x_2.

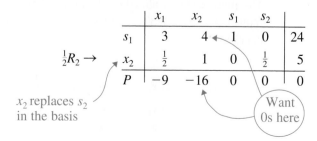

$\frac{1}{2}R_2 \to$

	x_1	x_2	s_1	s_2	
s_1	3	4	1	0	24
x_2	$\frac{1}{2}$	1	0	$\frac{1}{2}$	5
P	-9	-16	0	0	0

x_2 replaces s_2 in the basis

Want 0s here

The second step in the pivot operation is to subtract multiples of the *new* pivot row from the other rows to "zero out" the rest of the pivot column (the 4 and the -16). This is accomplished by the row operations listed on the left of the following tableau.

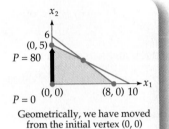

x_2
6
$(0, 5)$
$P = 80$
$(0, 0)$ $(8, 0)$ 10 x_1
$P = 0$

Geometrically, we have moved
from the initial vertex $(0, 0)$
to the vertex $(0, 5)$,
increasing P from 0 to 80.

	x_1	x_2	s_1	s_2		
$R_1 - 4R_2 \rightarrow$ s_1	1	0	1	-2	4	$\leftarrow s_1 = 4$
x_2	$\frac{1}{2}$	1	0	$\frac{1}{2}$	5	$\leftarrow x_2 = 5$
$R_3 + 16R_2 \rightarrow$ P	-1	0	0	8	80	$\leftarrow P = 80$

This completes the pivot operation. The basic variables have the values shown on the right, with the other variables set equal to zero (in particular, $x_1 = 0$). Observe that when $x_1 = 0$ and $x_2 = 5$ the objective function $P = 9x_1 + 16x_2$ *does* take the value of 80 as given in the lower right corner of the tableau.

All the steps we have just carried out—finding the pivot element, updating the basic variables on the left of the tableau, dividing to replace the pivot element by a 1, and zeroing out the rest of the column by subtracting multiples of the pivot row—constitute *one* pivot operation. You may want to carry out some of the calculations on scratch paper.

Notice that the tableau above still has a negative number (the -1) in the bottom row of the above tableau, meaning that there is more pivoting to be done.

➡ EXAMPLE 5 Pivoting a Second Time

Perform the pivot operation on the following tableau (from the end of Example 4):

	x_1	x_2	s_1	s_2	
s_1	1	0	1	-2	4
x_2	$\frac{1}{2}$	1	0	$\frac{1}{2}$	5
P	-1	0	0	8	80

Solution
The -1 in the bottom row means that the pivot column is column 1. The calculations on the right below show that the pivot row is row 1.

	x_1	x_2	s_1	s_2		
s_1	1	0	1	-2	4	$\frac{4}{1} = 4 \leftarrow$ Smallest $\leftarrow$ Pivot row
x_2	$\frac{1}{2}$	1	0	$\frac{1}{2}$	5	$\frac{5}{1/2} = 10$
P	-1	0	0	8	80	

Pivot column   Pivot element

The pivot entry is already a 1, so no division is needed (but we update the basis by replacing s_1 by x_1). The two row operations listed on the left below complete the pivot operation.

	x_1	x_2	s_1	s_2		
x_1 replaces $s_1 \rightarrow$ x_1	1	0	1	-2	4	$\leftarrow x_1 = 4$
$R_2 - \frac{1}{2}R_1 \rightarrow$ x_2	0	1	$-\frac{1}{2}$	$\frac{3}{2}$	3	$\leftarrow x_2 = 3$
$R_3 + R_1 \rightarrow$ P	0	0	1	6	84	$\leftarrow P = 84$

—

162 *Chapter 4: Linear Programming*

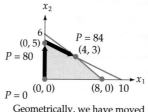

x_2

6
(0, 5)
$P = 84$
(4, 3)
$P = 80$

x_1

(0, 0)   (8, 0) 10
$P = 0$

Geometrically, we have moved from (0, 5) to the vertex (4, 3), increasing P from 80 to 84.

There are no negative numbers in the bottom row, so no more pivoting is possible—the objective function has been maximized. According to the box on page 159, we take the values from this final tableau: P from the bottom right corner, the basic variables from the right side, with the nonbasic variables set to 0.

The maximum value of P is 84 when $x_1 = 4$ and $x_2 = 3$.

Usually only the x-values are given in the solution, since only they were stated in the original problem. The nonbasic variables here take the values $s_1 = 0$ and $s_2 = 0$. In Example 7 we will see how to interpret the slack variables as "unused resources."

We continue to pivot until no more pivoting is possible, that is, until we arrive at a *final tableau*. How do we recognize and interpret a final tableau? Recall that choosing the smallest negative number in the bottom row was equivalent to choosing the variable that will increase the objective function most quickly. Therefore, if there is *no* negative number in the bottom row, then there is no variable that will increase the objective function, so P must be at its maximum value, and we read the solution from the tableau as we did in the preceding example. On the other hand, if there *is* a negative number in the bottom row (so that the objective function *can* be increased) but there is *no* pivot row, then there are no constraints limiting the objective function, which can therefore be increased arbitrarily, meaning that there is *no solution*.

Interpreting a Final Tableau

1. If all the numbers in the bottom row are positive or zero (so there is no pivot column), then the original problem *does* have a solution:
   a. The (basic) variables listed on the left of the rows take the values at the right ends of those rows.
   b. The (nonbasic) variables not listed on the left take the value 0.
   c. The maximum value of the objective function is in the lower right corner of the tableau.
2. If there *is* a negative number in the bottom row but every other element in its column is negative or zero (so there is a pivot column but no pivot row), then the original problem has *no solution*.

→ EXAMPLE 6 Interpreting a Final Tableau

Interpret the following final tableaux:

a.

	x_1	x_2	s_1	s_2	
s_1	2	0	1	3	4
x_2	5	1	0	6	7
P	8	0	0	9	10

b.

	x_1	x_2	s_1	s_2	
s_1	-2	0	1	3	4
x_2	0	1	0	6	7
P	-8	0	0	9	10

Solution

a. The first tableau has no negative numbers in the bottom row, so the original problem *does* have a solution. The basic variables listed on the left, s_1 and x_2, take the values on the right, 4 and 7; the other variables, s_2 and x_1, take the value 0. The value of P is in the lower right corner:

The maximum value of P is 10 when $x_1 = 0$ and $x_2 = 7$.

b. The second tableau *does* have a negative number in the bottom row, -8, so there *is* a pivot column (column 1). However, there is no pivot *row* (the ratios are $\frac{4}{-2}$, which is negative, and $\frac{7}{0}$, which is undefined). Therefore, there is *no solution*.

The Simplex Method

The simplex method proceeds by pivoting until a final tableau is reached and then interpreting it.

Simplex Method

To solve a standard maximum problem (page 155) by the simplex method:

1. Construct the initial tableau (see page 157).
2. Locate the pivot element (see pages 159–160) and go to step 3. If the tableau does not have a pivot element, go to step 4.
3. Perform the pivot operation (see pages 160–161) using the pivot element and return to step 2.
4. If the final tableau does not have a pivot column (see page 159), then the solution occurs at the vertex where the (basic) variables listed on the left of the tableau take the values at the right-hand ends of those rows; the other (nonbasic) variables take the value 0. The maximum value of the objective function is in the bottom right corner of the tableau.

   If the final tableau has a pivot *column* but no pivot *row* (see page 160), the problem has *no solution.*

→ EXAMPLE 7 A Manufacturing Problem with Many Variables

A pottery shop manufactures dinnerware in four different patterns by shaping the clay, decorating it, and then kiln-firing it. The numbers of hours required per place setting for the four designs are given in the following table, together with the expected profits. The shop employs two skilled workers to do the initial shaping and three artists to do the decorating, none of whom will work more than 40 hours each week. The kiln can be used no more than 55 hours per week. How many place settings of each pattern should be made this week to obtain the greatest possible profit?

	Classic	Modern	Art Deco	Floral
Shaping	2 hours	1 hour	4 hours	2 hours
Decorating	3 hours	1 hour	6 hours	4 hours
Kiln-firing	1 hour	1 hour	1 hour	1 hour
Expected profit	$10	$6	$9	$8

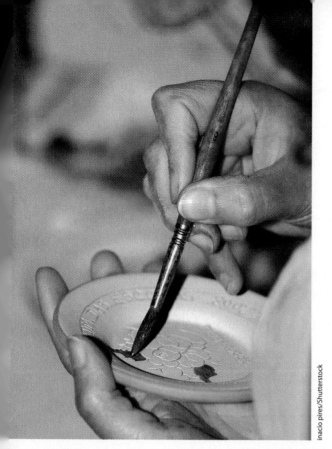

Solution

Let

$$x_1 = \left(\begin{array}{c} \text{Number of} \\ \text{Classic settings} \end{array} \right), \qquad x_2 = \left(\begin{array}{c} \text{Number of} \\ \text{Modern settings} \end{array} \right),$$

$$x_3 = \left(\begin{array}{c} \text{Number of} \\ \text{Art Deco settings} \end{array} \right), \quad \text{and} \quad x_4 = \left(\begin{array}{c} \text{Number of} \\ \text{Floral settings} \end{array} \right)$$

made during the week, so $x_1 \geq 0$, $x_2 \geq 0$, $x_3 \geq 0$, and $x_4 \geq 0$. From the last line in the table, the profit to be maximized is

$$P = 10x_1 + 6x_2 + 9x_3 + 8x_4.$$

The constraints on the time spent shaping, decorating, and firing come from the other numbers in the table, along with the time available for each:

$$2x_1 + x_2 + 4x_3 + 2x_4 \leq 80 \qquad \text{2 shapers @ 40 hours each}$$
$$3x_1 + x_2 + 6x_3 + 4x_4 \leq 120 \qquad \text{3 decorators @ 40 hours each}$$
$$x_1 + x_2 + x_3 + x_4 \leq 55 \qquad \text{Time kiln can be used}$$

The initial simplex tableau is

	x_1	x_2	x_3	x_4	s_1	s_2	s_3	
s_1	②	1	4	2	1	0	0	80
s_2	3	1	6	4	0	1	0	120
s_3	1	1	1	1	0	0	1	55
P	−10	−6	−9	−8	0	0	0	0

$\frac{80}{2} = 40$ ⟵ Smallest ratio ⟵ Pivot row

$\frac{120}{3} = 40$ (Tie for smallest ratio, so take top row)

$\frac{55}{1} = 55$

Pivot element

Smallest negative

Pivot column

When we pivot (on the 2 in column 1, row 1), the tableau becomes

x_1 replaces s_1 in basis

	x_1	x_2	x_3	x_4	s_1	s_2	s_3	
x_1	1	1/2	2	1	1/2	0	0	40
s_2	0	−1/2	0	1	−3/2	1	0	0
s_3	0	①/2	−1	0	−1/2	0	1	15
P	0	−1	11	2	5	0	0	400

$\frac{40}{1/2} = 80$

(Cannot use since $-\frac{1}{2}$ is < 0)

$\frac{15}{1/2} = 30$ ⟵ Smallest ratio, Pivot row

Pivot Element

Smallest negative

Pivot column

Pivoting again (on the 1/2 in column 2, row 3), we reach the final tableau:

	x_1	x_2	x_3	x_4	s_1	s_2	s_3	
x_1	1	0	3	1	1	0	-1	25
s_2	0	0	-1	1	-2	1	1	15
x_2	0	1	-2	0	-1	0	2	30
P	0	0	9	2	4	0	2	430

x_2 replaces s_3 in the basis ⟶

No negatives in bottom row
so final tableau

The basic variables take the values at the right ends of their rows, the nonbasic variables are zero, and the objective function value appears in the bottom right corner. Therefore, the maximum value of P is 430 when $x_1 = 25$, $x_2 = 30$, $x_3 = 0$, and $x_4 = 0$. The fact that $s_2 = 15$ means 15 hours of the available decorating time will not be needed.

Answer: The pottery shop should manufacture 25 place settings of Classic, 30 of Modern, and none of Art Deco and Floral.

➡ Solutions to Practice Problems

1. No. The -4 on the right of the second inequality violates the $b \geq 0$ condition in the box on page 155. (The -3 in the first inequality is not a violation: only the right-hand sides need to be nonnegative.)

2.

	x_1	x_2	s_1	s_2	s_3	
s_1	2	-3	1	0	0	24
s_2	5	2	0	1	0	20
s_3	3	2	0	0	1	12
P	-3	-5	0	0	0	0

➡ 4.3 Exercises

Construct the initial simplex tableau for each standard maximum problem.

1. Maximize $P = 8x_1 + 9x_2$

   Subject to $\begin{cases} 3x_1 + 2x_2 \leq 12 \\ 6x_1 + x_2 \leq 15 \\ x_1 \geq 0, \quad x_2 \geq 0 \end{cases}$

2. Maximize $P = 13x_1 + 7x_2$

   Subject to $\begin{cases} 4x_1 + 3x_2 \leq 12 \\ 5x_1 + 2x_2 \leq 20 \\ x_1 + 6x_2 \leq 12 \\ x_1 \geq 0, \quad x_2 \geq 0 \end{cases}$

For each simplex tableau, find the pivot element and carry out one complete pivot operation. If there is no pivot element, explain what the tableau shows about the solution of the original standard maximum problem.

3.

	x_1	x_2	s_1	s_2	
s_1	2	0	1	0	4
s_2	1	1	0	1	5
P	-7	-8	0	0	0

4.

	x_1	x_2	x_3	x_4	s_1	s_2	
s_1	4	2	6	2	1	0	12
s_2	3	1	2	1	0	1	8
P	-4	-5	6	-3	0	0	0

	x_1	x_2	x_3	x_4	s_1	s_2	s_3	
s_1	0	0	2	1	1	1	−1	10
x_2	0	1	1	1	0	1	0	15
x_1	1	0	−2	0	0	−1	1	10
P	0	0	0	2	0	1	3	90

5.

Solve each problem by the simplex method. (Exercise 6 can also be solved by the graphical method.)

6. Maximize $P = x_1 + 2x_2$

Subject to $\begin{cases} 3x_1 + x_2 \le 24 \\ x_1 + x_2 \le 14 \\ x_1 \ge 0, \quad x_2 \ge 0 \end{cases}$

7. Maximize $P = 6x_1 + 4x_2 + 5x_3$

Subject to $\begin{cases} x_1 - x_2 + x_3 \le 20 \\ x_1 + 2x_3 \le 10 \\ x_1 \ge 0, \quad x_2 \ge 0, \quad x_3 \ge 0 \end{cases}$

8. Maximize $P = 4x_1 + 3x_2 - 5x_3 + 6x_4$

Subject to $\begin{cases} x_1 + x_2 + x_4 \le 60 \\ x_1 + x_3 + x_4 \le 40 \\ x_1 + x_2 + x_3 \le 50 \\ x_1 \ge 0, \quad x_2 \ge 0, \quad x_3 \ge 0, \quad x_4 \ge 0 \end{cases}$

Formulate each situation as a linear programming problem by identifying the variables, the objective function, and the constraints. Be sure to state clearly the meaning of each variable. Check that the problem is a standard maximum problem and then solve it by the simplex method. State your final answer in terms of the original question.

9. *Production Planning* An automotive parts shop rebuilds carburetors, fuel pumps, and alternators. The numbers of hours to rebuild and then inspect and pack each part are shown in the following table, together with the number of hours of skilled labor available for each task. If the profit is $12 for each carburetor, $14 for each fuel pump, and $10 for each alternator, how many of each should the shop rebuild to obtain the greatest possible profit?

(hours)	Rebuilding	Inspection & Packaging
Carburetor	5	1
Fuel Pump	4	1
Alternator	3	0.5
Labor Available	200	45

10. *Agriculture* A farmer grows corn, peanuts, and soybeans on his 240-acre farm. To maintain soil fertility, the farmer rotates the crops and always plants at least as many acres of soybeans as the total acres of the other crops. Each acre of corn requires 2 days of labor and yields a profit of $150, each acre of peanuts requires 5 days of labor and yields a profit of $300, and each acre of soybeans requires 1 day of labor and yields a profit of $100. If the farmer and his family can put in at most 630 days of labor, how many acres of each crop should the farmer plant to obtain the greatest possible profit?

ATTENTION
NEED MORE PRACTICE? FIND MORE HERE:
CENGAGEBRAIN.COM

4.4 Standard Minimum Problems and Duality

In this section we will define and solve *standard minimum* linear programming problems. Rather than develop a new technique for minimum problems, we will find a relationship between minimum and maximum problems that allows us to solve one by solving the other.

Standard Minimum Problems

A *standard minimum problem* is a linear programming problem in which the objective function is to be minimized, all the coefficients in the objective function are nonnegative, the variables $y_1, y_2, \ldots y_n$ are all nonnegative, and the other constraints have $\geq$ inequalities.

Standard Minimum Problem

A problem is a *standard minimum problem* if it can be written:

Minimize an objective function

$$C = b_1 y_1 + b_2 y_2 + \cdots + b_n y_n \quad \text{with } nonnegative \quad b_1, b_2, \ldots, b_n$$

Subject to inequalities of the form

$$a_1 y_1 + a_2 y_2 + \cdots + a_n y_n \geq c$$

and

$$y_1 \geq 0, \quad y_2 \geq 0, \ldots, y_n \geq 0$$

We use C for the objective function to suggest cost, which is usually minimized, and y-variables to distinguish them from the x-variables used in maximum problems. Notice that in a standard minimum problem it is the *coefficients of the objective function* that must be nonnegative—the right-hand sides of the constraints may be of either sign.

→ EXAMPLE 1　Verifying Standard Minimum Form

Is the following a standard minimum problem?

$$\text{Minimize} \quad C = 24y_1 + 10y_2$$

$$\text{Subject to} \quad \begin{cases} 3y_1 + y_2 \geq 9 \\ 4y_1 + 2y_2 \geq 16 \\ y_1 \geq 0 \quad \text{and} \quad y_2 \geq 0 \end{cases}$$

Solution
The objective function is to be minimized, the coefficients 24 and 10 in the objective function are nonnegative, the variables are nonnegative, and the other inequalities all have $\geq$ inequalities, so this *is* a standard minimum problem.

→ Practice Problem 1

Is the following a standard minimum problem?

$$\text{Minimize} \quad C = 24y_1 - 4y_2$$

$$\text{Subject to} \quad \begin{cases} 2y_1 - 5y_2 \geq -10 \\ 3y_1 + 2y_2 \geq 6 \\ y_1 \geq 0 \quad \text{and} \quad y_2 \geq 0 \end{cases}$$

➤ Solution on page 176

The Dual of a Standard Minimum Problem

Given a standard minimum problem, we want to create from it a particular maximum problem, called the *dual problem* of the original, which will help us to solve the original minimum problem. For this purpose, recall that the *transpose* of a matrix turns every row into a column, or equivalently, every column into a row (see page 114). Denoting a transpose by a superscripted t, we have, for example:

$$\begin{pmatrix} 1 & 2 & 3 \\ 4 & 5 & 6 \end{pmatrix}^t = \begin{pmatrix} 1 & 4 \\ 2 & 5 \\ 3 & 6 \end{pmatrix}$$

To create the dual, we summarize the minimum problem by writing its numbers in a matrix (with the objective function numbers in the bottom row), transpose the matrix, and then reinterpret the matrix as a maximum problem (with $\leq$ inequalities).

➡ EXAMPLE 2   Finding a Dual Problem

Find the dual of the following problem:

$$\text{Minimize} \quad C = 24y_1 + 10y_2$$

$$\text{Subject to} \quad \begin{cases} 3y_1 + y_2 \geq 9 \\ 4y_1 + 2y_2 \geq 16 \\ y_1 \geq 0 \quad \text{and} \quad y_2 \geq 0 \end{cases}$$

Solution

Beginning with the original problem on the left below, we summarize, transpose, and rewrite as a maximum problem as shown directly below the original.

Minimize $C = 24y_1 + 10y_2$

Subject to $\begin{cases} 3y_1 + y_2 \geq 9 \\ 4y_1 + 2y_2 \geq 16 \\ y_1 \geq 0 \quad \text{and} \quad y_2 \geq 0 \end{cases}$ Summarize ➡ $\left(\begin{array}{cc|c} 3 & 1 & 9 \\ 4 & 2 & 16 \\ \hline 24 & 10 & 0 \end{array} \right)$ } Constraints

⟵ Objective function

Original ⟶

Dual ⟶

⬇ Transpose

Maximize $P = 9x_1 + 16x_2$

Subject to $\begin{cases} 3x_1 + 4x_2 \leq 24 \\ x_1 + 2x_2 \leq 10 \\ x_1 \geq 0 \quad \text{and} \quad x_2 \geq 0 \end{cases}$ ⬅ Rewrite $\left(\begin{array}{cc|c} 3 & 4 & 24 \\ 1 & 2 & 10 \\ \hline 9 & 16 & 0 \end{array} \right)$ } *New* constraints

⟵ *New* objective function

Be careful! The matrices on the right are *not* tableaux, so do not change the signs of the objective function numbers.

The dual of a standard minimum problem is a standard maximum problem. Since transposing the numbers a second time would just recover the original, we say that *either problem is the dual of the other.*

→ **Practice Problem 2**

Find the dual of the following problem:

Minimize $\quad C = 24y_1 + 20y_2 + 12y_3$

Subject to $\quad \begin{cases} 2y_1 + 5y_2 + 3y_3 \geq 3 \\ -3y_1 + 2y_2 + 2y_3 \geq 5 \\ y_1 \geq 0, \quad y_2 \geq 0, \quad \text{and} \quad y_3 \geq 0 \end{cases}$

➤ **Solution on page 176**

Practice Problem 2 shows that a problem and its dual need not have the same number of variables or the same number of constraints.

The two problems in Example 2 use only two variables and so can be solved by the graphical method of Section 4.2: drawing graphs, shading regions, finding vertices, and determining which vertex maximizes or minimizes the objective function. The graphical solutions of these two problems are shown below, with the solution to each being where the two lines intersect.

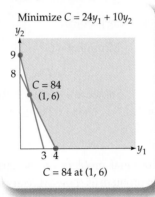

$C = 84$ at $(1, 6)$

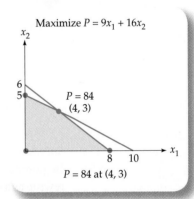

$P = 84$ at $(4, 3)$

Notice that although the problems are quite different, the maximum value of P in one is the same as the minimum value of C in the dual (as you can easily check by substituting the intersection points into the objective functions). That the minimum value of C in a standard minimum problem will *always* be the same as the maximum value of P in the dual maximum problem was a remarkable discovery and is called the Duality Theorem. The following statement of the theorem also shows how to read the solution of the minimum problem from the final tableau of the dual problem.

Duality Theorem

If a maximum problem has a solution, then its final tableau also displays the solution to the dual minimum problem in its bottom row: first the slack variables, then the variables, and finally the objective function, as shown below.

x_1	$\cdots$	x_n	s_1	$\cdots$	s_m	
$\vdots$	$\vdots$	$\vdots$	$\vdots$	$\vdots$	$\vdots$	
$\vdots$	$\vdots$	$\vdots$	$\vdots$	$\vdots$	$\vdots$	
t_1	$\cdots$	t_n	y_1	$\cdots$	y_m	C

$\underbrace{\text{Slack variables}} \quad \underbrace{\text{y-variables}} \quad \overset{\uparrow}{}$ Objective function

Note that for a minimum problem we use ts for the slack variables to distinguish them from the slacks in a maximum problem. Since dual problems are related by transposition, it is not surprising that the solution to the *minimum* problem appears in the last *row*, just as the solution to the *maximum* problem appears in the last *column*. If the maximum problem has no solution, then neither does the dual minimum problem.

For example, on page 169 we began with the following minimum problem and found its dual.

$$\text{Minimize} \quad C = 24y_1 + 10y_2$$

$$\text{Subject to} \quad \begin{cases} 3y_1 + y_2 \geq 9 \\ 4y_1 + 2y_2 \geq 16 \\ y_1 \geq 0 \quad \text{and} \quad y_2 \geq 0 \end{cases}$$

The dual maximum problem is the same one we solved in the previous section, obtaining the following final tableau on page 162:

	x_1	x_2	s_1	s_2	
x_1	1	0	1	-2	4
x_2	0	1	$-\frac{1}{2}$	$\frac{3}{2}$	3
P	0	0	1	6	84

Slacks $\quad$ y-variables $\quad$ Objective function

From the bottom row of this tableau we may read off the solution to the original minimum problem:

The minimum value of C is 84 when $y_1 = 1$ and $y_2 = 6$.

The slacks, which we usually ignore, are $t_1 = 0$ and $t_2 = 0$

This solution is the same as the graphical solution shown on the previous page.

➡ Practice Problem 3

Find the solution of the linear programming problem

$$\text{Minimize} \quad C = 255y_1 + 435y_2 + 300y_3 + 465y_4$$

$$\text{Subject to} \quad \begin{cases} 2y_1 + 3y_2 + 2y_3 + 4y_4 \geq 25 \\ y_1 + 3y_2 + 2y_3 + 2y_4 \geq 18 \\ 3y_1 + 4y_2 + 3y_3 + 2y_4 \geq 36 \\ y_1 \geq 0, \quad y_2 \geq 0, \quad y_3 \geq 0, \quad \text{and} \quad y_4 \geq 0 \end{cases}$$

from the fact that the final tableau of its dual maximum problem is

	x_1	x_2	x_3	s_1	s_2	s_3	s_4	
x_3	0	0	1	0	-2	3	0	30
x_1	1	0	0	1	3	-5	0	60
x_2	0	1	0	-1	0	1	0	45
s_4	0	0	0	-2	-8	12	1	75
P	0	0	0	7	3	1	0	3390

➤ Solution on page 176

We may summarize the entire process as follows:

Solution of a Standard Minimum Problem

To solve a standard minimum problem:

1. Construct the dual maximum problem (page 169).
2. Solve the dual maximum problem by the simplex method (page 164).
3. If there is a solution to the dual maximum problem, then there is a solution to the minimum problem: The values of the slacks, the variables, and the objective function appear (in that order) in the bottom row of the final tableau. If the maximum problem has no solution, then neither does the minimum problem.

→ **EXAMPLE 3**   Solving a Standard Minimum Problem

$$\text{Minimize} \quad C = 50y_1 + 60y_2$$

$$\text{Subject to} \quad \begin{cases} y_1 + 4y_2 \geq 20 \\ 2y_1 + 3y_2 \geq 30 \\ y_1 - y_2 \geq 5 \\ y_1 \geq 0 \quad \text{and} \quad y_2 \geq 0 \end{cases}$$

Solution

First we summarize, transpose, and write the dual:

$$\begin{pmatrix} 1 & 4 & | & 20 \\ 2 & 3 & | & 30 \\ 1 & -1 & | & 5 \\ \hline 50 & 60 & | & 0 \end{pmatrix} \quad \Longrightarrow \quad \begin{pmatrix} 1 & 2 & 1 & | & 50 \\ 4 & 3 & -1 & | & 60 \\ \hline 20 & 30 & 5 & | & 0 \end{pmatrix} \quad \Longrightarrow$$

$$\text{Maximize} \quad P = 20x_1 + 30x_2 + 5x_3$$

$$\text{Subject to} \quad \begin{cases} x_1 + 2x_2 + x_3 \leq 50 \\ 4x_1 + 3x_2 - x_3 \leq 60 \\ x_1 \geq 0, x_2 \geq 0, x_3 \geq 0 \end{cases}$$

The initial simplex tableau for the dual maximum problem is

	x_1	x_2	x_3	s_1	s_2	
s_1	1	2	1	1	0	50
s_2	4	3	-1	0	1	60
P	-20	-30	-5	0	0	0

Bottom row states that
$t_1 = -20, \quad t_2 = -30, \quad t_3 = -5,$
$y_1 = 0, \quad y_2 = 0, \quad \text{and} \quad C = 0$

To solve the dual maximum problem, we first pivot on the 3 in column 2 and row 2 to find the tableau

	x_1	x_2	x_3	s_1	s_2	
s_1	-5/3	0	5/3	1	-2/3	10
x_2	4/3	1	-1/3	0	1/3	20
P	20	0	-15	0	10	600

Bottom row states that
$t_1 = 20, \quad t_2 = 0, \quad t_3 = -15,$
$y_1 = 0, \quad y_2 = 10, \quad \text{and} \quad C = 600$

Then we pivot again on the 5/3 in column 3 and row 1 to reach the final tableau:

	x_1	x_2	x_3	s_1	s_2	
x_3	-1	0	1	3/5	$-2/5$	6
x_2	1	1	0	1/5	1/5	22
P	5	0	0	9	4	690

Bottom row states that
$t_1 = 5$, $t_2 = 0$, $t_3 = 0$,
$y_1 = 9$, $y_2 = 4$,   and $C = 690$

Answer: The minimum value is 690 when $y_1 = 9$ and $y_2 = 4$.

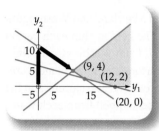

To compare this solution with the graphical method, we graph the feasible region of the minimum problem with y_1 on the horizontal axis and y_2 on the vertical axis. The path of nonfeasible points visited by the simplex tableaux of the dual maximum problem and leading to a feasible vertex is marked in bold.

The simplex method always begins at the origin and "pivots" to successively better vertices. Notice from the graph that in a minimum problem the origin is optimal (minimizing C to 0) but not feasible (not in the shaded region), in contrast to a maximum problem, where the origin is feasible but not optimal. That is, the simplex method *keeps the minimum problem optimal while making it feasible and keeps the maximum problem feasible while making it optimal.*

Matrix Form

The relationships between a minimum problem and its dual maximum problem and the initial and final tableaux may be stated simply in matrix form. For a minimum problem, we use Y for a column of y-variables and A^t in the coefficients in the constraints (since we used A in the maximum problem).

Standard Minimum Problem

Minimize $C = b^t Y$ $(b \geq 0)$

Subject to $\begin{cases} A^t Y \geq c \\ Y \geq 0 \end{cases}$

Dual Maximum Problem

Maximize $P = c^t X$

Subject to $\begin{cases} AX \leq b \quad (b \geq 0) \\ X \geq 0 \end{cases}$

Notice that the *objective* numbers and the *right-hand sides of the constraints* become interchanged when writing the dual, so the requirement that $b \geq 0$ applies to the objective numbers in the minimum problem and the right-hand sides of the constraints in the maximum problem. If the initial tableau for the maximum problem (on the left below) leads to a final tableau that solves the maximum problem (on the right below), then the minimum value is C with variables Y and slacks T taking values as shown in the bottom row.

Initial Tableau

$$\left(\begin{array}{c|cc|c} & X & S & \\ \hline S & A & I & b \\ \hline P & -c^t & 0 & 0 \end{array} \right)$$

If the solution exists

Final Tableau

$$\left(\begin{array}{c|cc|c} & X & S & \\ \hline & \vdots & \vdots & \vdots \\ \hline & T & Y & C \end{array} \right)$$

Mixed Constraints: A Transportation Problem

Although we have been careful to write our standard minimum problem with $\geq$ constraints, we do not mean to exclude the possibility of mixed constraints, some $\geq$ and some $\leq$. We simply multiply each $\leq$ inequality by -1 so it becomes a $\geq$ inequality. (Recall from page 168 that the minimum problem does not require that the right-hand sides of the inequalities be nonnegative.) The following example belongs to a general type of minimization problem with mixed constraints having the nice property that the pivot elements will all be 1s.

→ **EXAMPLE 4   A Transportation Problem**

A retail store chain has cartons of goods stored at warehouses in Maryland and Washington that must be distributed to its stores in Ohio and Louisiana. The cost to ship each carton from Maryland to Ohio is $6, from Maryland to Louisiana is $7, from Washington to Ohio is $8, and from Washington to Louisiana is $9. There are 300 cartons at the Maryland warehouse and 300 at the warehouse in Washington. If the Ohio stores need 200 cartons and the Louisiana stores need 300 cartons, how many cartons should be shipped from each warehouse to each state to incur the smallest shipping costs?

Solution

This is a minimization problem that requires four variables: one for the amount shipped from each warehouse to each state. Let

$$y_1 = \begin{pmatrix} \text{Cartons shipped} \\ \text{from MD to OH} \end{pmatrix}, \qquad y_2 = \begin{pmatrix} \text{Cartons shipped} \\ \text{from MD to LA} \end{pmatrix},$$

$$y_3 = \begin{pmatrix} \text{Cartons shipped} \\ \text{from WA to OH} \end{pmatrix}, \quad \text{and} \quad y_4 = \begin{pmatrix} \text{Cartons shipped} \\ \text{from WA to LA} \end{pmatrix}$$

Because each warehouse can ship no more than the number of cartons stored there and each state must receive at least the required number of cartons, this problem is the linear programming problem

Minimize $C = 6y_1 + 7y_2 + 8y_3 + 9y_4$ — Total shipping costs

Subject to
$$\begin{cases} y_1 + y_2 & \leq 300 \\ y_3 + y_4 & \leq 300 \\ y_1 + y_3 & \geq 200 \\ y_2 + y_4 & \geq 300 \\ y_1 \geq 0, \ y_2 \geq 0, \ y_3 \geq 0, \ y_4 \geq 0 \end{cases}$$

Have 300 in MD	
Have 300 in WA	
Need 200 in OH	
Need 300 in LA	
Nonnegativity	

We multiply the first two constraints by -1 to give them $\geq$ inequalities. This problem is then simply stated in matrix form:

$$\text{Minimize} \quad C = (6 \quad 7 \quad 8 \quad 9) \begin{pmatrix} y_1 \\ y_2 \\ y_3 \\ y_4 \end{pmatrix}$$

Subject to
$$\begin{cases} \begin{pmatrix} -1 & -1 & 0 & 0 \\ 0 & 0 & -1 & -1 \\ 1 & 0 & 1 & 0 \\ 0 & 1 & 0 & 1 \end{pmatrix} \begin{pmatrix} y_1 \\ y_2 \\ y_3 \\ y_4 \end{pmatrix} \geq \begin{pmatrix} -300 \\ -300 \\ 200 \\ 300 \end{pmatrix} \\ \\ \text{and} \quad \begin{pmatrix} y_1 \\ y_2 \\ y_3 \\ y_4 \end{pmatrix} \geq 0 \end{cases}$$

The dual maximum problem is

$$\text{Maximize} \quad P = (-300 \quad -300 \quad 200 \quad 300) \begin{pmatrix} x_1 \\ x_2 \\ x_3 \\ x_4 \end{pmatrix}$$

Subject to
$$\begin{cases} \begin{pmatrix} -1 & 0 & 1 & 0 \\ -1 & 0 & 0 & 1 \\ 0 & -1 & 1 & 0 \\ 0 & -1 & 0 & 1 \end{pmatrix} \begin{pmatrix} x_1 \\ x_2 \\ x_3 \\ x_4 \end{pmatrix} \leq \begin{pmatrix} 6 \\ 7 \\ 8 \\ 9 \end{pmatrix} \\ \\ \text{and} \quad \begin{pmatrix} x_1 \\ x_2 \\ x_3 \\ x_4 \end{pmatrix} \geq 0 \end{cases}$$

The initial simplex tableau is

	x_1	x_2	x_3	x_4	s_1	s_2	s_3	s_4	
s_1	-1	0	1	0	1	0	0	0	6
s_2	-1	0	0	1	0	1	0	0	7
s_3	0	-1	1	0	0	0	1	0	8
s_4	0	-1	0	1	0	0	0	1	9
P	300	300	-200	-300	0	0	0	0	0

We pivot at [column 4, row 2], [column 3, row 1], and [column 1, row 3] to reach the final tableau:

	x_1	x_2	x_3	x_4	s_1	s_2	s_3	s_4	
x_3	0	-1	1	0	0	0	1	0	8
x_4	0	-1	0	1	-1	1	1	0	9
x_1	1	-1	0	0	-1	0	1	0	2
s_4	0	0	0	0	1	-1	-1	1	0
P	0	100	0	0	0	300	200	0	3700
					y_1	y_2	y_3	y_4	C

Bottom row states that
$t_1 = 0$, $t_2 = 100$,
$t_3 = 0$, $t_4 = 0$,
$y_1 = 0$, $y_2 = 300$,
$y_3 = 200$, $y_4 = 0$,
and $C = 3700$

The Maryland warehouse should send nothing to Ohio and 300 cartons to Louisiana, while the Washington warehouse should send 200 cartons to Ohio and nothing to Louisiana to achieve the least shipping cost of $3700. The second slack shows that there will be 100 cartons unused in the Washington warehouse.

→ **Solutions to Practice Problems**

1. No. The negative coefficient -4 in the objective function violates the condition in the box on page 168. (The -10 on the right-hand side of the first inequality is *not* a violation; only the objective coefficients need to be nonnegative.)

2. $$\begin{pmatrix} 2 & 5 & 3 & | & 3 \\ -3 & 2 & 2 & | & 5 \\ 24 & 20 & 12 & | & 0 \end{pmatrix}^t \longrightarrow \begin{pmatrix} 2 & -3 & | & 24 \\ 5 & 2 & | & 20 \\ 3 & 2 & | & 12 \\ 3 & 5 & | & 0 \end{pmatrix} \longrightarrow$$

Maximize $P = 3x_1 + 5x_2$

Subject to
$$\begin{cases} 2x_1 - 3x_2 \leq 24 \\ 5x_1 + 2x_2 \leq 20 \\ 3x_1 + 2x_2 \leq 12 \\ x_1 \geq 0 \text{ and } x_2 \geq 0 \end{cases}$$

Notice that the original problem had three variables and two constraints, while the dual has two variables and three constraints. The numbers of variables and constraints will always interchange in this way because of the transposition.

3. The bottom row of the final tableau for the dual maximum problem displays the values for the slack variables $(t_1 = 0, \ t_2 = 0, \ t_3 = 0)$, the variables $(y_1 = 7, \ y_2 = 3, \ y_3 = 1, \ y_4 = 0)$, and the objective function $(C = 3390)$. The minimum value is 3390 when $y_1 = 7$, $y_2 = 3$, $y_3 = 1$, and $y_4 = 0$.

For each standard minimum problem construct the dual maximum problem and its initial simplex tableau.

1. Minimize $C = 60y_1 + 100y_2 + 300y_3$

   Subject to $\begin{cases} y_1 + 2y_2 + 3y_3 \geq 180 \\ 4y_1 + 5y_2 + 6y_3 \geq 120 \\ y_1 \geq 0, \quad y_2 \geq 0, \quad y_3 \geq 0 \end{cases}$

2. Minimize $C = 3y_1 + 20y_2$

   Subject to $\begin{cases} 3y_1 + 2y_2 \geq 150 \\ y_1 + 4y_2 \geq 100 \\ 3y_1 + 4y_2 \geq 228 \\ y_1 \geq 0, \quad y_2 \geq 0 \end{cases}$

Solve each standard minimum problem by finding the dual maximum problem and using the simplex method. (Exercises 3 through 5 can also be solved by the graphical method.)

3. Minimize $C = 15y_1 + 10y_2$

   Subject to $\begin{cases} y_1 + 2y_2 \geq 20 \\ 3y_1 - y_2 \geq 60 \\ y_1 \geq 0, \quad y_2 \geq 0 \end{cases}$

4. Minimize $C = 4y_1 + 5y_2$

   Subject to $\begin{cases} y_1 + y_2 \geq 10 \\ y_1 \geq 2 \\ y_2 \geq 3 \\ y_1 \geq 0, \quad y_2 \geq 0 \end{cases}$

5. Minimize $C = 2y_1 + y_2$

   Subject to $\begin{cases} -y_1 + y_2 \leq 20 \\ y_1 - y_2 \leq 20 \\ y_1 + y_2 \geq 10 \\ y_1 \geq 0, \quad y_2 \geq 0 \end{cases}$

6. Minimize $C = 10y_1 + 20y_2 + 10y_3$

   Subject to $\begin{cases} -y_1 + y_2 + y_3 \geq 50 \\ y_1 + y_2 - y_3 \geq 30 \\ y_1 \geq 0, \quad y_2 \geq 0, \quad y_3 \geq 0 \end{cases}$

7. Minimize $C = 20y_1 + 50y_2 + 30y_3$

   Subject to $\begin{cases} 2y_1 - y_2 + y_3 \leq 10 \\ y_1 + y_2 + y_3 \geq 30 \\ 2y_1 - y_2 + y_3 \geq 20 \\ y_1 \geq 0, \quad y_2 \geq 0, \quad y_3 \geq 0 \end{cases}$

Formulate each situation as a standard minimum linear programming problem by identifying the variables, the objective function, and the constraints. Be sure to state clearly the meaning of each variable. Solve it by finding the dual maximum problem and using the simplex method. State your final answer in terms of the original question.

8. **Nutrition** An athlete's training diet needs at least 44 more grams of carbohydrates, 12 more grams of fat, and 16 more grams of protein each day. A dietitian recommends two food supplements, Bulk-Up Bars (costing 48¢ each) and Power Drink (costing 45¢ per can), with nutritional contents (in grams) as given in the table. How much of each food supplement will provide the extra needed nutrition at the least cost?

(grams)	Bulk-Up Bar	Power Drink
Carbohydrates	4	3
Fat	1	1
Protein	2	1

9. **Agriculture** A soil analysis of a farmer's field showed that he needs to apply at least 3000 pounds of nitrogen, 2400 pounds of phosphoric acid, and 2100 pounds of potash. Plant fertilizer is labeled with three numbers giving the percentages of nitrogen, phosphoric acid, and potash. The local farm supply store sells 15-30-15 Miracle Mix for 15¢ per pound and a 10-5-5 store brand for 8¢ per pound. How many pounds of each fertilizer should the farmer buy to meet the needs of the field at the least cost?

10. **Transportation** A retail store chain has cartons of goods stored at warehouses in Kentucky and Utah that must be distributed to its stores in Kansas, Texas, and Oregon. Each carton shipped from the Utah warehouse costs $2 whether it goes to Kansas, Texas, or Oregon, but the cost to ship one carton from the Kentucky warehouse to Kansas is $2, to Texas is $4, and to Oregon is $5. There are 200 cartons at the Utah warehouse and 400 at the warehouse in Kentucky. If the Kansas stores need 200 cartons, the Texas stores need 300 cartons, and the Oregon stores need 100 cartons, how many cartons should be shipped from each warehouse to each state to incur the smallest shipping costs?

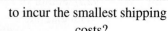

ATTENTION

NEED MORE PRACTICE? FIND MORE HERE:
CENGAGEBRAIN.COM

Chapter 5

Probability

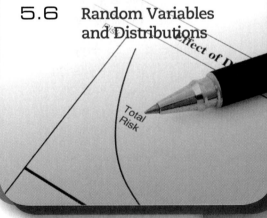

5.1 Sets, Counting, and Venn Diagrams

Probability is used frequently in everyday conversation, as in "you will probably get the job" or "it will probably rain today." In mathematics, however, we need to give a precise definition for probability, and we will see that it depends on a careful counting of all the possible events that can occur. With this in mind, we begin by reviewing sets and ways of counting their members.

Sets and Set Operations

A *set* is any well-defined collection of objects (also called *elements* or *members*). By "well-defined" we mean you can tell whether an object is in the set or not. For example, we may speak of *the set of all U.S. citizens*, since there are specific conditions for being a U.S. citizen. On the other hand, we cannot speak of *the set of all thin people*, since the word "thin" is not precisely defined.

We often use diagrams to represent sets. We use a rectangle for the *universal set*, denoted U, which is the set of all the elements that we are discussing. For example, depending on the question, the universal set might consist of all people, all customers of a company, all cars on the road, or any other collection of objects of interest. To represent sets within the universal set, we draw ellipses inside the rectangle containing the elements of the sets. Such drawings are called *Venn diagrams*, after the English logician John Venn (1843–1923), author of *Logic of Chance*. There are several operations on sets, which we illustrate with Venn diagrams.

The *intersection* of sets *A* and *B*, denoted $A \cap B$, is the set of all elements that are in *both A and B*.

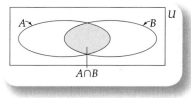

Intersection

Two sets *A* and *B* are *disjoint* if their intersection is empty (that is, if they have no elements in common).

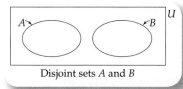

Disjoint sets *A* and *B*

Disjoint

The *union* of sets A and B, denoted $A \cup B$, is the set of all elements that are in *either* A *or* B (or both).

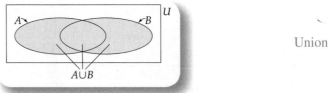

Union

The *complement* of a set A, denoted A^c, is the set of all elements *not* in A.

Complement

In general, the words "and," "or," and "not" translate into "intersection," "union," and "complement." We use the word "or" in the *inclusive* sense, meaning one possibility or the other *or both*. The complement of the universal set is the *empty* set, also called the *null* set, denoted $\varnothing$, which has no members. A *subset* of a set is a set (possibly empty) of elements from the original set. For a finite set A, we use the symbol $n(A)$ to mean the number of elements in A:

$$n(A) = \left(\begin{array}{c} \text{Number of} \\ \text{elements in } A \end{array} \right)$$

For example, $n(\varnothing) = 0$ because the empty set has no elements. From the previous Venn diagram, it is clear that the elements of the set A^c are precisely the elements of the universal set U that are *not* in A. Therefore:

Complementary Principle of Counting

$$n(A^c) = n(U) - n(A)$$

The number of elements in the *complement* of a set is the number of elements in the universal set minus the number of elements in the original set

We will find this principle useful when it is easier to count the elements in the complement rather than the elements in the set itself.

Addition Principle for Counting

How can we count the number of elements in the *union* of two sets without counting them one by one? In the following Venn diagram there are $n(A) = 7$ elements (dots) in A and $n(B) = 5$ elements in B.

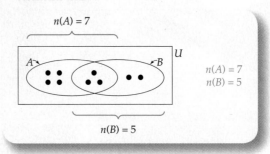

Altogether there are 9 elements in $A \cup B$ but adding $n(A) = 7$ and $n(B) = 5$ gives 12, not the correct total of 9 for the union. The diagram shows the trouble: Adding $n(A)$ and $n(B)$ counts the elements in the middle, $A \cap B$, *twice*—once in A and a second time in B. To correct this double-counting, we must subtract the number of elements in $A \cap B$ from the total, obtaining the general rule:

Addition Principle of Counting

$$n(A \cup B) = n(A) + n(B) - n(A \cap B)$$

The number of elements in the *union* of two sets is the number of elements in one plus the number of elements in the other minus the number of elements in *both*

Of course, if A and B are disjoint, then $A \cap B$ is empty, so $n(A \cap B) = 0$, giving a simpler addition principle for disjoint sets:

$$n(A \cup B) = n(A) + n(B) \qquad \text{for } A, B \text{ disjoint}$$

➡ EXAMPLE 1 Cars in a Parking Lot

A mall parking lot has 300 cars, of which 150 have alarm systems, 200 have sound systems, and 90 have both alarm and sound systems.

a. How many cars have an alarm system or a sound system?
b. How many have neither?

Solution
Let A be the set of cars with alarm systems and S be the set of cars with sound systems. Starting with the Venn diagram on the left, we enter the numbers successively, beginning with the intersection:

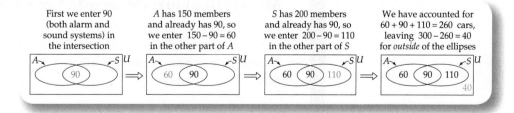

The 40 *outside* of the ellipses means that 40 cars have *neither* system.

Answer:

260 cars have either an alarm system or a sound system. $\qquad$ 60 + 90 + 110
40 cars have neither an alarm system nor a sound system. $\qquad$ 300 − 260

Alternatively, since "or" means "union," we could find $n(A \cup S)$ by the Addition Principal of Counting, using $n(A) = 150$, $n(S) = 200$, and $n(A \cap S) = 90$:

$$n(A \cup S) = n(A) + n(S) - n(A \cap S) = 150 + 200 - 90 = 260$$

$$\underbrace{}_{150} \quad \underbrace{}_{200} \quad \underbrace{}_{90}$$

Addition principle of counting

Again we see that 260 cars have an alarm or a sound system, so $300 - 260 = 40$ have neither.

Hint: With Venn diagrams you usually begin at the "inside" and work "out."

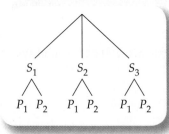

Picsfive/Shutterstock

➡ Practice Problem 1

A survey of insurance coverage in 300 metropolitan businesses revealed that 150 offer their employees dental insurance, 150 offer vision coverage, and 100 offer both dental and vision coverage. How many of these businesses offer their employees dental or vision insurance? How many offer neither?

➤ **Solution on page 185**

The Multiplication Principle for Counting

Suppose you are choosing an outfit to wear, and you may choose any one of 3 shirts, S_1, S_2, or S_3, and either of 2 pairs of pants, P_1 or P_2. If the 3 shirts can be combined freely with the 2 pairs of pants, there are $3 \cdot 2 = 6$ different possible outfits, namely $S_1 P_1$, $S_1 P_2$, $S_2 P_1$, $S_2 P_2$, $S_3 P_1$, and $S_3 P_2$. The possible choices are shown in the *tree diagram* on the left, indicating a first choice of a shirt, then "branching" from each shirt to the second choice of a pair of pants, ending in 6 "leaves" at the bottom. More generally, we have the following *multiplication principle*, which will be very useful throughout this chapter.

Multiplication Principle for Counting

If two choices are to be made, and there are m possibilities for the first choice and n possibilities for the second choice, and if any first choice can be combined with any second choice, then the two choices together can be made in $m \cdot n$ ways.

The multiplication principle can also be proved by enumerating all the possibilities, as in the following example.

→ EXAMPLE 2   Counting Different Products

A toy company makes red, green, blue, and yellow plastic cars, trucks, and planes. How many different kinds of toys do they make?

Solution

Let the set of colors be $C = \{\text{red, green, blue, yellow}\}$,   and let the set of shapes be $S = \{\text{car, truck, plane}\}$.   Each possible toy is described by its color and shape:

	Red	**Green**	**Blue**	**Yellow**
Car	(red, car)	(green, car)	(blue, car)	(yellow, car)
Truck	(red, truck)	(green, truck)	(blue, truck)	(yellow, truck)
Plane	(red, plane)	(green, plane)	(blue, plane)	(yellow, plane)

This rectangular table contains all possible combination of colors and shapes, and the number of boxes is found by multiplying length times width: There are $4 \cdot 3 = 12$ possible toys that can be made.

A convenient way of using the multiplication principle is to imagine making up a typical combination of the two choices. For instance, in the preceding example there were 4 choices for the color and 3 choices for the shape, giving

$$(\underline{\qquad}, \underline{\qquad})$$

$$\begin{array}{cc} \uparrow & \uparrow \\ 4 & 3 \\ \text{choices} & \text{choices} \\ \text{(color)} & \text{(shape)} \end{array}$$

$4 \cdot 3 = 12$ possible toys

The Multiplication Principle for Counting generalizes to *more* than two choices.

Generalized Multiplication Principle for Counting

If k choices are to be made, and there are m_1 possibilities for the first choice, m_2 possibilities for the second choice, m_3 possibilities for the third choice, and so on down to m_k possibilities for the kth choice, and if the choices can be combined in any way, then the k choices can be made together in $m_1 \cdot m_2 \cdot \cdots \cdot m_k$ ways.

→ EXAMPLE 3   Counting Parking Permits

A parking permit displays an identification code consisting of a letter (A to Z) followed by two digits (0 to 9). How many different permits can be issued?

Solution

Since each identification code consists of three symbols, we need to fill three blanks:

(———— , ———— , ————) $26 \cdot 10 \cdot 10 = 2600$ ways

    ↑   ↑   ↑

   26   10   10

choices   choices   choices

(letter)   (digit)   (digit)

Therefore, 2600 different permits can be issued.

➡ Practice Problem 2

A computer password consists of two letters (A to Z) followed by three digits (0 to 9). How many different passwords are there?

> ➤ **Solution on next page**

The Number of Subsets of a Set

Given a set, how many subsets does it have? For example, the set $\{a, b\}$ has 4 subsets: $\{a\}$, $\{b\}$, $\{a, b\}$, and $\varnothing$. (We consider the set itself and the empty set $\varnothing$ to be subsets.) To form these subsets we had two choices: *include or exclude a* and *include or exclude b.* Two choices (for *a* and *b*) with each made in two possible ways (*include* or *exclude*) gives a total of $2 \cdot 2$ choices, for 4 subsets (as we saw).

(———— , ————) $2 \cdot 2 = 4$ subsets

     ↑   ↑

Include *a*?   Include *b*?

(yes/no)   (yes/no)

In this same way we can count the number of subsets of *any* set. A set with *n* members means *n* choices: *include* or *exclude the first member, include* or *exclude the second member,* and so on. Making *n* choices, each in 2 possible ways (*include* or *exclude*), means a total of 2^n choices, for 2^n subsets.

Number of Subsets of a Set

A set with *n* elements has 2^n subsets.

➡ EXAMPLE 4   **Finding the Subsets of a Set**

List all subsets of the set $\{a, b, c\}$ and verify that there are $2^3 = 8$ of them.

Solution

The subsets are

$$\{a\}, \quad \{b\}, \quad \{c\}, \quad \{a, b\}, \quad \{a, c\}, \quad \{b, c\}, \quad \{a, b, c\}, \quad \text{and} \quad \varnothing$$

Indeed, there are 8 subsets (including the set itself and the empty set).

→ **EXAMPLE 5** **Counting Subsets**

A restaurant offers pizza with mushrooms, peppers, onions, pepperoni, and sausage. How many different types of pizza can be ordered?

Solution
The set of toppings has 5 members, so there are $2^5 = 32$ possible subsets, and so 32 different pizzas. (Do you see which of these subsets corresponds to "plain" pizza?)

→ **Solutions to Practice Problems**

1. Let D be the set of businesses offering dental insurance, and let V be those offering vision insurance. Then, using a Venn diagram:

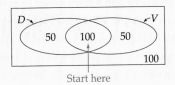

Start here

Or, using the Addition Principle:

$$n(D \cup V) = n(D) + n(V) - n(D \cap V)$$
$$= 150 + 150 - 100 = 200$$

Two hundred businesses offer dental or vision insurance. One hundred offer neither.

2. Since $26 \cdot 26 \cdot 10 \cdot 10 \cdot 10 = 676,000$, there are 676,000 different passwords.

2 letters   3 digits

→ **5.1 Exercises**

Find each number using the following Venn diagram.

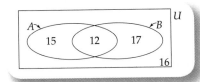

1. $n(A)$
2. $n(B)$
3. $n(U)$
4. $n(A \cap B)$
5. $n(A \cup B)$
6. $n(A^c)$
7. $n(B^c)$
8. $n(A \cap B^c)$
9. $n(A^c \cap B)$
10. $n(A \cup B^c)$

11. Given that $n(A) = 20$, $n(B) = 10$, $n(A \cap B) = 6$, and $n(U) = 40$, fill in the four regions in the Venn diagram.

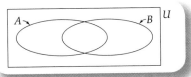

12. **Parking Permits** A parking permit sticker displays an identification code consisting of a letter (from A, B, and C) followed by a digit (from 1, 2, 3, and 4). How many different permits can be issued?

13. Lacrosse The sophomore lacrosse team has 24 players, of whom 10 played defense last year, 12 played offense, and 5 played both defense and offense, while the rest of the players did not play last year. How many members of the team played last year?

14. Computer Passwords How many eight-symbol computer passwords can be formed using the letters A to J and the digits 2 to 6?

15. False Testimony A witness in a trial testified that he searched 20 cars, of which 12 were convertibles, 15 had out-of-state plates, and 5 were convertibles with out-of-state plates. Explain why the witness was not telling the truth.

5.2 Permutations and Combinations

In this section we will develop two very useful formulas for counting various types of choices, known as *permutations* and *combinations*, and then apply these formulas to a wide variety of problems. We begin by describing *factorial notation.*

Factorials

Products of successive integers from a number down to 1, such as $5 \cdot 4 \cdot 3 \cdot 2 \cdot 1$, are called *factorials*. We denote factorials by exclamation points, so the preceding product would be written 5! (read "5 factorial"). Formally:

Factorials

For any positive integer n,

$$n! = n(n-1) \cdot \cdots \cdot 1$$

n factorial is the product of the integers from n down to 1

$$0! = 1$$

Zero factorial is 1

The next example shows that some factorial expressions are most easily found by using cancellation *before* evaluating the factorials.

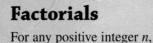

EXAMPLE 1 Calculating Factorials

Find: **a.** 4! **b.** $\dfrac{7!}{6!}$ **c.** $\dfrac{6!}{3!}$ **d.** $\dfrac{100!}{99!}$

Solution

a. $4! = 4 \cdot 3 \cdot 2 \cdot 1 = 24$

b. $\dfrac{7!}{6!} = \dfrac{7 \cdot 6 \cdot 5 \cdot 4 \cdot 3 \cdot 2 \cdot 1}{6 \cdot 5 \cdot 4 \cdot 3 \cdot 2 \cdot 1} = \dfrac{7 \cdot \cancel{6} \cdot \cancel{5} \cdot \cancel{4} \cdot \cancel{3} \cdot \cancel{2} \cdot \cancel{1}}{\cancel{6} \cdot \cancel{5} \cdot \cancel{4} \cdot \cancel{3} \cdot \cancel{2} \cdot \cancel{1}} = 7$

When finding quotients of factorials, look for cancellation

c. $\dfrac{6!}{3!} = \dfrac{6 \cdot 5 \cdot 4 \cdot 3 \cdot 2 \cdot 1}{3 \cdot 2 \cdot 1} = \dfrac{6 \cdot 5 \cdot 4 \cdot \cancel{3} \cdot \cancel{2} \cdot \cancel{1}}{\cancel{3} \cdot \cancel{2} \cdot \cancel{1}} = 6 \cdot 5 \cdot 4 = 120$

d. $\dfrac{100!}{99!} = \dfrac{100 \cdot 99 \cdot \cdots \cdot 1}{99 \cdot \cdots \cdot 1} = 100$

Canceling $99 \cdot \cdots \cdot 1$

Parts (b) and (d) of the above example are instances of the formula $\dfrac{n!}{(n-1)!} = n$. For $n = 1$ this becomes $\dfrac{1!}{0!} = 1$, and multiplying each side by 0! gives $1 = 0!$, which is why we define 0! to be 1.

We will use factorials to count the number of ways that objects can be *ordered*.

Permutations

How many different orderings are there for the letters a, b, and c? We may list the orderings as *abc*, *acb*, *bac*, *bca*, *cab*, and *cba*, so there are 6. Each of these orderings is called a *permutation*. Instead of listing them all, we could observe that there are 3 ways of choosing the first letter, 2 ways of choosing the second (because one letter was "used up" in the first choice), and 1 way of choosing the last (whichever is left), so by the multiplication principle there are $3 \cdot 2 \cdot 1 = 6$ possible orderings, just as we found before. How many permutations are there of n distinct objects? By the same reasoning, the answer is $n(n-1) \cdot \cdots \cdot 1 = n!$.

jymsts/Shutterstock

→ EXAMPLE 2   Counting Batting Orders

How many different batting orders are there for a 9-player baseball team?

Solution

The first batter can be any one of the 9 players, the second batter can be any one of the remaining 8, next can be any one of the remaining 7, and so on down to the last batter who will be the only one left. By the multiplication principle we should multiply together the numbers of ways of making each choice, so the number of batting orderings will be

$$9! = 9 \cdot 8 \cdot 7 \cdot 6 \cdot 5 \cdot 4 \cdot 3 \cdot 2 \cdot 1 = 362{,}880$$

→ Practice Problem 1

You have six different tasks to do today. In how many different orders can they be done?   **➤ Solution on page 191**

How many different orderings are there for just *the first 3 batters* on the 9-player team? Clearly any one of the 9 players can bat first, then any one of the remaining 8, then any one of the remaining 7, for a total of $9 \cdot 8 \cdot 7$ orderings (by the multiplication principle). In general, if we have *n* distinct objects and want to count all possible orderings of some *r* of them, there are *n* choices for the first, $n - 1$ choices for the second, $n - 2$ choices for the third, down to $n - r + 1$ choices for the *r*th. (Notice that taking *r* of them means leaving $n - r$, so the last one kept is the preceding one, $n - r + 1$.) Such orderings are called *permutations* and can be counted using the following formula.

Permutations

The number of permutations (ordered arrangements) of *n* distinct objects taken *r* at a time is

$$\overbrace{{}_nP_r = n \cdot (n - 1) \cdot \cdots \cdot (n - r + 1)}^{r \text{ factors}}$$

Product of *r* numbers from *n* down

Alternative notations for ${}_nP_r$ are $P_{n,r}$ P_r^n and $P(n, r)$.

We define ${}_nP_0 = 1$ since there is exactly one way of taking zero objects from *n*, namely, taking nothing.

➡ **EXAMPLE 3   Counting Nonsense Words**

How many five-letter "nonsense" words (that is, strings of letters without regard to meaning) can be made from the letters A to Z with no letter repeated?

Solution
Words are ordered arrangements of letters, so we use the permutation formula with $n = 26$ and $r = 5$.

$${}_{26}P_5 = 26 \cdot 25 \cdot 24 \cdot 23 \cdot 22 = 7{,}893{,}600$$

Product of 5 numbers from 26 down

There are 7,893,600 five-letter nonsense words with distinct letters.

Instead of using the permutation formula, we may think of building each word by filling in 5 blanks with distinct letters and multiplying the numbers of choices:

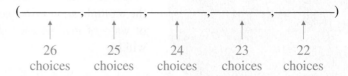

This gives $26 \cdot 25 \cdot 24 \cdot 23 \cdot 22 = 7{,}893{,}600$ nonsense words, just as before.

➡ **EXAMPLE 4   Counting License Plates**

If a car license plate consists of three distinct letters followed by three distinct digits, how many different license plates are possible?

Solution

For three letters with no repeats we have $_{26}P_3 = 26 \cdot 25 \cdot 24$, and for three digits with no repeats we have $_{10}P_3 = 10 \cdot 9 \cdot 8$. By the multiplication principle, the number of possible license plates is the product of these numbers:

$$\underbrace{26 \cdot 25 \cdot 24}_{_{26}P_3} \cdot \underbrace{10 \cdot 9 \cdot 8}_{_{10}P_3} = 11{,}232{,}000$$

➜ Practice Problem 2

Jurors at an art exhibition must select the first-, second-, and third-place winners from an exhibition of 35 paintings. In how many different ways can the winning paintings be chosen? ➤ Solution on page 191

Combinations

Sometimes we want to count choices *where order does not matter.* Such choices are called *combinations.* For example, a committee of three people is the same committee regardless of the order in which they were chosen. How do we count the number of combinations? We begin with an example.

➜ EXAMPLE 5   Counting Committees

A group of 8 student representatives wants to choose a committee of 3 to write a petition to the administration asking for better internet access. How many different committees are possible?

56

Solution

If we were considering the order in which the committee members were chosen, then there would be $8 \cdot 7 \cdot 6$ ways. However, a committee consisting of Alice, Bob, and Cathy is exactly the same as the committee of Bob, Cathy, and Alice—the order in which they were chosen makes no difference. Any particular committee of 3 people can be chosen in $3 \cdot 2 \cdot 1 = 6$ different orders, all making the same committee. Therefore, to eliminate multiple countings of the same committee we divide the $8 \cdot 7 \cdot 6$ orderings by $3 \cdot 2 \cdot 1$, obtaining for the number of different committees:

$$\frac{8 \cdot 7 \cdot 6}{3 \cdot 2 \cdot 1} = 56 \qquad \text{Number of committees from 8 of size 3}$$

This number of ways of choosing from 8 a committee of 3 is called the *number of combinations of 8 things taken 3 at a time*, written $_8C_3$, sometimes read "8 choose 3." If we were to multiply the numerator of the above fraction by $5 \cdot 4 \cdot 3 \cdot 2 \cdot 1$, it would "complete" the factorial into 8!. Therefore, multiplying numerator and denominator by $5 \cdot 4 \cdot 3 \cdot 2 \cdot 1$ gives an alternate way expression for $_8C_3$:

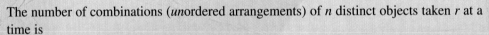

$$\frac{8 \cdot 7 \cdot 6}{3 \cdot 2 \cdot 1} = \frac{8 \cdot 7 \cdot 6 \cdot 5 \cdot 4 \cdot 3 \cdot 2 \cdot 1}{3 \cdot 2 \cdot 1 \cdot 5 \cdot 4 \cdot 3 \cdot 2 \cdot 1} = \frac{8!}{3! \cdot 5!} \qquad \begin{array}{l}\text{Multiplying top} \\ \text{and bottom by 5!}\end{array}$$

The expressions on the left and the right above are each equal to $_8C_3$: The left side is easier to calculate, but the right side simpler to write (using factorials). Replacing the 8 and 3 by *any* positive integers, we obtain the following general formulas.

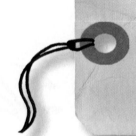

Combinations

The number of combinations (*un*ordered arrangements) of n distinct objects taken r at a time is

$$_nC_r = \frac{n \cdot (n - 1) \cdot \cdots \cdot (n - r + 1)}{r \cdot (r - 1) \cdot \cdots \cdot 1} \qquad \begin{array}{l}\longleftarrow r \text{ numbers beginning with } n \\ \longleftarrow r!\end{array}$$

$$= \frac{n!}{r!(n - r)!} \qquad \text{In factorial form}$$

The second formula comes from the first by multiplying the numerator and denominator by $(n - r) \cdot \cdots \cdot 1 = (n - r)!$. We define $_nC_0 = 1$ since there is clearly one way to choose zero objects from n, namely, taking nothing. Alternative notations for $_nC_r$ are $C_{n,r}$, C_r^n, $C(n, r)$, and $\binom{n}{r}$, which are also called *binomial coefficients*.

Notice that 8 things taken 3 at a time is the same as 8 things taken 5 at a time:

$$_8C_3 = \frac{8 \cdot 7 \cdot 6}{3 \cdot 2 \cdot 1} = 56$$

$$\qquad\qquad\qquad\qquad\qquad\qquad\qquad \text{Same value}$$

$$_8C_5 = \frac{8 \cdot 7 \cdot 6 \cdot 5 \cdot 4}{5 \cdot 4 \cdot 3 \cdot 2 \cdot 1} = 56$$

There is a simple reason for this, which may be explained in terms of committees: Whenever you choose a committee of 3 from 8 people you are, in a sense, also selecting another committee, *the 5 left behind.* Any different choice of the 3 results in a different 5 being left behind. Therefore, from 8 people there must be exactly as many committees of 3 as committees of 5. This is also clear from the factorial formulas, since $_8C_3 = \frac{8!}{3!\,5!}$ and $_8C_5 = \frac{8!}{5!\,3!}$ differ only in the order of the denominators. As before, replacing 8 and 3 by *any* positive integers n and r shows that from n people there are exactly as many committees of r people as committees of $n - r$ people (in symbols: $_nC_r = {_nC_{n-r}}$). This relationship is useful in reducing calculation; for example, given $_8C_5$ it is easier to calculate it as $_8C_3$, as we saw in the calculations on the bottom of the previous page.

➔ EXAMPLE 6   Counting Permutations and Combinations

A student club has 15 members.
a. How many ways can a president, vice president, and treasurer be chosen?
b. How many ways can a committee of three members be chosen?

Solution
Each question involves choosing three members from the club, but for the officers we want *ordered* arrangements (the order determines the offices: the president is listed first, the vice president second, and the treasurer third), while for the committee we want *unordered* arrangements. Thus part (a) asks for permutations, part (b) for combinations.

a. For the officers: $_{15}P_3 = 15 \cdot 14 \cdot 13 = 2730$ 3 numbers from 15 down

b. For the committee: $_{15}C_3 = \dfrac{15 \cdot 14 \cdot 13}{3 \cdot 2 \cdot 1} = 455$ Permutations divided by 3!

There are 2730 different ways of choosing the president, vice president, and treasurer, and 455 ways of choosing the committee.

➔ Practice Problem 3

A college business major can also minor in computer science by taking any 7 courses from an approved list of 10 courses. How many different collections of courses will satisfy the requirements for the computer science minor?

➤ **Solution below**

➔ Solutions to Practice Problems

1. There are $6! = 6 \cdot 5 \cdot 4 \cdot 3 \cdot 2 \cdot 1 = 720$ different orders.

2. Since the winners must be selected in order (first, second, and third place), there are $_{35}P_3 = 35 \cdot 34 \cdot 33 = 39{,}270$ different ways of choosing the winners.

3. Since the courses can be taken in any order, there are $_{10}C_7$ different ways. It is easier to calculate this as $_{10}C_3 = \frac{10 \cdot 9 \cdot 8}{3 \cdot 2 \cdot 1} = 120$ different ways to fulfill a minor in computer science.

→ 5.2 Exercises

Calculate each factorial or quotient of factorials.

1. *a.* 2!   *b.* 6!

2. $\dfrac{10!}{7!}$

Determine each number of permutations.

3. $_6P_3$

4. $_8P_1$

Find each number of combinations.

5. $_6C_2$

6. $_7C_3$

7. **Telephone Numbers** A telephone number consists of seven digits, and the first digit cannot be a zero or a one. How many telephone numbers have no repeated digits?

8. **Computer Passwords** A computer password is to consist of four alphanumeric characters with no repeats. (An alphanumeric character is a letter from A to Z or a digit from 0 to 9.) How many such passwords are there? How many are there if the letter O and the digit 0 are excluded to avoid confusion?

9. **Books** You have 2 science books, 3 language books, and 4 philosophy books. In how many ways can they be arranged on the shelf if all books of the same subject are to be together?

10. **Basketball Teams** A junior high girls' basketball team is to consist of 5 players. How many different teams can the manager select from a roster of 12 girls?

11. **Test Taking** A test consists of 12 questions, from which each student chooses 10 to answer. How many choices are possible?

5.3   Probability Spaces

Some experiments always have the same outcome, while others involve "chance" or "random" effects that produce a variety of outcomes. In this section we will consider such "chance" experiments and show that, despite their random nature, we can draw many useful conclusions about their results. We restrict ourselves to experiments that have a finite number of possible outcomes.

Random Experiments and Sample Spaces

A *random* experiment is one that, when repeated under identical conditions, may produce different outcomes. Each repetition of the experiment is a called a *trial*, and each result is an *outcome*. The set of all possible outcomes, exactly one of which must occur, is called the *sample space* for the experiment. The following table lists some simple random experiments and sample spaces.

Random Experiment	Sample Space	
a. Flip a coin	$\{H, T\}$	H means Heads, T means Tails
b. Flip a coin twice	$\{(H, H), (H, T), (T, H), (T, T)\}$	(H, T) means Heads then Tails
c. Choose a card from a deck and observe its suit	$\{\spadesuit, \heartsuit, \diamondsuit, \clubsuit\}$	Spades, hearts, diamonds, clubs
d. Roll a die	$\{1, 2, 3, 4, 5, 6\}$	

There may be more than one possible sample space for a given random experiment. For example, in the experiment of choosing a card [experiment (c) above] the sample space could be all 52 cards (the ace of spades to the king of clubs). In general, we will choose the *simplest* sample space that allows us to observe what we are interested in [which in experiment (c) is just the suit]. The only restriction is that exactly one of the outcomes in the sample space must occur whenever the experiment is performed. In general, if the possible outcomes are represented as $e_1, e_2, \ldots, e_n,$ then the sample space is the *set* of these possible outcomes $\{e_1, e_2, \ldots, e_n\}.$

Events

An *event* is a *subset* of the sample space.

Vit Kovalcik/Shutterstock

→ EXAMPLE 1 Rolling a Die

For the experiment of rolling a die [experiment (d) above], represent the following events as subsets of the sample space $\{1, 2, 3, 4, 5, 6\}$:

a. Event E: *Rolling evens.*

b. Event F: *Rolling 5 or better.*

Solution

a. $E = \{2, 4, 6\}$ Rolling evens
b. $F = \{5, 6\}$ Rolling 5 or better

→ Practice Problem 1

For the experiment of flipping a coin twice [experiment (b) above], represent the event S: *same face* as a subset of the sample space $\{(H, H), (H, T), (T, H), (T, T)\}$.

➤ Solution on page 200

Probabilities of Possible Outcomes

Having identified the possible outcomes of an experiment, we now assign to each of them a probability from 0 (impossible) to 1 (certain). How do we assign probabilities? This depends on our knowledge of the experiment. For example, if there are three possible outcomes and each seems equally likely, then we assign probability $\frac{1}{3}$ to each; if there are 10 outcomes that seem equally likely, then we assign probability $\frac{1}{10}$ to each. In general,

Equally Likely Outcomes

If each of the n possible outcomes in the sample space S is equally likely to occur, then we assign probability $\frac{1}{n}$ to each.

→ EXAMPLE 2 Assigning Equal Probabilities

a. A coin is said to be *fair* if "heads" and "tails" are equally likely. Therefore, for a fair coin we assign probabilities $P(H) = \frac{1}{2}$ and $P(T) = \frac{1}{2}$.

b. A die is said to be *fair* if each of its six faces is equally likely. Therefore, for a fair die we assign $P(1) = P(2) = P(3) = P(4) = P(5) = P(6) = \frac{1}{6}$.

c. A card randomly drawn from a standard deck is equally likely to be any one of the 52 cards. Therefore, the probability of drawing any particular card (such as the queen of hearts) is $\frac{1}{52}$.

From now on when we speak of coins or dice, we will assume that they are fair, unless stated otherwise.

→ EXAMPLE 3 Assigning Equal Probability Using Combinations

A student club has 15 members. If a 3-member fund-raising committee is selected at random, what is the probability that the committee will consist of Bob, Sue, and Tim?

Solution
Since there are $_{15}C_3 = \frac{15 \cdot 14 \cdot 13}{3 \cdot 2 \cdot 1} = 455$ different 3-member committees that can be selected from the 15 members and each committee is equally likely, the probability that any one particular committee is selected is $\frac{1}{455}$.

Not all outcomes are equally likely. For example, long-term weather records for an area might indicate that the probability that it will rain on any given day is $P(R) = \frac{1}{10}$ so that the probability of no rain is $P(N) = \frac{9}{10}$. Probabilities must be assigned on the basis of experience and knowledge of the basic experiment, but they are subject to two conditions: Each probability must be between 0 and 1 (inclusive), and the probabilities must add to 1.

→ EXAMPLE 4 Assigning Unequal Probabilities

The arrow of the spinner on the right can point to any one of three regions labeled A, B, and C. If the probability of pointing to a region is proportional to its area, find the probability of pointing to each of the areas A, B, and C.

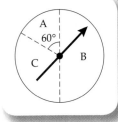

Solution

Let the sample space be $S = \{A, B, C\}$, representing the events that the spinner lands in regions A, B, or C. Clearly, region A is $\frac{60°}{360°} = \frac{1}{6}$ of the circle, region B is $\frac{1}{2}$ of the circle, and region C is the remaining $\frac{1}{3}$ of the circle. Therefore,

$$P(A) = \tfrac{1}{6} \qquad P(B) = \tfrac{1}{2} \qquad P(C) = \tfrac{1}{3}$$

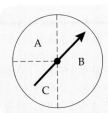

➡ Practice Problem 2

Suppose the spinner on the left is spun. What are the probabilities of the three outcomes?

➤ **Solution on page 200**

Probabilities of Events

We have assigned probabilities to events that consist of exactly one outcome. How do we find the probabilities of other events? We simply add up the probabilities of all the outcomes in the event.

Probability Summation Formula

$$P(E) = \sum_{\text{All } e_i \text{ in } E} P(e_i)$$

The probability of an event is the sum of the probabilities of the possible outcomes in the event

Σ is the Greek capital letter sigma, which is equivalent to our capital S, and stands for "sum." From this formula, since the sample space S contains *all* the outcomes,

$$P(S) = 1 \qquad\qquad \text{Something } must \text{ happen}$$

and, since the empty set contains *no* outcomes,

$$P(\varnothing) = 0 \qquad\qquad \text{"Nothing" cannot happen}$$

➡ EXAMPLE 5   Finding the Probability of an Event

When rolling a pair of dice, what is the probability of rolling *doubles*?

Solution

For two dice, the sample space is

$$
\begin{aligned}
S = \{ \ &(1,1), \ (1,2), \ (1,3), \ (1,4), \ (1,5), \ (1,6),\\
&(2,1), \ (2,2), \ (2,3), \ (2,4), \ (2,5), \ (2,6),\\
&(3,1), \ (3,2), \ (3,3), \ (3,4), \ (3,5), \ (3,6),\\
&(4,1), \ (4,2), \ (4,3), \ (4,4), \ (4,5), \ (4,6),\\
&(5,1), \ (5,2), \ (5,3), \ (5,4), \ (5,5), \ (5,6),\\
&(6,1), \ (6,2), \ (6,3), \ (6,4), \ (6,5), \ (6,6) \ \}
\end{aligned}
$$

Each of the 36 outcomes is equally likely, so each has probability $\frac{1}{36}$. Rolling doubles means rolling one of the outcomes $(1, 1)$, $(2, 2)$, $(3, 3)$, $(4, 4)$, $(5, 5)$, or $(6, 6)$, each having probability $\frac{1}{36}$, so we add $\frac{1}{36}$ six times:

$$\underbrace{\frac{1}{36} + \frac{1}{36} + \cdots + \frac{1}{36}}_{6 \text{ times}} = \frac{6}{36} = \frac{1}{6}$$

Probability of rolling doubles

In the preceding example, we could instead have taken the number of "favorable" outcomes (the 6 doubles) divided by the *total* number of possible outcomes (36) to get $\frac{6}{36} = \frac{1}{6}$, the same answer as before. Although this method is sometimes stated as a basic rule of probability, it is important to remember that it holds only when the outcomes are *equally likely*.

Probability for Equally-Likely Outcomes

$$P(E) = \frac{n(E)}{n(S)}$$

← Number of outcomes "favorable to" (or in) E
← Total number of outcomes

Probability That an Event Does *Not* Occur

An event E is a subset of the sample space, so the event that E does *not* occur, the event E^c, is the complement of E. Since E and E^c are disjoint and contain all outcomes, we must have $P(E) + P(E^c) = 1$. Solving this equation for $P(E^c)$ gives the probability of the complement:

Complementary Probability

$$P(E^c) = 1 - P(E)$$

The probability that an event does *not* occur is 1 minus the probability that it *does* occur

→ **EXAMPLE 6**   **Quality Control**

A shipment of 100 memory chips contains 4 that are defective. The company that ordered them has a policy of testing 3 of the chips and, if any are defective, rejecting the entire shipment. What is the probability that the shipment is rejected?

Solution
The shipment will be rejected if among the three chosen chips there are 1, 2, or 3 defectives. It is much easier to find the probability of the *complementary* event, that the ship-

ment is *accepted* (the number of defectives is 0). For no defectives, all three tested chips must be selected from the 96 good ones in the shipment, and this can occur in $_{96}C_3$ ways. Since the total number of ways that the sample of three can be selected is $_{100}C_3$, and all choices are equally likely, the probability that the shipment passes inspection is

$$\frac{_{96}C_3}{_{100}C_3} = \frac{96 \cdot 95 \cdot 94}{100 \cdot 99 \cdot 98} \approx 0.884$$

Numerator and denominator are both divided by 3!, which cancels and so may be omitted

This is the probability that the shipment is *accepted*, so the probability that it is *rejected* is one minus this number: The shipment will be rejected with probability $1 - 0.884 = 0.116$, or about 12%.

→ **Practice Problem 3**

For the spinner in the Example 4, the probability of the spinner pointing to region A was $\frac{1}{6}$. What is the probability of it *not* pointing to region A?

➤ **Solution on page 200**

Probability Space

A sample space and an assignment of probabilities to the possible outcomes make up a *probability space*. To summarize:

Probability Space

A *probability space* is a sample space $S = \{e_1, e_2, \ldots, e_n\}$ of possible outcomes $e_1, e_2, \ldots, e_n$ together with their probabilities $P(e_1), P(e_2), \ldots, P(e_n)$ satisfying two conditions:

1. $0 \leq P(e_i) \leq 1$ for each outcome e_i in S

   Probabilities are between 0 and 1

2. $P(e_1) + P(e_2) + \cdots + P(e_n) = 1$

   Probabilities sum to 1

An *event E* is a subset of S and has probability

$$P(E) = \sum_{\text{All } e_i \text{ in } E} P(e_i)$$

Sum of the probabilities of the event's outcomes

In particular, $P(S) = 1$, $P(\varnothing) = 0$, and $P(E^c) = 1 - P(E)$.

Addition Rule for Probability

On page 181 we found that the number of elements in the union of two sets is given by the formula $n(A \cup B) = n(A) + n(B) - n(A \cap B)$, with the subtraction at the end done to avoid double-counting the elements that are in both A and B. For exactly the same reason, to find the probability of a union $A \cup B$, we add the probabilities of events A and B but then must subtract the probability of the intersection (to avoid double-counting outcomes in both events):

Addition Rule for Probability

$$P(A \cup B) = P(A) + P(B) - P(A \cap B)$$

The probability of a *union* is the sum of the probabilities minus the probability of the intersection

For disjoint events (also called *mutually exclusive events*), the intersection is empty and so has probability zero, leading to a simpler addition rule:

Addition Rule for Disjoint Events

$$P(A \cup B) = P(A) + P(B)$$

For *disjoint* events, the probability of the *union* is the sum of the probabilities

➡️ **EXAMPLE 7**   **Probability of a Union**

If you roll two dice, what is the probability of rolling doubles or a sum of at least 11?

Solution

Let A be the event of rolling doubles and B the event of rolling a sum of at least 11 (that is, 11 or 12). The sample space and these events may be diagrammed as follows.

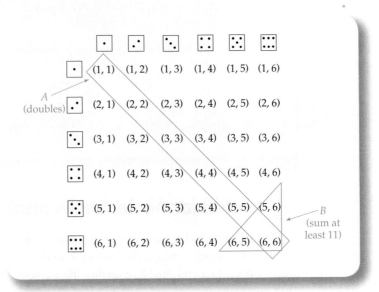

The diagram shows that the probability of doubles is $P(A) = \frac{6}{36}$, and that the probability of a sum of at least 11 is $P(B) = \frac{3}{36}$. The outcome $(6, 6)$ satisfies both conditions, so $P(A \cap B) = \frac{1}{36}$. Therefore:

$$P(A \cup B) = P(A) + P(B) - P(A \cap B) \qquad \text{Addition rule}$$

$$\underbrace{}_{6/36} \quad \underbrace{}_{3/36} \quad \underbrace{}_{1/36}$$

$$= \frac{6}{36} + \frac{3}{36} - \frac{1}{36} = \frac{8}{36} = \frac{2}{9}$$

The probability of doubles or a sum of at least 11 is $\frac{2}{9}$, or about 22%.

Notice that the subtraction of $P(A \cap B)$ prevents the outcome $(6, 6)$ from being counted twice, once in A and once in B.

For events that are disjoint, we may use the simpler Addition Rule for Disjoint Events.

➡ EXAMPLE 8   Finding the Probability of a Disjoint Union

If you roll two dice, what is the probability of rolling a sum of 10 or 11?

Solution

Let A be the event that the sum is 10 and B the event that it is 11. Clearly the sum cannot be two different numbers at the same time, so the events A and B are *disjoint* and we may use the simpler Addition Rule for Disjoint Events. The sum is 10 for the three outcomes $(4, 6)$, $(5, 5)$, and $(6, 4)$, giving $P(A) = \frac{3}{36}$, and it is 11 for the two outcomes $(5, 6)$ and $(6, 5)$, so $P(B) = \frac{2}{36}$. Therefore:

$$P(A \cup B) = P(A) + P(B) = \frac{5}{36} \qquad \begin{array}{l}\text{Using the Addition Rule}\\\text{for Disjoint Events}\end{array}$$

$$\underbrace{}_{3/36} \quad \underbrace{}_{2/36}$$

The probability of rolling a sum of 10 or 11 is $\frac{5}{36}$ or about 14%.

The disjoint events A: sum is 10 and B: sum is 11 are shown in the following diagram, again showing that the probability is $\frac{5}{36}$.

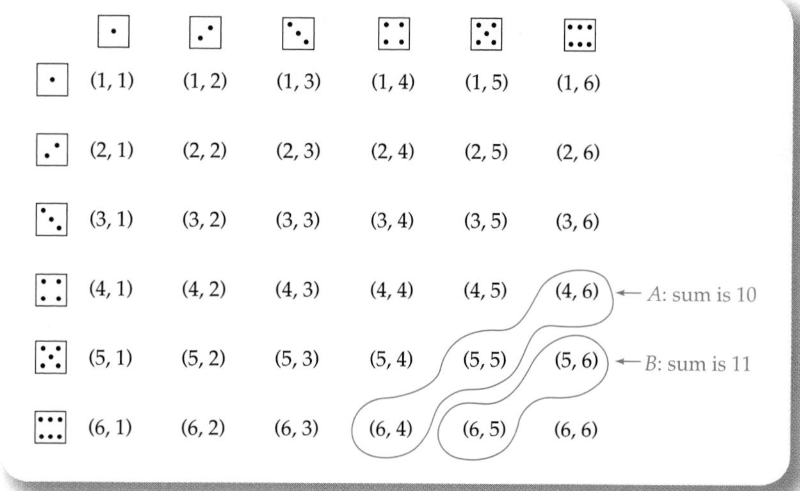

→ 5.3 Exercises

1. **Committees** Find the sample space for a committee of two chosen from Alice, Bill, Carol, and David. Then find the sample space if both sexes must be represented. (*Use initials.*)

2. **Marbles** A box contains three marbles, one red, one green, and one blue. A first marble is chosen, its color is recorded, and then it is replaced in the box and a second marble is chosen, and its color is recorded. Find the sample space. Then find the sample space if the first marble is *not* replaced before the second is chosen.

3. **Dice** Two dice are rolled. E is the event that the sum is even, F is the event of rolling at least one six, and G is the event that the sum is eight. List the outcomes for the following events:

   a. $E \cap F$ *b.* $E^c \cap G$

4. **Committees** If a committee of 3 is to be chosen at random from a class of 12 students, what is the probability of any particular committee being selected? What if the committee is to consist of a president, a vice president, and a treasurer?

5. **Political Contributions** In a town, 38% of the citizens contributed to the Republicans, 42% contributed to the Democrats, and 12% contributed to both. What percentage contributed to neither party?

6. **Surveys** A college survey claimed that 63% of students took English composition, 48% took calculus, 15% took both, and 10% took neither. Show that these figures cannot be correct.

7. **Marbles** A box contains 4 red and 8 green marbles. You reach in and remove 3 marbles all at once. Find the probability that these 3 marbles:

   a. are all red.
   b. are all of the same color.

8. **Elevator Stops** An elevator has 5 people and makes 7 stops. What is the probability that no two people get off on the same floor?

9. **Senate Committees** The U.S. Senate consists of 100 members, 2 from each state. A committee of 8 senators is formed. What is the probability that it contains at least one senator from your state?

10. **Keys** You carry six keys in your pocket, two of which are for the two locks on your front door. You lose one key. What is the probability that you can get into your house through the front door?

5.4 Conditional Probability and Independence

We often ask about the probability of one event *given another*. For example, a card player may want to know the probability of being dealt a third ace given that the first two cards were aces. A smoker might ask about the probability of developing cancer given that one continues to smoke. Such questions involve *conditional probability*, which is the subject of this section. Conditional probability will then lead to a discussion of independence.

Conditional Probability

When two dice (always fair unless stated otherwise) are rolled, there are $6 \cdot 6 = 36$ possible outcomes in the sample space:

$$S = \{(1, 1), (1, 2), \dots, (2, 1), (2, 2), \dots, (6, 6)\}$$

See also page 195

→ EXAMPLE 1   Finding a Conditional Probability

What kind of probability best describes the game of Rochambeau, or rock–paper–scissors?

Brooke Fuller/Shutterstock

When rolling two dice, what is the probability that the first is a 3 given that the sum is 5?

Solution

Let B be the event that the sum is 5, so $B = \{(1, 4), (2, 3), (3, 2), (4, 1)\}$. B contains four equally likely outcomes, and the first roll is a 3 in only one, $(3, 2)$, so the probability that the first is a 3 (event A) given that the sum is 5 is one out of four, which we write as

$$P(A \text{ given } B) = \frac{1}{4}$$

A: first is 3
B: sum is 5

In this example we found the conditional probability by restricting ourselves to a new *smaller* sample space corresponding to the "given" event B. The *conditional* probability was then the probability of A *in* B, that is, the probability of $A \cap B$ relative to the *restricted* sample space B. This leads to the following definition of conditional probability.

Conditional Probability

For events A and B with $P(B) > 0$, the conditional probability of A given B is

$$P(A \text{ given } B) = \frac{P(A \cap B)}{P(B)}$$

The probability of A given B is the probability of *both* divided by the probability of the *second*

We assume $P(B) > 0$ to avoid zero denominators. The conditional probability of A given B may also be written $P(A \mid B)$. If we show the events in a Venn diagram with area representing probability, then the conditional probability is the ratio of the intersection relative to the "given" area.

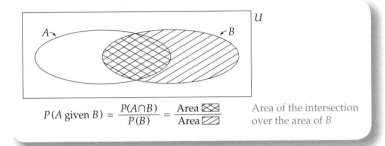

$$P(A \text{ given } B) = \frac{P(A \cap B)}{P(B)} = \frac{\text{Area} \boxtimes}{\text{Area} \boxtimes}$$ Area of the intersection over the area of B

We may check the result of Example 1 by using the above formula:

$$P(A \cap B) = \frac{1}{36}$$ The first is 3 *and* the sum is 5 means just (3, 2)

$$P(B) = \frac{4}{36} = \frac{1}{9}$$ The four outcomes were listed in Example 1

Then, according to the formula,

$$P(A \text{ given } B) = \frac{P(A \cap B)}{P(B)} = \frac{\frac{1}{36}}{\frac{1}{9}} = \frac{9}{36} = \frac{1}{4}$$ Same answer as before

The *unconditional* probability of getting a 3 on the first roll is $\frac{1}{6}$ (all six numbers are equally likely), so conditioning on the sum being 5 changed the probability from $\frac{1}{6}$ to $\frac{1}{4}$.

There are many other ways to express *the probability of A given B* using different words. A few of them are:

- the probability that A happens if B happens
- the probability that A occurs knowing that B occurs
- if B happens, the probability that A happens
- supposing that B occurs, the probability that A occurs

A *standard deck* of 52 playing cards contains four *suits* (spades, hearts, diamonds, and clubs), each of which contains three *face cards* (jack, queen, and king), cards numbered 2 through 10, and an *ace*. A *hand* of cards is a selection of cards from the deck.

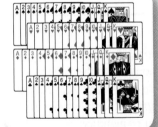

➡ EXAMPLE 2   Finding Probabilities of Hands of Cards

You are playing cards with a friend, and each of you has been dealt 5 cards at random. If you have no face cards, what is the probability that your friend doesn't either?

Solution

Let A be the event that your friend has no face cards, and let B be the event that you have no face cards. The event $A \cap B$ means that *neither* of you has face cards, and since the deck has 12 face cards, this means that the first 10 cards dealt were from the 40 non-face cards.

$$P(A \cap B) = \frac{_{40}C_{10}}{_{52}C_{10}} \approx 0.0536$$

Ways of choosing 10 from the 40 nonface cards ←

← Ways of choosing 10 from 52

Using a calculator

$$P(B) = \frac{_{40}C_5}{_{52}C_5} \approx 0.2532$$

Ways of choosing 5 from the 40 nonface cards ←

← Ways of choosing 5 from 52

Therefore,

$$P(A \text{ given } B) = \frac{P(A \cap B)}{P(B)} \approx \frac{0.0536}{0.2532} \approx 0.212$$

The probability that your friend has no face cards given that you don't have any is about 21%.

Notice that the probability of having no face cards decreases from 25% to 21% when you include the information that you have none. Can you give an intuitive reason why this information should *decrease* the probability?

We could also solve this problem by looking directly at the restricted sample space: Given that you have 5 nonface cards, your friend's cards come from the remaining 47 cards, 12 of which are face cards and 35 of which are not. Therefore,

$$P(A \text{ given } B) = \frac{_{35}C_5}{_{47}C_5} \approx 0.212$$

5 from the remaining 35 nonface cards ←

← 5 from the remaining 47

Same answer as before

We will see that this is true in general: You can solve conditional probability problems either by using the formula or by looking at the restricted sample space. You should use whichever way seems easier for that problem, but preferably use both ways to check that they agree.

The Product Rule for Probability

The conditional probability formula $P(A \text{ given } B) = \dfrac{P(A \cap B)}{P(B)}$ multiplied through by $P(B)$ gives $P(B) \cdot P(A \text{ given } B) = P(A \cap B)$. Reversing sides and interchanging A and B gives the following formula.

Product Rule for Probability

4x5 FILM

$$P(A \cap B) = P(A) \cdot P(B \text{ given } A)$$

The probability of $A \cap B$ is the probability of A times the probability of B *given* A

This rule is very useful when a conditional probability is known.

➡ EXAMPLE 3 Using the Product Rule

You need to be somewhere in half an hour, and your friend has borrowed your car. If your friend arrives soon (you give this a 50–50 chance), the probability that you will be on time is 90%. Otherwise, you will have to walk, with only a 60% chance of being on time. What is the probability that you arrive in your car and on time?

Solution

Much depends on whether your friend arrives soon, so that will be the "given" event, *A: Your friend arrives soon.* Then *B* is the other event, *B: You are on time.* The 90% is the probability of being on time *if* (or *given that*) your friend arrives soon, so $P(B \text{ given } A) = 0.90.$ We are asked for the probability that you arrive in your car *and* on time, meaning that *both* event *A* and *B* occurs. Therefore, by the Product Rule for Probabilities:

$$P(A \cap B) = P(A) \cdot P(B \text{ given } A)$$
$$= \underbrace{0.5} \cdot \underbrace{0.9} = 0.45$$
$$\,\, P(A)\,\, P(B \text{ given } A)$$

A: Friend arrives soon
B: You are on time

The probability that you arrive on time and in your car is 45%.

Suppose instead we were asked for the probability of *arriving on time* (either by car or on foot). We will use a *tree diagram* to represent the different ways, similar to our use of tree diagrams to find different outcomes in Section 5.1.

➡ EXAMPLE 4 Using a Tree Diagram

For the situation in Example 3, find the probability that you arrive on time.

Solution

Again, much depends upon your friend's arrival, so we first "branch" on whether your friend arrives soon (event *A*, having probability 0.5) or not (event A^c, also having complementary probability 0.5). The second-stage branching represents whether you arrive on time (event *B*) or not (B^c), with the given (conditional) probabilities depending on the first-stage branching. Be sure you understand the correct placement of the probabilities given in the problem, 0.90 and 0.60, along with their complements 0.10 and 0.40. The probabilities at the end of *two* branchings are the *products* of the probabilities along those branches.

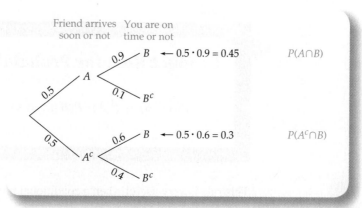

Arriving on time (by car or on foot) corresponds to the two branches ending in B. We multiply out those probabilities and add the results, $0.45 + 0.30 = 0.75$. Answer:

You have a 75% chance of arriving on time.

Notice that the uppermost branches give the 0.45 that we found in Example 3. In general, a tree diagram gives all possible probabilities that can be found from the given information.

> **Practice Problem 1**

Use the tree diagram in Example 4 to answer the following questions.
  a. What is the probability of arriving late and in your car?
  b. What is the probability of arriving late? ➤ **Solutions on page 208**

Notice several things about such tree diagrams:

- The branches from any point must have probabilities that add to 1.
- The probabilities at the right-hand end of a branch are found by multiplying along the branch.
- The probabilities written on the second-stage branches are *conditional* probabilities, so we are indeed using the Product Rule for Probability.
- There can be any number of branches per stage and any number of stages.
- There is no need to multiply out all the products, only the ones you need.

➤ **EXAMPLE 5   Sampling without Replacement**

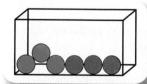

A box contains three blue marbles, two green marbles, and one red marble. A marble is randomly chosen (and not replaced) and then a second marble is chosen. What is the probability that the second marble is green?

Solution
We branch on the color of the first marble, using B, G, and R for the events that it is blue, green, or red. Of the six marbles, 3 are blue, 2 are green, and 1 is red, so the probabilities on the first-stage branching are $\frac{3}{6} = \frac{1}{2}$, $\frac{2}{6} = \frac{1}{3}$, and $\frac{1}{6}$, as shown in the following diagram. For the second-stage branching, if a blue is chosen first, then the remaining 5 are 2 blue, 2 green, and 1 red, leading to the probabilities $\frac{2}{5}, \frac{2}{5}, \frac{1}{5}$ for the top part of the second stage choice. The other second-stage probabilities are calculated similarly and depend on the color that was removed in the first stage.

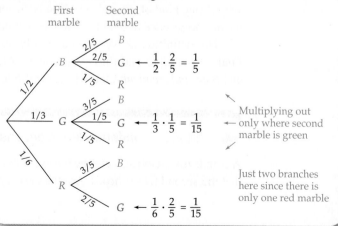

Multiplying out and adding up the branches that end in green:

The probability that the second marble is green is $\frac{1}{5} + \frac{1}{15} + \frac{1}{15} = \frac{5}{15} = \frac{1}{3}$.

> **Practice Problem 2**
>
> Use the diagram in Example 5 to find the probability that the second marble is blue.
>
> ➤ **Solution on page 208**

Independent Events

Roughly speaking, two events are said to be *independent* if one has nothing to do with the other, so that the occurrence of one has no bearing on the occurrence of the other. In terms of conditional probability, this means that for independent events A and B, we have $P(A$ given $B) = P(A)$. Using the definition of conditional probability, this equation becomes

$$\frac{P(A \cap B)}{P(B)} = P(A) \qquad \text{For } P(B) \neq 0$$

Multiplying each side by $P(B)$ gives the following equivalent condition, which we take as the *definition* of independence.

Independent Events

Events A and B are *independent* if

$$P(A \cap B) = P(A) \cdot P(B)$$

The probability of the intersection is the *product* of the probabilities

Events that are not independent are *dependent*.

Be careful! Independent does not mean the same thing as *disjoint* or *mutually exclusive*. If events A and B have positive probabilities and are disjoint, then $P(A \cap B) = 0$ but $P(A) \cdot P(B)$ is positive, so they cannot be independent. Intuitively, disjointness is a very strong kind of *dependence*: The occurrence of one of two disjoint events guarantees the *non*occurrence of the other, so they cannot be independent.

The definition above gives a simple test for the independence of events A and B: Find $P(A \cap B)$ and find the product $P(A) \cdot P(B)$. If the results are *equal*, then A and B are *independent*; otherwise, A and B are *dependent*.

➤ **EXAMPLE 6** **Independent Coin Tosses**

A coin is tossed twice. Let A be the event that the first toss is heads, and let B be the event that the second toss is heads. Are the events A and B independent?

Alexey Stiop/Shutterstock

Solution

The sample space consists of four equally likely events:

$$S = \{(H, H), (H, T), (T, H), (T, T)\}$$

so

$$A = \{(H, H), (H, T)\}, \quad B = \{(H, H), (T, H)\}, \quad A \cap B = \{(H, H)\}$$

$$P(A) = \frac{2}{4} = \frac{1}{2}, \quad P(B) = \frac{2}{4} = \frac{1}{2}, \quad P(A \cap B) = \frac{1}{4}$$

Since $P(A) \cdot P(B) = \frac{1}{2} \cdot \frac{1}{2} = \frac{1}{4}$ and $P(A \cap B) = \frac{1}{4}$ are equal, the events A and B *are* independent.

If $P(A \cap B)$ and $P(A) \cdot P(B)$ had *not* been equal, the events would have been *dependent*. That successive coin tosses are independent should come as no surprise, and this fact is sometimes expressed by saying that the coin has "no memory." Thus, each toss is a new experiment with "no influence from the past."

The concept of independence can be extended to more than two events.

Many Independent Events

A collection $E_1, E_2, \dots, E_m$ of events is *independent* if any subcollection of them satisfies the multiplication formula:

$$P(E_i \cap E_j \cap \cdots \cap E_k) = P(E_i) \cdot P(E_j) \cdot \cdots \cdot P(E_k)$$

The probability of the intersection is the *product* of the probabilities

In practice, the most important uses of independence do not involve *proving* it but *assuming* it.

➡ EXAMPLE 7   Quality Control

A computer consists of a logic board, three memory chips, a screen, and an I/O (input–output) module. If these individual components work with probabilities 0.98, 0.99 (for each memory chip), 0.96, and 0.97, respectively, what percentage of the computers should be expected to work?

Solution

Assuming independence, we multiply the probabilities:

$$0.98 \cdot 0.99^3 \cdot 0.96 \cdot 0.97 \approx 0.885$$

About 89% of the assembled computers should be expected to work.

---

1. *a.* From the branches corresponding to A and B^c, $0.50 \cdot 0.10 = 0.05$.

   b. From the branches ending in B^c, $0.05 + 0.50 \cdot 0.40 = 0.25$.   (We could also have found this as 1 minus the probability of being on time, which we found in Example 4.)

2. Adding up the products along branches ending in blue:

$$\tfrac{1}{2} \cdot \tfrac{2}{5} + \tfrac{1}{3} \cdot \tfrac{3}{5} + \tfrac{1}{6} \cdot \tfrac{3}{5} = \tfrac{1}{2}.$$

→ **5.4 Exercises**

Use the given values to find:

   a. $P(A$ given $B)$ *b.* $P(B$ given $A)$

1. $P(A) = 0.6$, $P(B) = 0.4$, $P(A \cap B) = 0.2$

2. $P(A) = 0.4$, $P(B) = 0.5$, $P(A \cup B) = 0.6$

3. *Marbles* A box contains 4 white, 2 red, and 4 black marbles. One marble is chosen at random, and it is not black. Find the probability that it is white.

4. *Gender* Your friend has two children, and you know that at least one is a girl. What is the probability that both are girls? (Assume that girls and boys are equally likely.)

5. *Choosing Courses* You will take either a basket-weaving course or a philosophy course, depending on what your advisor decides. You estimate that the probability of your getting an A in basket weaving is 0.95, while in philosophy it is 0.70. However, the chances of your advisor choosing the basket-weaving course is only 20%, while there is an 80% chance that he will put you in the philosophy course. What is the probability that you end up with an A?

6. *Driving* Suppose that 70% of drivers are "careful" and 30% are "reckless." Suppose further that a careful driver has a 0.1 probability of being in an accident in a given year, while for a reckless driver the probability is 0.3. What is the probability that a randomly selected driver will have an accident within a year?

7. *Independence* For the experiment of tossing a coin twice, find whether events

    A: heads on the first toss
    B: different results on the two tosses

are independent or dependent.

8. *Class Attendance* Two students are registered for the same class and attend independently of each other, student A 90% of the time and student B 70% of the time. The teacher remembers that on a given day at least one of them is in class. What is the probability that student A was in class that day?

9. *Dice* Three dice are rolled. Find the probability of getting:

   a. all sixes
   b. all the same outcomes
   c. all different outcomes

10. *Longevity* Based on *Life Tables*, in a population of 100,000 females, 89.8% will live to age 60 and 57.1% will live to age 80. Given that a woman is 60, what is the probability that she will live to age 80?

5.5 Bayes' Formula

Sometimes we want to *reverse the order* in conditional probability. For example, the probability of developing cancer given that one smokes is different from the probability of being a smoker given that one develops cancer. The first might be of interest to a smoker and the second to a medical researcher. In this section we will develop Bayes' formula to reverse the order in conditional probability.

Bayes' Formula

In general, $P(A \text{ given } B)$ will not be the same as $P(B \text{ given } A)$.

> ### ➡ Practice Problem 1
>
> Explain the difference between
>
> $$P(\text{you are rich given that you win the lottery})$$
>
> and
>
> $$P(\text{you win the lottery given that you are rich})$$
>
> ➤ Solution on page 212

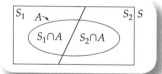

Suppose the sample space S is divided into two parts, S_1 and S_2, that are disjoint and whose union is the whole sample space $S_1 \cup S_2 = S$. These sets then partition any event A into two parts, the part in S_1 and the part in S_2, as shown in the diagram on the left. The probability of A can then be found by adding the probabilities of the two parts:

$$P(A) = P(S_1 \cap A) + P(S_2 \cap A)$$

The definition of conditional probability $P(S_1 \text{ given } A) = \dfrac{P(S_1 \cap A)}{P(A)}$ with the denominator replaced by the above expression gives

$$P(S_1 \text{ given } A) = \frac{P(S_1 \cap A)}{P(S_1 \cap A) + P(S_2 \cap A)}$$

Using the Product Rule for Probability (see page 203) in the numerator and twice in the denominator, we may write this in the following form:

$$P(S_1 \text{ given } A) = \frac{P(S_1) \cdot P(A \text{ given } S_1)}{P(S_1) \cdot P(A \text{ given } S_1) + P(S_2) \cdot P(A \text{ given } S_2)}$$

This formula generalizes by replacing S_1 and S_2 by any *partition* of S into sets $S_1, S_2, \ldots, S_n$ that are pairwise disjoint and whose union is the entire sample space, $S_1 \cup S_2 \cup \cdots \cup S_n = S$:

Bayes' Formula

$$P(S_1 \text{ given } A) = \frac{P(S_1) \cdot P(A \text{ given } S_1)}{P(S_1) \cdot P(A \text{ given } S_1) + \cdots + P(S_n) \cdot P(A \text{ given } S_n)}$$

Thomas Bayes (1702–1761), an English nonconformist minister, was also an amateur mathematician and a defender of the new subject of calculus when it was attacked as illogical and incorrect. His mathematical writings were not published during his lifetime; the one containing "Bayes' formula" was not published until three years after his death.

Petr Vaclavek/Shutterstock (speech bubble) and stevanovic.igor/Shutterstock (grid)

The sum in the denominator has n terms, one for each of the sets $S_1, S_2, \ldots, S_n$. Notice that the order in the conditional probability is reversed: S_1 *given* A on the left and A *given* S_1 on the right. Bayes' formula in this form is needed in only the most complicated cases. In fact, we will use tree diagrams to carry out the calculations in Bayes' formula.

→ EXAMPLE 1   Medical Screening

Suppose that a medical test for a disease is 90% accurate for both those who have the disease and for those who don't. Suppose, furthermore, that only 5% of the population has this disease. Given that a person tests positive for the disease (meaning that the test says that the person has the disease), what is the probability that the person actually has the disease? Would this test be useful for widespread medical screening?

Solution

The probabilities depend on whether a person has the disease (event D, with probability 0.05) or not (event D^c, with probability 0.95), which represents the first-stage branching. The second-stage branching is according to whether the person tests positive (T) or not (T^c), with probabilities based on the test being 90% accurate.

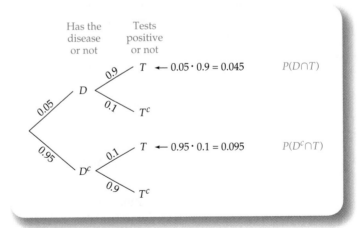

The probability that a randomly selected person has the disease given that the person tests positive is $P(D \text{ given } T)$,   which, by the definition of conditional probability, is

$$P(D \text{ given } T) = \frac{P(D \cap T)}{P(T)}$$

For the numerator we take the branch though D and T (with probability 0.045), and for the denominator we add the two branches ending in T:

$$P(D \text{ given } T) = \frac{0.045}{0.045 + 0.095} = \frac{0.045}{0.14} \approx 0.32$$

Therefore, the probability that a person actually has the disease given that the person tests positive for it is only about 32%. In other words, 68% of those testing positive in fact do not have the disease and are being needlessly worried by a false diagnosis, so this test should *not* be used for widespread medical screening.

➡ **EXAMPLE 2   Assessing Voting Patterns**

Registered voters in Marin County are 45% Democratic, 30% Republican, and 25% Independent. In the last election for county supervisor, 70% of the Democrats voted, as did 80% of the Republicans and 90% of the Independents. What is the probability that a randomly selected voter in this election was a Democrat?

Solution

Since the information is stated in terms of the political parties, they represent the first-stage branching $(D, R, \text{or } I)$, with the second-stage represented by voting or not voting $(V \text{ or } V^c)$. The probabilities come from the given values together with their complements.

The probability that a randomly selected voter was a Democrat is $P(D \text{ given } V)$, which is

$$P(D \text{ given } V) = \frac{P(D \cap V)}{P(V)}$$

Using the definition of conditional probability

For the numerator $P(D \cap V)$ we take the branch through D and V (the uppermost branch), and for the denominator $P(V)$ we take the sum of all of the V branches:

$$\frac{0.315}{0.315 + 0.24 + 0.225} = \frac{0.315}{0.780} \approx 0.404$$

The probability that a voter in the election was a Democrat is about 40%.

➡ **Practice Problem 2**

Using the diagram in Example 2, what is the probability that a randomly selected voter in the election was an Independent?

➤ **Solution below**

Notice that as required for Bayes' formula, the three parties, Democrat, Republican, and Independent, do not overlap (the sets are *disjoint*) and include all voters (their probabilities *add to 1*). If the probabilities had not added to 1, we would have had to add in another group ("Others") with the remaining probability.

➡ **Solutions to Practice Problems**

1. *P* (you are rich given that you win the lottery) should be very high since you certainly will be rich if you win the lottery.

P (you win the lottery given that you are rich) should be very small since your chances of winning the lottery are very small regardless of how much money you have.

2. Using the notation from the solution to Example 2, we have

$$P(I \text{ given } V) = \frac{P(I \cap V)}{P(V)} = \frac{0.225}{0.315 + 0.24 + 0.225} = \frac{0.225}{0.780} \approx 0.288$$

The probability that a voter in the last election was an Independent is about 29%.

➡ **5.5 Exercises**

1. *Voting* In a town, 60% of the citizens are Republicans and 40% are Democrats. In the last election, 55% of the Republicans voted and 65% of the Democrats voted. If a voter is randomly selected, what is the probability that the person is a Republican?

2. *Medical Testing* A new test is developed to test for a certain disease, giving "positive" or "negative" results to indicate that the person does or does not have the disease. For a person who actually *has* the disease, the test will give a positive result with probability 0.95 and a negative result with probability 0.05 (a so-called "false negative"). For a person who does *not* have the disease, the test will give a positive result with probability 0.05 (a "false positive") and a negative result with probability 0.95. Furthermore, only one person in 1000 actually has this disease. If a randomly selected person is given

the test and tests positive, what is the probability that the person actually has the disease?

3. *Manufacturing Defects* A computer chip factory has three machines, A, B, and C, for producing the memory chips. Machine A produces 50% of the factory's chips, machine B produces 30%, and machine C produces 20%. It is known that 3% of the chips produced by machine A are defective, as are 2% of chips produced by machine B and 1% of the chips from machine C. If a randomly selected chip from the factory's output is found to be defective, what is the probability that it was produced by machine B?

4. *Airline Hijacking* The "hijacker profile" developed by the Federal Aviation Administration fits 90% of hijackers and only 0.05% of legitimate passengers.

Based on historical data, assume that only 30 of 300 million passengers are hijackers. What is the probability that a person who fits the profile is actually a hijacker?

5. *Random Drug Testing* In 1986 the Reagan administration issued an executive order allowing agency heads to subject all employees to urine tests for drugs. Suppose that the test is 95% accurate both in identifying drug users and in clearing nonusers.

Suppose further that 1% of employees use drugs. What is the probability that a person who tests positive is *not* a drug user?

5.6 Random Variables and Distributions

Often a random experiment results in a *number*, such as the sum on the faces of two dice or your winnings in a lottery. Such numerical quantities that depend on chance events are called *random variables* and are the central objects of probability and statistics. In this section (and also in the following chapter) we will introduce some of the random variables that have proved most useful in applications.

Random Variables

A random variable is an assignment of a number to each outcome in the sample space. We will use capital letters such as X and Y for random variables.

→ **EXAMPLE 1** Defining a Random Variable

A coin is tossed four times, and

$$X = \begin{pmatrix} \text{Number} \\ \text{of heads} \end{pmatrix}$$

Find the possible values for X and the outcomes corresponding to each of its possible values.

Solution

The number of heads in four tosses can be 0, 1, 2, 3, or 4, so these are the possible values for X. The sample space consists of the 16 sequences of Hs and Ts shown on the left.

Since X is the number of heads, for any particular outcome we can find its value by counting Hs. For example, (H, T, H, H) gives $X = 3$. The following table lists the possible values of X and the outcomes for which it takes those values (as you should check by "counting heads").

Values of X	Outcomes
$X = 0$	(T, T, T, T)
$X = 1$	$(H, T, T, T), (T, H, T, T), (T, T, H, T), (T, T, T, H)$
$X = 2$	$(H, H, T, T), (H, T, H, T), (H, T, T, H), (T, H, H, T), (T, H, T, H), (T, T, H, H)$
$X = 3$	$(H, H, H, T), (H, H, T, H), (H, T, H, H), (T, H, H, H)$
$X = 4$	(H, H, H, H)

We can find the probability that X takes any particular value, such as $X = 2$, by adding up the probabilities of the outcomes corresponding to that value. The *probability distribution* of a random variable is the collection of these probabilities for its various values.

Random Variable

A *random variable* X is an assignment of a number to each element in the sample space. The *probability distribution* of the random variable X is the collection of all probabilities $P(X = x)$ for each possible value x.

➡ **EXAMPLE 2** Finding a Probability Distribution

A coin is tossed four times. Find and graph the probability distribution for

$$X = \left(\begin{matrix} \text{Number} \\ \text{of heads} \end{matrix}\right)$$

Solution

Using the table in Example 1, $X = 0$ occurs only for the outcome (T, T, T, T), which is 1 out of 16 equally likely outcomes, so $P(X = 0) = \frac{1}{16}$. The event $X = 1$ corresponds to 4 outcomes in the table, so $P(X = 1) = \frac{4}{16} = \frac{1}{4}$. The other probabilities $P(X = 2)$, $P(X = 3)$, and $P(X = 4)$ are similarly found by counting outcomes and dividing by 16, giving the probabilities in the following table. These probabilities are graphed on the left, the height of each bar being the probability that X takes the value at the bottom of the bar.

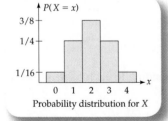
Probability distribution for X

x	0	1	2	3	4
$P(X = x)$	$\frac{1}{16}$	$\frac{1}{4}$	$\frac{3}{8}$	$\frac{1}{4}$	$\frac{1}{16}$

Observe that the *most likely* number of heads in 4 tosses is 2 and that the least likely are the extreme values 0 and 4, just as you might expect. The probabilities add to 1: $\frac{1}{16} + \frac{1}{4} + \frac{3}{8} + \frac{1}{4} + \frac{1}{16} = \frac{16}{16} = 1$, so *the area under the graph is 1* (since each bar has width 1).

→ **Practice Problem 1**

Your dog has a litter of four pups. Assuming that each pup is equally likely to be male or female, which is more likely: two of each sex or 3–1 split (3 of one sex and 1 of the other)? [*Hint*: Use Example 2.]

➤ **Solution on page 220**

→ **EXAMPLE 3**   Raffle Winnings

To raise money, a children's hospital sells 400 raffle tickets for $100 each. First prize is a $3000 Florida vacation for two, second prize is a $1000 credit at the town's supermarket, and the five third prizes are $100 "dinner for two" gift certificates at a local restaurant. Find the probability distribution of the value of a raffle ticket.

Solution

If X represents the value of a ticket, then its possible values are $3000, 1000, 100$, and 0, with probabilities

$$P(X = 3000) = \tfrac{1}{400} \qquad \text{One \$3000 prize among 400 tickets}$$

$$P(X = 1000) = \tfrac{1}{400} \qquad \text{One \$1000 prize}$$

$$P(X = 100) = \tfrac{5}{400} \qquad \text{Five \$100 prizes}$$

$$P(X = 0) = \tfrac{393}{400} \qquad \text{The other 393 tickets are worth nothing}$$

Expected Value

On average, how much is each ticket worth? We divide the total value of all the tickets, $3000 + 1000 + 100 \cdot 5 + 0 \cdot 393$, by the number of tickets, 400. The result can be written $3000 \cdot \tfrac{1}{400} + 1000 \cdot \tfrac{1}{400} + 100 \cdot \tfrac{5}{400} + 0 \cdot \tfrac{393}{400}$. Looking back at the above example, this is just the values of X multiplied by their probabilities and added. This leads to the following definition of the expected value of a random variable.

Expected Value

A random variable X taking values $x_1, x_2, \ldots, x_n$ with probabilities $p_1, p_2, \ldots, p_n$ has expected value:

$$E(X) = x_1 \cdot p_1 + x_2 \cdot p_2 + \cdots + x_n \cdot p_n \qquad \text{Sum of the possible values times their probabilities}$$

The expected value is also called the *expectation* or the *mean* of the random variable (or of the probability distribution) and is sometimes denoted by μ (pronounced "mu," the Greek letter *m*). Using this definition, we may calculate the expected value of the raffle ticket in the preceding example by multiplying the possible values times their probabilities and adding:

$$E(X) = 3000 \cdot \tfrac{1}{400} + 1000 \cdot \tfrac{1}{400} + 100 \cdot \tfrac{5}{400} + 0 \cdot \tfrac{393}{400} = \$11.25$$

This expected value represents your average winnings per ticket. Since the tickets cost $100, you are on average losing $88.75 on each ticket. To put this in a more positive light, this latter figure represents your true generosity to the hospital.

→ **Practice Problem 2**

If the prizes in Example 3 were changed to one $3000 first prize, two $1000 second prizes, and ten $100 third prizes, what would be the expected value of a raffle ticket?

➤ Solution on page 220

→ **EXAMPLE 4**   **Expected Dice Winnings**

You roll a die and win any *even* amount and lose any *odd* amount (in dollars). (That is, if you roll 2, you win $2, and if you roll 3, you lose $3.) What is the expected value of your winnings?

Solution

The expected value of $X = \begin{pmatrix} \text{Your} \\ \text{winnings} \end{pmatrix}$ is the sum of the values multiplied by their probabilities:

$$E(X) = -1 \cdot \tfrac{1}{6} + 2 \cdot \tfrac{1}{6} - 3 \cdot \tfrac{1}{6} + 4 \cdot \tfrac{1}{6} - 5 \cdot \tfrac{1}{6} + 6 \cdot \tfrac{1}{6} = \tfrac{3}{6} = \tfrac{1}{2}$$

The expected value of your winnnings is half a dollar, or 50¢.

This answer of $\tfrac{1}{2}$ means that if you play the game many times, your *average winnings per play* will be about half a dollar. The expected value is positive, so this game is *favorable* to you. If the expected value were zero, the game would be *fair*, and if it were negative, the game would be *unfavorable* to you. The expected value can be interpreted as *the fair price for playing the game*, meaning that paying that amount before each play would make the game *fair*.

Note that if the values are equally likely, then the expected value is just the arithmetic average of the values. Generally, the probabilities will not all be equal, so the expected value will be a *weighted average*, weighting each value by its probability. If none of the probabilities is zero, the expected value will lie somewhere between the highest and lowest values of the random variable.

→ **Practice Problem 3**

Find the expected value of one roll of a die.

➤ Solution on page 220

hunta/Shutterstock

→ **EXAMPLE 5** Concert Insurance

A concert promoter is planning an outdoor concert and estimates a profit of $100,000 if there is no rain but only $10,000 if it does rain. He can take out an insurance policy costing $15,000 that will pay $100,000 if it rains. The weather bureau estimates the chances of rain at 10%. Based on expected value, should he buy the insurance?

Solution

First we calculate the profit under both conditions.

	Without Insurance	With Insurance	
No rain	100,000	$100,000 - 15,000 = 85,000$	← Subtracting the cost of insurance
Rain	10,000	$10,000 - 15,000 + 100,000 = 95,000$	← and adding the insurance payment

The expected values are

Without insurance: $100,000 \cdot \frac{9}{10} + 10,000 \cdot \frac{1}{10} = 91,000$

With insurance: $85,000 \cdot \frac{9}{10} + 95,000 \cdot \frac{1}{10} = 86,000$

Each value times its probability

The expected value *without* insurance is higher, so on this basis he should skip the insurance. (However, in view of the relatively small difference in expected profits, he might decide to take the insurance after all to reduce the risk.)

Binomial Distribution

Suppose a biased coin has probability p of coming up heads (and therefore probability $1 - p$ of tails) on any toss. Since successive tosses are independent, the probability of a particular succession of outcomes would be just the product of the probabilities for each toss. For example, the probability of the outcome $H T H H T$ is $p \cdot (1 - p) \cdot p \cdot p \cdot (1 - p)$, which simplifies to $p^3(1 - p)^2$. Clearly, any other particular ordering of 3 heads and 2 tails would have this same probability. How many such orderings are there? Choosing the three tosses on which heads will occur (so the other two tosses are automatically tails) can be done in exactly ${}_5C_3 = \frac{5 \cdot 4 \cdot 3}{3 \cdot 2 \cdot 1}$ ways (using the combinations formula from page 190). Therefore, the probability of exactly 3 heads in 5 tosses is ${}_5C_3 \, p^3(1 - p)^2$.

We can generalize this result to any number n of tosses with k heads and $n - k$ tails to obtain the formula ${}_nC_k \, p^k(1 - p)^{n-k}$. Furthermore, we spoke of heads and tails only for familiarity. The same ideas apply to any experiment that has just two possible outcomes, *success* and *failure*, occurring with probabilities p and $1 - p$, respectively, with successive repetitions being independent. Such repeated experiments with fixed probabilities are called *Bernoulli trials*, and the number of successes in Bernoulli trials is called a *binomial random variable*.

Who's Bernoulli?

Bernoulli trials are named after the Swiss mathematician James Bernoulli (1654–1705), who first recognized the importance of such random variables. His main work was Ars Conjectandi (Art of Conjecturing).

Robyn Mackenzie/Shutterstock

Binomial Distribution

For independent repetitions of an experiment with probability p of success on each trial, the probability that in n trials the number X of successes will be k is

$$P(X = k) = {}_nC_k\, p^k(1 - p)^{n-k}$$

n = number of trials
k = number of successes $(0 \le k \le n)$
p = probability of success

X is called a *binomial random variable with parameters n and p*. Its expected value or mean is $E(X) = n \cdot p$.

The formula for the expected value may be made intuitively reasonable as follows: If a coin with probability of heads $\frac{1}{3}$ is tossed 12 times, you should expect about a third of them, $12 \cdot \frac{1}{3} = 4$,   to be heads, and this is just the formula $E(X) = n \cdot p$. Recall that the number of combinations ${}_nC_k$ may be calculated from the formula ${}_nC_k = \frac{n \cdot (n-1) \cdot \cdots \cdot (n-k+1)}{k!} = \frac{n!}{k!(n-k)!}$.   We may use the binomial distribution in any situation with independent trials having probability p of success, where the word *success* may mean whatever outcome we choose.

→ **EXAMPLE 6**   **A Binomial Distribution**

Let X be the number of heads in four tosses of a coin. Find the probability distribution and expected value of X.

Solution

Coin tossing is merely Bernoulli trials, so we want the probability distribution of a binomial random variable with $n = 4$ and $p = \frac{1}{2}$.

$$P(X = 0) = 1 \cdot \left(\tfrac{1}{2}\right)^0 \cdot \left(\tfrac{1}{2}\right)^4 = \tfrac{1}{16} \qquad\qquad {}_4C_0 p^0 (1-p)^4$$
$$P(X = 1) = 4 \cdot \left(\tfrac{1}{2}\right)^1 \cdot \left(\tfrac{1}{2}\right)^3 = \tfrac{4}{16} = \tfrac{1}{4} \qquad\qquad {}_4C_1 p^1 (1-p)^3$$
$$P(X = 2) = 6 \cdot \left(\tfrac{1}{2}\right)^2 \cdot \left(\tfrac{1}{2}\right)^2 = \tfrac{6}{16} = \tfrac{3}{8} \qquad\qquad {}_4C_2 p^2 (1-p)^2$$
$$P(X = 3) = 4 \cdot \left(\tfrac{1}{2}\right)^3 \cdot \left(\tfrac{1}{2}\right)^1 = \tfrac{4}{16} = \tfrac{1}{4} \qquad\qquad {}_4C_3 p^3 (1-p)^1$$
$$P(X = 4) = 1 \cdot \left(\tfrac{1}{2}\right)^4 \cdot \left(\tfrac{1}{2}\right)^0 = \tfrac{1}{16} \qquad\qquad {}_4C_4 p^4 (1-p)^0$$

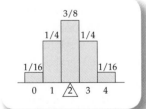

These five probabilities, graphed on the left, make up the probability distribution for X, the number of heads in four tosses. For the expected value, we use the formula in the preceding box:

$$E(x) = n \cdot p = 4 \cdot \tfrac{1}{2} = 2$$

We found exactly this distribution in Example 2 (page 214). Here we found it using the binomial probability formula, and there we found it from basic principles. The results agree. It should come as no surprise that the expected number of heads in four tosses of a fair coin is two. This "mean" of 2 is the "balance point" of the probability distribution graph, as shown above on the left.

→ **Practice Problem 4**

Let X be the number of heads in six tosses of a coin. Find $P(X = 3)$ and $E(X)$.

➤ **Solution on next page**

→ **EXAMPLE 7**   **Employee Retention**

A restaurant manager estimates the probability that a newly hired waiter will still be working at the restaurant six months later is only 60%. For the five new waiters just hired, what is the probability that at least four of them will still be working at the restaurant in six months?

Solution

Assuming that the waiters decide independently of one another, their decisions make up five Bernoulli trials. Counting a waiter who stays as a "success," the number X of waiters who stay is a binomial random variable with $n = 5$ and $p = 0.6$. We want $P(X \geq 4)$, which means $P(X = 4 \text{ or } 5) = P(X = 4) + P(X = 5)$.

$$P(X = 4) + P(X = 5) = {}_5C_4(0.6)^4(0.4)^1 + {}_5C_5(0.6)^5(0.4)^0 \approx 0.337$$

$$\underbrace{}_{5} \qquad \underbrace{}_{1}$$

The probability that at least four of the new waiters will stay for six months is only about 34%.

→ **EXAMPLE 8**   **Money-Back Guarantees**

A manufacturer of compact disks (CDs) for computer storage sells them in packs of 10 with a "double your money back" guarantee if more than one CD is defective. If each CD is defective with a 1% probability independently of the others, what proportion of packages will require refunds?

Solution

If X is the number of defective CDs in a package, then X is a binomial random variable with $n = 10$ and $p = 0.01$. Rather than calculating the probability of a refund (2 or more defectives), we find the complementary probability:

$$P(X = 0) + P(X = 1) = {}_{10}C_0(0.01)^0(0.99)^{10} + {}_{10}C_1(0.01)^1(0.99)^9 \approx 0.996$$

$$\underbrace{}_{1} \qquad \underbrace{}_{10}$$

Subtracting this answer from 1 (since it is the complementary probability) gives 0.004. Therefore, the company will have to provide refunds for only about 0.4% of its packs.

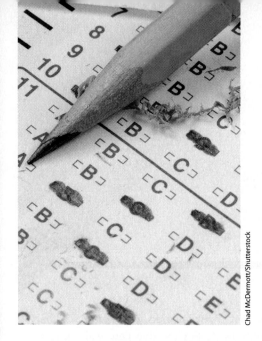
Chad McDermott/Shutterstock

→ EXAMPLE 9   Multiple-Choice Tests

A multiple-choice test consists of 25 questions, each of which has 5 possible answers. What is the expected number of correct answers that would result from random guessing?

Solution

The 25 questions are 25 Bernoulli trials, and random guessing among the 5 possible answers for each question gives a probability $p = \frac{1}{5}$ of success for each. Therefore, the expected number of correct answers is

$$E(X) = 25 \cdot \tfrac{1}{5} = 5 \qquad\qquad E(X) = n \cdot p$$

This is why multiple-choice tests are often *rescaled* by only counting the number of correct answers above a certain expected number. In this case, the grade would be based on the number of correct answers beyond the first 5.

→ Solutions to Practice Problems

1. The sample space of all gender orders (such as F, M, M, F) is just like that for four coin tosses on pages 213–214 but with different letters. Example 2 says that two of each gender has probability $\frac{3}{8}$ while a 3–1 split (which can occur in two ways) has probability $\frac{1}{4} + \frac{1}{4} = \frac{1}{2}$. Therefore, three of one gender and one of the other is more likely than two of each.

2. $E(X) = 3000 \cdot \frac{1}{400} + 1000 \cdot \frac{2}{400} + 100 \cdot \frac{10}{400} + 0 \cdot \frac{387}{400} = 15$.
The expected value is \$15.

3. $E(X) = 1 \cdot \frac{1}{6} + 2 \cdot \frac{1}{6} + 3 \cdot \frac{1}{6} + 4 \cdot \frac{1}{6} + 5 \cdot \frac{1}{6} + 6 \cdot \frac{1}{6} = \frac{21}{6} = 3.5$.
This answer shows that the expected value may not be one of the possible values of the random variable.

4. $P(X = 3) = {}_6C_3(\frac{1}{2})^3(\frac{1}{2})^3 = \frac{6 \cdot 5 \cdot 4}{3 \cdot 2 \cdot 1}(\frac{1}{2})^6 = 20 \cdot \frac{1}{64} = \frac{5}{16}$

$E(X) = n \cdot p = 6 \cdot \frac{1}{2} = 3$

→ 5.6 Exercises *(A calculator will be helpful.)*

1. **Coins** A coin is tossed three times, and X is the number of heads. Find and graph the probability distribution of X.

2. **Dice** A die is rolled, and you win \$2 if it comes up odd, you lose \$12 if it comes up 2, and you win \$3 if it comes up 4 or 6. Find and graph the probability distribution of your winnings.

3. **Mean** Find the mean of the random variable in Exercise 1.

4. **Mean** Find the mean of the random variable in Exercise 2.

5. **Raffle Tickets** One thousand raffle tickets are sold, and there is one first prize worth $2000, one second prize worth $250, and 20 third prizes worth $50 each. Find the expected value of a ticket.

6. **Sales** An automobile salesperson predicts that a person walking into a showroom will buy a $30,000 car with probability $\frac{1}{10}$, a $25,000 car with probability $\frac{1}{5}$, a $20,000 car with probability $\frac{3}{10}$, and otherwise buy nothing. What is the expected value of this sales opportunity?

7. **Product Quality** A company manufactures products of which 2% have hidden defects. If you buy 10 of them, what is the expected number that are defective?

8. **ESP** A person claims to have ESP (extrasensory perception) and calls 7 out of 10 tosses of a fair coin correctly. What is the probability of doing at least that well by guessing randomly?

9. **Multiple-Choice Tests** A multiple-choice test consists of 20 questions, each of which has 4 possible answers. What is the expected number of correct answers that would result from random guessing?

10. **Airline Safety** An airplane has four independent engines and each operates with probability 0.99. If it takes at least two engines to land safely, what is the probability of a safe landing? What if the two engines must be on opposite sides of the plane?

Chapter 6

Statistics

6.1 Random Samples and Data Organization

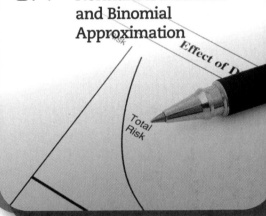

Often you have to make decisions based on incomplete information. For example, to find out how many Americans watch a television program you can't ask everyone, and to find out how long a lightbulb lasts a company can't test all its bulbs, particularly if it wants to have any left to sell. These and many other "real world" problems depend on taking samples and analyzing the data. Statistics is the branch of mathematics concerned with the collection, analysis, and interpretation of numerical data.

Random Samples

A *statistical population* is the collection of data that you are studying. Examples of statistical populations are the ages of citizens of the United States, the size of bank accounts in Illinois, and the brands of automobiles in California. Although in principle it is possible to measure the entire population, considerations of time and cost usually dictate that we measure only part of the population. For that purpose we use a *random sample* to represent the population.

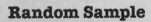

Random Sample

A *random sample* is a selection of members of the population satisfying two requirements:
1. Every member of the population is equally likely to be included in the sample.
2. Every possible sample of the same size from the population is equally likely to be chosen.

A sample that does not meet these criteria is not representative and cannot be used to infer characteristics of the entire population. For example, a telephone survey to homes at 10 A.M. on a weekday would miss most employed people and so could not produce valid information about work skills. Random samples can be chosen using random number tables or by other methods, and we will assume in this chapter that the data we are analyzing are from a random sample.

Bar Chart

A *bar chart* provides a visual summary of data with the number of times each value appears corresponding to the length of the bar for that value. The bars may be drawn vertically or horizontally, and they may have spaces between them.

→ **EXAMPLE 1** **Constructing a Bar Chart**

The colors of a random sample of thirty cars in the Country Corners Mall parking lot were recorded using the following scheme: 1—blue, 2—green, 3—white, 4—yellow, 5—brown, and 6—black. Construct a bar chart for this data {3, 5, 2, 4, 1, 5, 1, 1, 1, 2, 4, 1, 6, 1, 6, 5, 6, 3, 1, 1, 1, 1, 1, 6, 3, 1, 2, 2, 4, 2}.

Solution
We first tally the frequency of each color and then sketch the bar chart.

Color	Tally	Frequency
1	ⅲ⅏ ⅲ⅏ ‖	12
2	ⅲ⅏	5
3	‖‖	3
4	‖‖	3
5	‖‖	3
6	‖‖‖	4

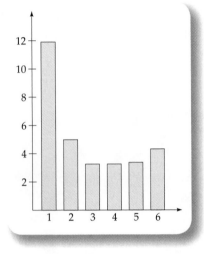

Notice that the most popular color by far is blue—about as popular as any three of the other colors combined. This observation is immediately obvious from the bar chart but is much more difficult to see from the list of data.

Maksim Toome/Shutterstock

→ **Practice Problem**

Ten vehicles passing through a turnpike toll booth were classified as 1 for a car, 2 for a bus, and 3 for a truck, giving data {1, 3, 2, 3, 1, 1, 1, 1, 3, 1}. Construct a bar chart for this data. Which type of vehicle was the most common? Which was the least common?   ➤ **Solution on next page**

Histogram

A *histogram* is a type of bar chart in which the sides of the bars touch and the *width* of the rectangles has meaning. The data are divided into *classes*, usually numbering from 5 to 15, one for each bar, and each class has the same *width*:

$$\begin{pmatrix} \text{Class} \\ \text{width} \end{pmatrix} \approx \frac{\begin{pmatrix} \text{Largest} \\ \text{data value} \end{pmatrix} - \begin{pmatrix} \text{Smallest} \\ \text{data value} \end{pmatrix}}{\begin{pmatrix} \text{Number} \\ \text{of classes} \end{pmatrix}}$$

Round the class width *up* so that all data will be covered

We then make a tally of the number of data values that fall into each class. Some histograms use the convention that any value on the boundary between two classes belongs to the upper class, while others are designed so that no data value falls on a class boundary.

→ EXAMPLE 2   Constructing a Histogram

When FirstAmerica National Bank set up its website for online banking, they asked users to fill out a questionnaire. Those who said that they would use online banking at least twice a week listed their ages as follows: {34, 10, 53, 50, 80, 38, 39, 31, 52, 41, 46, 46, 41, 69, 73, 57, 40, 52, 47, 68, 22, 33, 51, 65, 23, 47, 64, 45, 26, 74}.   Construct a histogram for this data and determine whether online banking is most popular with young, middle-aged, or elderly users.

Solution

Scanning the data, we see that the smallest and largest ages are 10 and 80. Therefore, if we choose 6 classes, we obtain a class width of $\frac{80-10}{6} \approx 11.7$,   which we round up to 12. Since we rounded up, we may choose our first class to start just below the lowest data point of 10, say at 9.5. We then add successively the class width of 12 to get the other class boundaries.

Class Boundaries	Tally	Frequency
9.5–21.5	|	1
21.5–33.5	⊬⊣⊣	5
33.5–45.5	⊬⊣⊣ ||	7
45.5–57.5	⊬⊣⊣ ⊬⊣⊣	10
57.5–69.5	||||	4
69.5–81.5	|||	3

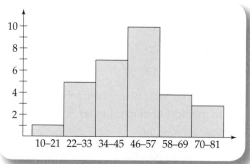

The histogram shows that FirstAmerica's website is most popular with middle-aged customers (in their 40s and 50s).

We may label the base of each bar with either the range of its data values or the class boundaries. Different numbers of bars or class boundaries will give different histograms.

How do histograms compare to the graphs of probability distributions that we drew in Chapter 5? They are very similar, but in Chapter 5 the heights of the bars were *probabilities*, with the heights adding up to 1. Here the heights of the bars are *frequencies*, with the heights adding to equal the total number of data points.

→ Solution to Practice Problem

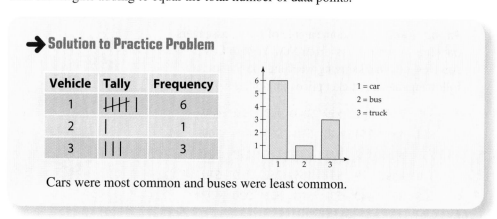

Cars were most common and buses were least common.

→ 6.1 Exercises

Construct a bar chart of the data in each situation.

1. **Marital Status** A random sample of shoppers in the Kingston Valley Mall counted 39 "never married," 97 "married," 16 "widowed," and 8 "divorced" individuals.

2. **State Legislators** A random sample of candidates running for election to the Nevada legislature counted 4 business executives, 7 real estate developers, 18 lawyers, 1 doctor, 3 certified public accountants, and 2 retirees.

For each situation construct a histogram and use it to answer the question. (For Exercises 5 through 8, histograms may vary.)

3. **Shipping Weights** A random sample of twenty packages shipped from the Holiday Values Corporation mail order center found the following weights (in ounces). (Use 6 classes of width 8 starting at 12.5.)

29	21	60	23	39	24	33	49	37	17
13	27	56	21	30	43	34	24	18	13

Based on your histogram, which type of boxes are in greatest demand: lighter weight, midweight, or heavier weight?

4. **Websites** A random sample of twenty-eight student accounts at the campus Academic Computing Center showed the following number of website "hits" between 9 P.M. and midnight last Wednesday (use 8 classes of width 10 starting at 10.5):

38	29	16	35	69	63	13
58	11	58	34	73	45	57
70	16	45	37	39	88	69
47	30	40	12	47	20	61

Based on your histogram, which two classes have the highest number of hits?

5. **Airline Tickets** A random sample of forty passengers traveling economy class from New York to Los Angeles last Thanksgiving weekend found the following one-way ticket prices (in dollars):

425	236	481	559	258	473	522	480
544	451	244	440	530	565	445	569
255	515	412	377	453	518	490	301
379	439	229	531	465	252	219	510
565	477	440	551	518	525	230	437

Based on your histogram, are more ticket prices grouped at the lower end, the middle, or the upper end of the range of costs?

6. **Law Practice** A random sample of twenty-one attorneys hired within the past year at the U.S. Department of Justice found the following numbers of hours worked last week:

48	44	50	49	44	45	51
50	57	36	53	44	48	54
60	52	52	39	47	49	57

Based on your histogram, are more hours grouped at the lower end, the middle, or the upper end of the workweeks?

7. **Medical Insurance** A random sample of twenty private practice physicians in the Lehigh Valley found the following numbers of insurance plans accepted by each:

16	21	9	23	19	23	20	23	18	19
19	29	20	21	20	23	20	14	11	19

Based on your histogram, are the numbers of plans roughly equally distributed over the range or grouped in one area?

8. **Automobile Accidents** A random sample of twenty drivers involved in accidents in Columbia County during August showed the following ages:

23	16	75	18	19
74	17	21	73	27
61	18	72	18	44
16	75	29	17	24

Based on your histogram, are the accidents roughly evenly distributed across age groups? Give an explanation for the general shape of the histogram.

ATTENTION
NEED MORE PRACTICE? FIND MORE HERE:
CENGAGEBRAIN.COM

6.2 Measures of Central Tendency

We often hear of *averages*, from grade point averages to batting averages. Given a collection of data, we often use the average as a representative or typical value. In this section we will see that not all types of data are appropriately summarized by an average, and sometimes another kind of "typical" value is more representative. To avoid confusion with casual meanings of "average," these different kinds of typical values are called *measures of central tendency*.

Mode

The *mode* of a collection of data is the *most frequently occurring* value. If all the values in a data set occur the same number of times, there is no mode, while if only a few of the values occur the same maximal number of times, there are several modes.

➡ EXAMPLE 1   Finding Modes

Find the mode for each collection of numbers.

a. $\{1, 2, 2, 2, 2, 3, 3, 4, 4, 4, 10\}$

b. $\{1, 2, 2, 2, 2, 3, 3, 4, 4, 4, 4\}$

c. $\{2, 2, 2, 3, 3, 3, 4, 4, 4\}$

Solution

a. Since 2 appears more often than any other value, the mode is 2.

b. Since both 2 and 4 occur four times and every other number occurs fewer times, the modes are 2 and 4.

c. Since each number occurs the same number of times, there is no mode.

The following are bar graphs for these collections of data. Notice that the presence of one number that is very different from the others, such as the 10 in the first graph, has no effect on the mode: If the 10 were replaced by 100 or even 1000, the mode would not change.

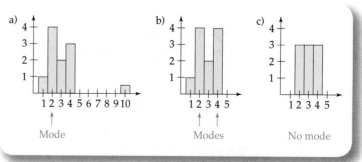

If a collection of data has just one mode (as in the first graph), it is said to be *unimodal* and if it has two modes (as in the second), it is said to be *bimodal*.

➡ Practice Problem 1

Find the mode of   {17, 20, 20, 13, 10, 17, 10, 20}.

➤ Solution on page 231

Median

The *median* is the *middle* value of a list of data values when sorted in ascending or descending order. If there is an even number of data values, the median is the arithmetic average of the two middle values (their sum divided by 2).

➡ EXAMPLE 2   Finding Medians

Find the median of each data set.

a. {1, 3, 4, 7, 11, 18, 39}

b. {1, 3, 4, 7, 11, 18, 1000}

c. {26, 16, 10, 6, 4, 2}

d. {62, 56, 52, 53, 58, 64, 68, 67, 60, 61, 57, 58, 57, 63, 55}

Solution

a. Since these seven numbers are arranged in ascending order, the median is the fourth number, 7, because three numbers are smaller and three are larger.

b. Just as in part (a), the median is 7. Notice that replacing the largest data value in the previous data set with a much larger value did not change the median.

c. Since these six values are arranged in descending order, the median is 8, the number halfway between the two middle values of 10 and 6. Thus the median need not be one of the data values.

d. Since these fifteen values are not in ascending or descending order, we first order them:

$$\{52, 53, 55, 56, 57, 57, 58, 58, 60, 61, 62, 63, 64, 67, 68\}$$

The median of these is the middle value, 58, since there are seven data values no larger and seven values no smaller.

➡ Practice Problem 2

Find the median of the values   {4, 5, 5, 6, 7, 8}.

➤ Solution on page 231

Mean

The *mean* is the usual arithmetic average found by summing the values and dividing by the number of them:

Mean

The *mean* $\bar{x}$ of the n values $\quad x_1, x_2, \ldots, x_n \quad$ is

$$\bar{x} = \frac{1}{n}(x_1 + x_2 + \cdots + x_n)$$

The mean of a collection of numbers is equivalent to the mean of a random variable that is equally likely to be any of the numbers. The mean of a data set is sometimes called the *sample mean* to distinguish it from the mean of a probability distribution.

→ **EXAMPLE 3** **Finding Means**

Find the mean of each data set.

a. $\{5, 16, 14, 1, 4, 20\}$

b. $\{5, 16, 14, 1, 4, 200\}$

Solution

a. $\bar{x} = \frac{1}{6}(5 + 16 + 14 + 1 + 4 + 20) = \frac{1}{6} \cdot 60 = 10$

b. $\bar{x} = \frac{1}{6}(5 + 16 + 14 + 1 + 4 + 200) = \frac{1}{6} \cdot 240 = 40$

The bar over x indicates *mean*

Notice that the data set in part (b) is the same as in part (a) except that the last number was changed from 20 to 200, with the result that the mean changed substantially. Thus, unlike the mode and the median, the mean is sensitive to extreme values. The mean is where the histogram of the values would just balance.

From the data in Example 3

These calculations show that when finding the mean there is no need to arrange the numbers in any particular order and that the mean need *not* be one of the original data values.

Mean, Median, and Mode

We have three measures of central tendency: the mean, the median, and the mode. Each has advantages and disadvantages, and one may be more appropriate than the others as a "representative" value for a particular data set.

- The *mean* has the advantage of being easy to calculate and takes all the values into account, but it can be heavily influenced by extremes (see Example 3).

- The *median* is not influenced by extremes and is usually found by first sorting the data, which can be difficult for a large data set.

- The *mode* is easy to find and is not influenced by extremes. However, a data set may not have a mode, or it may have several.

Dmitry Telegin/Shutterstock

The following histograms show several possible relationships between the mean, the median, and the mode.

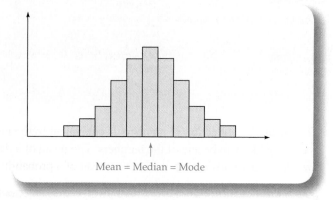

Mean = Median = Mode

For data that are symmetric about a single mode, the mean, the median, and the mode will all be the same.

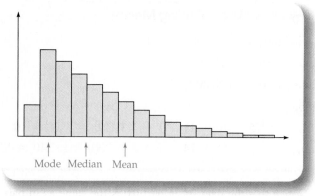

Mode   Median   Mean

For a *skewed* data set, the mode can occur at an early peak and the mean may be severely affected by extreme values.

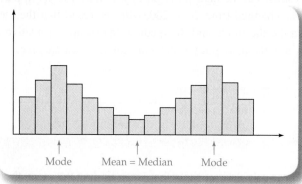

Mode   Mean = Median   Mode

For *bimodal* data, the modes may be quite far from the mean and median.

For an example of a histogram like that shown above, think of the graph of the number of automobile accidents for drivers of different ages: Teenage drivers have frequent accidents, then the number decreases with age, bottoming out at ages in the 40s and 50s, and finally increases again for elderly drivers with another peak in the top age bracket. That is, such a histogram would look like the one above but with the two peaks exactly at the ends (see Exercise 8 on page 226).

➡ Practice Problem 3

For an extreme case of skewing, imagine that Bill Gates (the richest man in the United States, with an estimated wealth of $60 billion in 2010) lived in a town along with 5999 penniless people.

a. What would be the average (mean) wealth of people in the town?

b. What would be the median wealth?

c. What would be the mode?

d. Which of these seems the most "representative" measure of central tendency for people in the town?

➤ **Solutions below**

➔ **EXAMPLE 4** **Baseball Salaries**

In the 1995 baseball strike it was disclosed that the mean salary of major league baseball players was $1.2 million, while the median salary was only $500,000. What does this say about the distribution of baseball salaries?

Solution

If the *middle* salary is $500,000, then for the mean to be higher, there must be a few very large salaries. That is, a few baseball players must be receiving enormously high salaries to skew the mean that much. In such cases, the median is usually considered a more representative value for most players than the mean.

➔ **Solutions to Practice Problems**

1. Since 20 occurs three times and no other value occurs this often, the mode is 20.

2. Since these six values are arranged in ascending order, the median is 5.5, the number halfway between the two middle values of 5 and 6.

3. *a.* $\bar{x} = \dfrac{\$60 \text{ billion}}{6000} = \$10 \text{ million}$

   b. The median is $0.

   c. The mode is $0.

   d. The median and mode seem most representative.

➔ **6.2 Exercises**

Find the mode, median, and mean of each data set.

1. {6, 17, 12, 10, 15, 16, 5, 8, 9, 14, 9}

2. {15, 14, 5, 7, 5, 14, 9, 7, 7, 14, 7, 14, 12}

3. {19, 11, 10, 19, 18, 12, 19, 10, 14, 12, 13, 17, 20, 19, 15}

4. {8, 15, 10, 14, 13, 14, 5, 10, 7, 9, 13, 14, 15, 14, 15, 19, 13, 12, 8}

5. {19, 9, 15, 8, 18, 15, 12, 22, 16, 19, 15, 19, 20, 24, 7, 14, 21, 19, 20, 5, 19}

6. **Emergency Services** A random sample from the records of the Farmington Volunteer Ambulance Corps found the following twenty response times (in minutes) to 911 calls received during February:

14	9	25	14	12	16	5	20	20	25
20	24	12	7	19	12	6	13	24	18

7. **Volunteer Recycling** A random sample of thirty volunteers helping sort newspapers, aluminum, and glass at the Parkerville Regional Recovery Center found the following hours contributed last week:

10	7	13	11	7	11	10	5	9	14
14	12	12	5	10	12	19	10	7	10
9	21	13	10	7	18	10	15	10	12

8. **Unemployment** A random sample of nineteen newly unemployed workers in Wilmington last August reported the following times (in weeks) until they found new jobs:

12	6	9	9	11	14	8	15	12	13
9	8	9	7	9	5	6	13	15	

9. **Chicken Sandwiches** *Consumer Reports* listed the calories for chicken sandwiches as follows. Looking at the data, can you see why the mean is higher than the median?

Type	Calories
Chick-fil-A	290
Wendy's Grilled Fillet	300
KFC Tender Roast	350
Arby's Grilled Deluxe	420
Wendy's Fillet (breaded)	430
McDonald's McGrill	450
KFC Original (breaded)	450
Burger King BK Broiler	500
McDonald's (breaded)	550
Burger King (breaded)	660
Boston Market	750

10. **Basketball Salaries** The salaries of the Los Angeles Lakers for the 2009–2010 season were as follows. By looking at the data, can you see why the mean is so much higher than the median?

Player	Salary ($ millions)
Kobe Bryant	23.0
Pau Gasol	16.5
Andrew Bynum	12.5
Lamar Odom	7.5
Ron Artest	5.9
Adam Morrison	5.3
Derek Fisher	5.0
Sasha Vujacic	5.0
Luke Walton	4.8
Shannon Brown	2.0
Jordan Farmar	1.9
Didier Ilunga-Mbenga	1.0
Josh Powell	1.0

ATTENTION
NEED MORE PRACTICE? FIND MORE HERE!
CENGAGEBRAIN.COM

6.3 Measures of Variation

It is helpful to summarize a given data set by a few numbers as we did with random variables in the previous chapter. For a *representative* or *typical* value, we use one of the measures of central tendency (mode, median, or mean, whichever is most appropriate). But how do we measure whether the data are grouped tightly or spread widely about this central value? In this section we discuss three ways of describing the spread of the values.

Range

The simplest measure of the spread of a data set is the *range*, which is the difference between the smallest and largest data values.

➤ EXAMPLE 1   Finding Ranges

Find the range of each data set.

a. $\{1, 5, 6, 7, 7, 7, 8, 15\}$

b. $\{1, 2, 10, 10, 12, 13, 15\}$

Solution

a. The largest value is 15 and the smallest is 1, so the range is $15 - 1 = 14$.

b. Again, the range is $15 - 1 = 14$.   Notice that the range is *not* sensitive to the distribution of values between the two extremes.

➤ Practice Problem 1

Find the range of the data values $\{15, 17, 27, 12, 30, 15, 10, 27, 25, 29\}$.

➤ Solution on page 237

Box-and-Whisker Plot

A *box-and-whisker plot* provides a quick way to visualize how data values are distributed between the largest and smallest by graphically displaying a *five-point summary* of the data. The *minimum* is the smallest value, the *maximum* is the largest, and we have already discussed the *median*. The *first quartile* is the median of the data values below the median (but not including the median), and the *third quartile* is the median of the data values above the median (but not including the median). The median may then be called the *second quartile*. The box extends from the first quartile to the third quartile with a vertical bar at the median. The whiskers extend from the box out to the minimum and maximum data values.

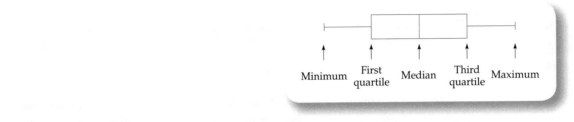

Notice that the distance between the extreme ends of the whiskers is the *range* of the data and that 50% of the data is contained in the box. The length of the box, from the first quartile to the third quartile, is sometimes called the *inter-quartile range*. In general, the quartiles will not be evenly spaced.

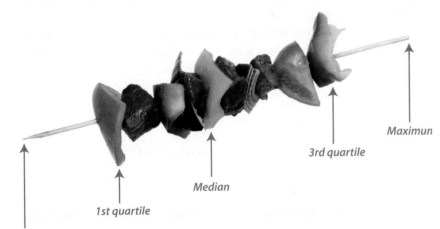

→ **EXAMPLE 2**   **Constructing a Box-and-Whisker Plot**

Make a five-point summary and draw the box-and-whisker plot for the data values {20, 37, 65, 77, 78, 79, 81, 82, 83, 85, 87, 90}.

Solution

Since these values are already in order, the minimum is 20 and the maximum is 90. Since there are twelve values, the median is halfway between the sixth value (79) and the seventh (81), so the median is 80. The values below the median are the six values {20, 37, 65, 77, 78, 79}, and the first quartile is the median of these values (that is, the value halfway between 65 and 77), so the first quartile is 71. Similarly, the values above the median are the six values {81, 82, 83, 85, 87, 90}, and the third quartile is the median 84 because it is the value halfway between 83 and 85. The five-point summary of this data set and the corresponding box-and-whisker plot are shown below.

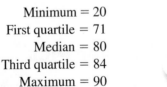

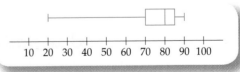

Minimum = 20
First quartile = 71
Median = 80
Third quartile = 84
Maximum = 90

The box-and-whisker plot shows clearly that the bottom quarter of the values are spread rather thinly from 20 to 71, whereas the entire top half is concentrated between 80 and 90 (with a fourth of the values in the narrow range between 80 and 84). Such clustering is much easier to see from the box-and-whisker plot than from simply looking at the data.

➡️ **Practice Problem 2**

Make a five-point summary and draw the box-and-whisker plot for the data values
{8, 10, 13, 14, 15, 16, 17, 19, 23, 24, 29}.   From your plot, which quarter is the
most tightly clustered?

➤ **Solution on page 237**

Interpreting Box-and-Whisker Plots

How do we interpret box-and-whisker plots? Consider an example.

➡️ **EXAMPLE 3**   Drawing and Interpreting Box-and-Whisker Plots

The following five-point summaries are based on a survey of annual incomes (in thousands of dollars) for those with three different levels of education: high school only, some college, and a bachelor's degree.* Draw the box-and-whisker plot for each and interpret the results.

High school only:	9.6, 16.2, 23.0, 32.6, 50.1
Some college:	10.8, 19.1, 26.8, 37.5, 57.5
Bachelor's degree:	12.5, 24.9, 35.8, 51.1, 83.6

Solution

For each five-point summary (minimum, first quartile, median, third quartile, and maximum), we plot the five points and draw the boxes and whiskers as follows.

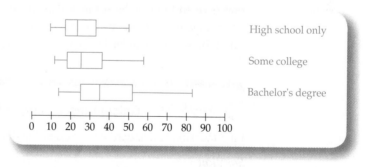

How do we interpret these plots? The first thing to look at is the median (the middle bar): Notice that it changes very little from "high school" to "some college" but then shifts significantly for "bachelor's degree," indicating that there is little income benefit from college unless you complete it. Then look at the boxes (the middle 50% of each group): As with the median, there is only a slight change from "high school" to "some college" and then a much larger change for "bachelor's degree." Notice that the minimum points (the left end of the left whisker) change very little (possibly because in any category some will leave employment), while the maximum increases much more dramatically (indicating a much greater earning potential for college graduates).

In each plot the right half of the box (the 25% above the mean) is wider than the left half (the 25% below the mean) and the right whisker is longer than the left whisker. This

* These and other data in this section are based on the *Current Population Survey* of the Bureau of Labor Statistics and the Bureau of the Census.

means that the distribution is more spread out to the right or *skewed to the right*, indicating that there are a few unusually high incomes. (If the data set were symmetric, the distances from the median to the quartiles and the extremes would be the same on either side.)

Clearly one can get much more information from comparing box-and-whisker plots than from just comparing medians.

Sample Standard Deviation

A third way of measuring variation is the *sample standard deviation*, which estimates the typical variation of the data values from the (sample) mean of the data.

Sample Standard Deviation

The *sample standard deviation* of the n data values $x_1, x_2, \ldots, x_n$ is

$$s = \sqrt{\frac{(x_1 - \bar{x})^2 + \cdots + (x_n - \bar{x})^2}{n - 1}} \qquad \bar{x} = \text{mean}$$

The differences between the values and the mean are *squared*, $(x_i - \bar{x})^2$, so positive differences don't cancel with negative ones. Since squaring gives an answer in *square units* (like "dollars squared"), we take the square root of the result to return to the original units. For n values we divide by $n - 1$ because the n numbers are related: Their sum divided by n is the mean $\bar{x}$. Therefore, any one value can be determined from the others together with the mean, leaving only $n - 1$ independent values in the sum. We call it the *sample* standard deviation to remind us of this.

➡ **EXAMPLE 4** **Calculating a Sample Standard Deviation**

Find the sample standard deviation of the data values $\{9, 13, 16, 18\}$.

Solution
Since there are four values, $n = 4$. First we must find the mean:

$$\bar{x} = \tfrac{1}{4}(9 + 13 + 16 + 18) = \tfrac{1}{4}(56) = 14$$

Then

$$s = \sqrt{\frac{(9 - 14)^2 + (13 - 14)^2 + (16 - 14)^2 + (18 - 14)^2}{4 - 1}}$$

$$= \sqrt{\frac{25 + 1 + 4 + 16}{3}} = \sqrt{\frac{46}{3}} \approx 3.9 \qquad s = \sqrt{\frac{(x_1 - \bar{x})^2 + \cdots + (x_n - \bar{x})^2}{n - 1}}$$

The sample standard deviation s is (approximately) 3.9.

→ **Solutions to Practice Problems**

1. The largest value is 30 and the smallest is 10, so the range is $30 - 10 = 20$.

2. The minimum is 8 and the maximum is 29.

    Since there are eleven values, the median is 16, the sixth value.

    From the first five values,   {8, 10, 13, 14, 15},   the first quartile is 13.

    From the last five values,   {17, 19, 23, 24, 29},   the third quartile is 23.

    From this five-point summary, the box-and-whisker plot is

    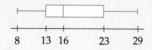

    The second quartile is most clustered.

3. First, find the mean: $\bar{x} = \frac{1}{4}(6 + 7 + 12 + 15) = \frac{1}{4}(40) = 10$.   Then

$$s = \sqrt{\frac{(6 - 10)^2 + (7 - 10)^2 + (12 - 10)^2 + (15 - 10)^2}{4 - 1}}$$

$$= \sqrt{\frac{16 + 9 + 4 + 25}{3}} = \sqrt{\frac{54}{3}} = \sqrt{18} \approx 4.2$$

→ **6.3 Exercises**

Find the range of each data set.

1. {21, 44, 48, 52, 83}

2. {5, 25, 6, 9, 7, 7, 21, 19, 23, 16}

Make a five-point summary and draw the box-and-whisker plot for each data set.

3. {3, 10, 12, 14, 17, 21, 23}

4. {8, 9, 11, 12, 13, 14, 15, 19, 26}

Find the sample standard deviation of each data set.

5. {9, 2, 17, 13, 14}

6. {22, 19, 11, 2, 20, 4}

In Exercises 7 through 9, analyze the data by finding the range, the five-point summary, and the sample standard deviation and then drawing the box-and-whisker plot.

7. *Stock Prices* Last Thursday was a very active trading day for shares of the new DiNextron technology stock. A random sample of twenty purchase transactions found the following share selling prices (in dollars):

16	12	17	11	12	7	12	17	11	20
10	11	14	17	16	14	25	9	17	5

8. **Airline Cancellations** A random sample of thirty airplane departures from Dallas International found the following numbers of "no shows" at the boarding gates:

2	6	8	4	10	5	10	8	7	5
10	10	9	6	11	9	6	6	10	8
11	6	7	13	12	8	9	5	9	10

9. **Trial Schedules** A random sample from the clerk's files at the Marksburg District Court found the following jail waiting times (in weeks) between arraignment and trial (or plea bargain) for twenty individuals charged with felonies:

11	35	15	33	20	21	2	21	23	24
18	16	9	25	12	20	20	25	21	33

10. **Income and Gender** The following five-point summaries are based on a survey of annual incomes (in thousands of dollars). The first summary is for men, and the second is for women. Draw the box-and-whisker plots for each. Interpret the results.

Men's income:	9.6, 18.4, 28.8, 40.6, 74.1
Women's income:	8.4, 15.0, 21.5, 31.7, 53.5

6.4 Normal Distributions and Binomial Approximation

The normal distribution, with its famous bell-shaped curve, is perhaps the most important of all probability distributions. The *central limit theorems* proved by Carl F. Gauss (1777–1855) and other mathematicians during the nineteenth century showed that many quantities, such as errors in measurement or means of random samples, follow the normal distribution. Today it is used to predict everything from sizes of newborn babies to stock market fluctuations to retirement benefits. After we examine some of its most important features, we will see how it is used in applications and examine its relation to the binomial distribution.

Discrete and Continuous Random Variables

The random variables we considered in Chapter 5 take values like 0, 1, 2, ... that are *separated* or *discrete* and so are called *discrete random variables*. There are other random variables whose values can be *any number in an interval* and are called *continuous random variables*. Examples of continuous random variables are your (exact) weight at birth and the time until a lightbulb burns out, since these will most likely

not be whole numbers but may be any number (in some reasonable interval). For discrete random variables, we found probabilities of events by adding probabilities of outcomes, which can be interpreted geometrically as adding areas of bars of probability distributions that change in jumps (see page 214). Analogously, for continuous random variables, we will find probabilities by calculating areas under curves.

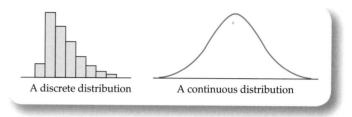

A discrete distribution   A continuous distribution

The most famous of all continuous distributions is the *normal distribution*, which we now discuss.

Normal Distribution

The normal distribution depends on two constants or *parameters*, its *mean* μ, which may be any number, and its *standard deviation* σ (pronounced "sigma," the Greek letter "s"), which must be positive. The parameter σ measures the *spread* of the distribution away from its mean, and it is analogous to the sample standard deviation of the previous section. The square of the standard deviation is called the *variance*. The normal distribution is given by the following formula, although we will not use it explicitly in what follows.

Normal Probability Distribution

The normal probability distribution with mean μ and standard deviation σ is given by the curve

$$f(x) = \frac{1}{\sigma\sqrt{2\pi}} e^{-\frac{1}{2}\left(\frac{x-\mu}{\sigma}\right)^2} \qquad \text{for } -\infty < x < \infty$$

A random variable with this distribution is called a *normal random variable* with mean μ and standard deviation σ.

The curve has a central peak at μ and then falls back symmetrically on either side to approach the *x*-axis. Different values of μ change the location of the peak, as is shown below by three normal distributions with different means.

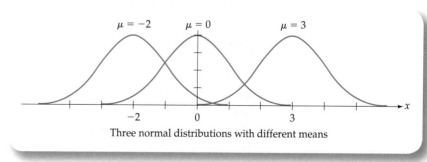

Three normal distributions with different means

Alexander Motrenko

The standard deviation σ measures how the values spread away from the mean. The distribution will be higher and more peaked if σ is small, and lower and more rounded if σ is large.

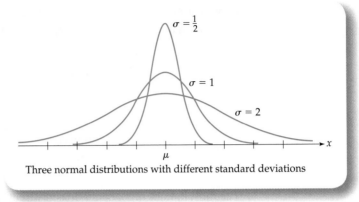

Three normal distributions with different standard deviations

As the distribution becomes "shorter" it also becomes "wider," and the area under the curve always stays at 1, although we will not prove this fact. The peaked shape of the normal distribution concentrates most of the probability (or area) near the center at μ. About 68% of the area under the normal curve is within one standard deviation of the mean, about 95% is within two standard deviations, and more than 99% is within three standard deviations. Notice also that the curve rises or falls most steeply at one standard deviation from the mean.

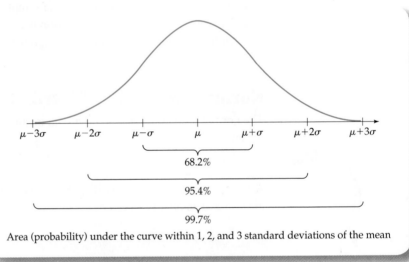

Area (probability) under the curve within 1, 2, and 3 standard deviations of the mean

The probability that a normal random variable has its value between any two given x-values corresponds to the *area* under the normal curve between those x-values. The probability is most easily found using a graphing calculator, but it can also be found using tables of values of the normal distribution.

z-Scores

The normal distribution with *mean* 0 and *standard deviation* 1 has special significance and is called the *standard* normal distribution (the word *standard* indicating mean 0 and standard deviation 1). The letter z is traditionally used for the variable in standard normal calculations. Since the mean is zero, the highest point on the curve is at $z = 0$, and the curve is symmetric about this value.

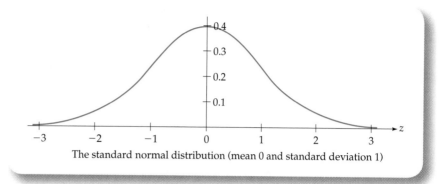

The standard normal distribution (mean 0 and standard deviation 1)

From the second diagram on the previous page with $\mu = 0$ and $\sigma = 1$ we have the following facts about the standard normal distribution: More than

68% of its probability is between -1 and $+1$

95% of its probability is between -2 and $+2$

99% of its probability is between -3 and $+3$

To change any x-value for a normal distribution with mean μ and standard deviation σ into the corresponding z-value for the *standard* normal distribution, we subtract the mean and then divide by the standard deviation. The resulting number is called the *z-score*, and this process is called *standardizing*.

z-Score

$$z = \frac{x - \mu}{\sigma}$$

Subtracting the mean μ and dividing by the standard deviation σ

The z-score gives the number of standard deviations the value is from the mean.

➡ EXAMPLE 1   Finding z-Scores for Heights of Men

The heights of American men are approximately normally distributed with mean $\mu = 68.1$ inches and standard deviation $\sigma = 2.7$ inches. Find the z-scores corresponding to the following heights (in inches) and interpret the results:

a. 73.5   **b.** 62.7

Solution

a. $z = \dfrac{73.5 - 68.1}{2.7} = \dfrac{5.4}{2.7} = 2$ Using $z = (x - \mu)/\sigma$ with $\mu = 68.1$ and $\sigma = 2.7$

Therefore, a height of 73.5 inches is 2 standard deviations *above* the mean.

b. $z = \dfrac{62.7 - 68.1}{2.7} = \dfrac{-5.4}{2.7} = -2$

Therefore, a height of 62.7 inches is two standard deviations *below* the mean.

Since the standard normal distribution has more than 95% of its probability between -2 and $+2$, these results mean that more than 95% of men have heights between 62.7 inches and 73.5 inches (5 feet 2.7 inches and 6 feet 1.5 inches).

➤ Practice Problem 1

The weights of American women are approximately normally distributed with mean 134.7 pounds and standard deviation 30.4 pounds. Find the z-scores for the following weights and interpret your answers:

a. 165.1 lb   *b.* 104.3 lb   ➤ Solutions on page 246

On page 248 is a brief table for the normal distribution, giving the probability that a standard normal random variable takes a value between 0 and any given positive number. For example, to find $P(0 \leq Z \leq 1.24)$ (in words, the probability that a standard normal random variable has a value between 0 and 1.24), we locate in the table the row headed **1.2** and the column headed **0.04** (the second decimal place), and $P(0 \leq Z \leq 1.24)$ is the number where this row and column intersect, **0.3925**.

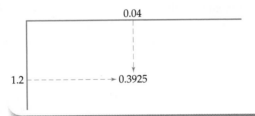

In the table on page 248, the row headed **1.2** and the column headed **0.04** intersect at the table value **0.3925**

Therefore,

$$P(0 \leq Z \leq 1.24) \approx 0.3925$$

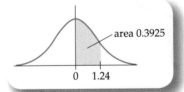

Probabilities for other intervals can be found by adding and subtracting such areas and using the symmetry of the normal curve, as the following examples illustrate.

➤ EXAMPLE 2   Finding a Probability for a Normal Random Variable

Using the mean and standard deviation for the heights of men given in Example 1, find the proportion of men who are between 5 feet 9 inches and 6 feet tall.

Solution

Converting to inches, we want the probability of a man's height being between 69 inches and 72 inches. We must convert these numbers to the corresponding z-scores for a *standard* normal random variable by subtracting the mean and dividing by the standard deviation:

$$x = 69 \quad \text{corresponds to} \quad z = \frac{69 - 68.1}{2.7} \approx 0.33$$

Using $z = \dfrac{x - \mu}{\sigma}$

$$x = 72 \quad \text{corresponds to} \quad z = \frac{72 - 68.1}{2.7} \approx 1.44$$

with $\mu = 68.1$ and $\sigma = 2.7$

5'11"

0'0"

tan4ikk/Shutterstock

Using these values, we then want $P(0.33 \leq Z \leq 1.44)$, which is equivalent to the shaded area in the first of the following graphs, which is equal to the difference between the next two areas on the right. These two areas (or probabilities) are found from the table on page 248, with the calculations shown below.

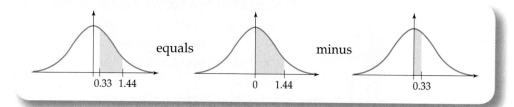

equals   minus

0.33 1.44   0   1.44   0.33

$$P(0.33 \leq Z \leq 1.44) \quad = \quad P(0 \leq Z \leq 1.44) \quad - \quad P(0 \leq Z \leq 0.33)$$
$$= \quad 0.4251 - 0.1293 = 0.2958$$

From the table
on page 248

Therefore, about 30% of American men are between 5 feet 9 inches and 6 feet tall.

→ Practice Problem 2

Using the mean and standard deviation for weights of women given in Practice Problem 1, find the proportion of women who weigh between 130 and 150 pounds.

➤ Solution on page 247

The Normal and Binomial Distributions

As we saw on page 218, the binomial distribution has a kind of "bell" shape, with a peak at the expected value $\mu = np$ and falling on both sides to very small probabilities farther away from the mean.

$n = 20, p = 0.4$ $n = 15, p = 0.5$ $n = 25, p = 0.7$

Several binomial distributions with different values for n and p

Since both the binomial and the normal distributions have similar "mound" shapes, could they be related? In the eighteenth century, Abraham de Moivre (1667–1754) and Pierre-Simon Laplace (1749–1827) discovered and proved that for any choice of p between 0 and 1, the binomial distribution *approaches* the normal distribution as n becomes large. This fundamental fact is known as the *de Moivre–Laplace theorem*.

We may use the de Moivre–Laplace theorem to approximate binomial distributions by the normal distribution.

Normal Approximation to the Binomial

Let X be a binomial random variable with parameters n and p. If $np > 5$ and $n(1 - p) > 5$, then the distribution of X is approximately normal with mean $\mu = np$ and standard deviation $\sigma = \sqrt{np(1 - p)}$.

Approximating the *discrete* binomial distribution by the *continuous* normal distribution requires an adjustment: To have the same width for a "slice" of the normal distribution as for the binomial, we adopt the convention that the binomial probability $P(X = x)$ corresponds to the area under the normal distribution curve from $x - \frac{1}{2}$ to $x + \frac{1}{2}$. This is called the *continuity correction*.

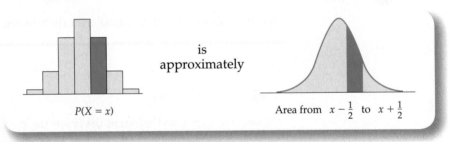

$P(X = x)$ is approximately   Area from $x - \frac{1}{2}$ to $x + \frac{1}{2}$

→ EXAMPLE 3   Normal Approximation of a Binomial Probability

Estimate $P(X = 12)$ for a binomial random variable X with $n = 25$ and $p = 0.6$ using the corresponding normal distribution.

Solution

Since

$$np = 25 \cdot 0.6 = 15 \quad \text{and} \quad n(1 - p) = 25 \cdot 0.4 = 10$$

are both greater than 5, we may use the normal distribution with

$$\mu = np = 25 \cdot 0.6 = 15$$

and

$$\sigma = \sqrt{np(1 - p)} = \sqrt{25 \cdot 0.6 \cdot 0.4} = \sqrt{6} \approx 2.45$$

By the continuity correction we interpret the event $X = 12$ as $11.5 \le X < 12.5$ to include numbers that would round to 12. Converting these x-values into z-scores, we find that

$$x = 11.5 \quad \text{corresponds to} \quad z = \frac{11.5 - 15}{2.45} \approx -1.43$$

$$x = 12.5 \quad \text{corresponds to} \quad z = \frac{12.5 - 15}{2.45} \approx -1.02$$

Using $z = \dfrac{x - \mu}{\sigma}$ with $\mu = 15$ and $\sigma = 2.45$

The probability $P(-1.43 \le Z \le -1.02)$ is represented below by the shaded area in the first graph, which, by symmetry, is equivalent to the second graph, which in turn is equivalent to the difference between the third and fourth graphs. The calculation with the probabilities found from the normal table is shown below.

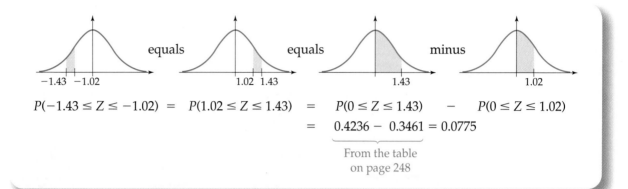

$$P(-1.43 \le Z \le -1.02) \;=\; P(1.02 \le Z \le 1.43) \;=\; P(0 \le Z \le 1.43) \;-\; P(0 \le Z \le 1.02)$$
$$= \; \underbrace{0.4236 \;-\; 0.3461}_{\substack{\text{From the table} \\ \text{on page 248}}} = 0.0775$$

The required probability is approximately 0.078.

The normal approximation of 0.0775 is indeed very close to the actual value calculated from the binomial probability formula as $_{25}C_{12}(0.6)^{12}(1-0.6)^{13} \approx 0.076$.

→ EXAMPLE 4   Management MBAs

At a major Los Angeles accounting firm, 73% of the managers have MBA degrees. In a random sample of 40 of these managers, what is the probability that between 27 and 32 will have MBAs?

Solution

Because each manager either has or does not have an MBA, presumably independently of one another, the question asks for the probability that a binomial random variable X with $n = 40$ and $p = 0.73$ satisfies $27 \le X \le 32$. Since both $np = 40 \cdot 0.73 = 29.2$ and $n(1-p) = 40 \cdot 0.27 = 10.8$ are greater than 5, this probability can be approximated as the area under the normal distribution curve with mean $\mu = np = 40 \cdot 0.73 = 29.2$ and standard deviation $\sigma = \sqrt{np(1-p)} = \sqrt{40 \cdot 0.73 \cdot 0.27} \approx 2.81$ from $26\frac{1}{2}$ to $32\frac{1}{2}$ (again using the continuity correction to include values that would round to between 27 and 32). We convert the x-values to z-scores:

$$x = 26.5 \quad \text{corresponds to} \quad z = \frac{26.5 - 29.2}{2.81} \approx -0.96 \qquad \text{Using } z = \frac{x - \mu}{\sigma}$$

$$x = 32.5 \quad \text{corresponds to} \quad z = \frac{32.5 - 29.2}{2.81} \approx 1.17 \qquad \begin{array}{l} \text{with } \mu = 29.2 \\ \text{and } \sigma = 2.81 \end{array}$$

The probability $P(-0.96 \le Z \le 1.17)$ is represented by the shaded area in the first graph, which is equivalent to the sum of the next two areas, with the calculation shown on the next page.

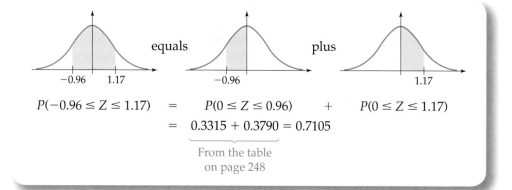

$$P(-0.96 \leq Z \leq 1.17) = P(0 \leq Z \leq 0.96) + P(0 \leq Z \leq 1.17)$$

$$= 0.3315 + 0.3790 = 0.7105$$

From the table
on page 248

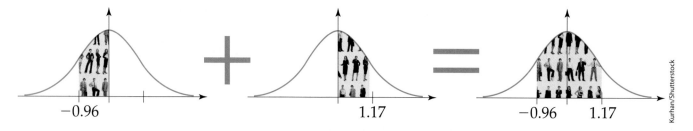

Kurhan/Shutterstock

The probability that the number of MBAs will be between 27 and 32 is 0.711, or about 71%.

The exact answer to the preceding example, from summing the binomial distribution from 27 to 32, is 71.5%, so the normal approximation is very accurate.

➜ Practice Problem 3

A brand of imported DVR is known to have defective timing mechanisms in 8% of the units imported last April. If Jerry's Discount Electronics received a shipment of 80 of these DVRs, what is the probability that 10 or more are defective?

➤ **Solution on next page**

➜ Solutions to Practice Problems

1. Using $\mu = 134.7$ and $\sigma = 30.4$:

   a. $z = \dfrac{165.1 - 134.7}{30.4} = \dfrac{30.4}{30.4} = 1$

   b. $z = \dfrac{104.3 - 134.7}{30.4} = \dfrac{-30.4}{30.4} = -1$

   Interpretation: More than 68% of women have weights between about 104 and 165 pounds.

2. We first change the weights into z-scores using $z = \frac{x - \mu}{\sigma}$ with $\mu = 134.7$ and $\sigma = 30.4$:

$$x = 130 \quad \text{corresponds to} \quad z = \frac{130 - 134.7}{30.4} \approx -0.15$$

$$x = 150 \quad \text{corresponds to} \quad z = \frac{150 - 134.7}{30.4} \approx 0.50$$

Using these values, we then want $P(-0.15 \leq Z \leq 0.50)$, which is equivalent to the shaded area shown in the graph on the left below, which is equal to the *sum* of the following two shaded areas. The two areas are found from the table on page 248, with the calculation shown below.

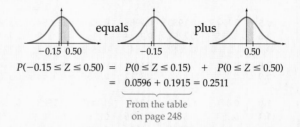

$$P(-0.15 \leq Z \leq 0.50) = P(0 \leq Z \leq 0.15) + P(0 \leq Z \leq 0.50)$$
$$= \underbrace{0.0596 + 0.1915}_{\substack{\text{From the table} \\ \text{on page 248}}} = 0.2511$$

Therefore, about 25% of American women weigh between 130 and 150 pounds.

3. Since $np = 80 \cdot 0.08 = 6.4$ and $n(1 - p) = 80 \cdot 0.92 = 73.6$ are each greater than 5, we may use the normal distribution with $\mu = np = 80 \cdot 0.08 \approx 6.4$ and $\sigma = \sqrt{np(1 - p)} = \sqrt{80 \cdot 0.08 \cdot 0.92} \approx 2.43$ to approximate $P(X \geq 10)$ as the area under the normal curve from 9.5 to 80.5 (corresponding to all 80 being defective, and again including rounding). Converting the x-values into z-scores using $z = \frac{x - \mu}{\sigma}$ with $\mu = 6.4$ and $\sigma = 2.43$:

$$x = 9.5 \quad \text{corresponds to} \quad z = \frac{9.5 - 6.4}{2.43} \approx 1.28$$

$$x = 80.5 \quad \text{corresponds to} \quad z = \frac{80.5 - 6.4}{2.43} \approx 30.49$$

The probability $P(1.28 \leq Z \leq 30.49)$ is represented by the shaded area on the left, which is equivalent to the difference between the two areas on the right with the calculation shown below.

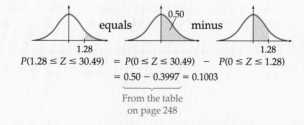

$$P(1.28 \leq Z \leq 30.49) = P(0 \leq Z \leq 30.49) - P(0 \leq Z \leq 1.28)$$
$$= \underbrace{0.50 - 0.3997}_{\substack{\text{From the table} \\ \text{on page 248}}} = 0.1003$$

The probability of at least 10 defective DVRs is (about) 0.10, or 10%.

Area under the standard normal distribution from 0 to z

x	0.00	0.01	0.02	0.03	0.04	0.05	0.06	0.07	0.08	0.09
0.0	0.0000	0.0040	0.0080	0.0120	0.0160	0.0199	0.0239	0.0279	0.0319	0.0359
0.1	0.0398	0.0438	0.0478	0.0517	0.0557	0.0596	0.0636	0.0675	0.0714	0.0754
0.2	0.0793	0.0832	0.0871	0.0910	0.0948	0.0987	0.1026	0.1064	0.1103	0.1141
0.3	0.1179	0.1217	0.1255	0.1293	0.1331	0.1368	0.1406	0.1443	0.1480	0.1517
0.4	0.1554	0.1591	0.1628	0.1664	0.1700	0.1736	0.1772	0.1808	0.1844	0.1879
0.5	0.1915	0.1950	0.1985	0.2019	0.2054	0.2088	0.2123	0.2157	0.2190	0.2224
0.6	0.2258	0.2291	0.2324	0.2357	0.2389	0.2422	0.2454	0.2486	0.2518	0.2549
0.7	0.2580	0.2612	0.2642	0.2673	0.2704	0.2734	0.2764	0.2794	0.2823	0.2852
0.8	0.2881	0.2910	0.2939	0.2967	0.2996	0.3023	0.3051	0.3078	0.3106	0.3133
0.9	0.3159	0.3186	0.3212	0.3238	0.3264	0.3289	0.3315	0.3340	0.3365	0.3389
1.0	0.3413	0.3438	0.3461	0.3485	0.3508	0.3531	0.3554	0.3577	0.3599	0.3621
1.1	0.3643	0.3665	0.3686	0.3708	0.3729	0.3749	0.3770	0.3790	0.3810	0.3820
1.2	0.3849	0.3869	0.3888	0.3907	0.3925	0.3944	0.3962	0.3980	0.3997	0.4015
1.3	0.4032	0.4049	0.4066	0.4082	0.4099	0.4115	0.4131	0.4147	0.4162	0.4177
1.4	0.4192	0.4207	0.4222	0.4236	0.4251	0.4265	0.4279	0.4292	0.4306	0.4319
1.5	0.4332	0.4345	0.4357	0.4370	0.4382	0.4394	0.4406	0.4418	0.4429	0.4441
1.6	0.4452	0.4463	0.4474	0.4484	0.4495	0.4505	0.4515	0.4525	0.4535	0.4545
1.7	0.4554	0.4564	0.4573	0.4582	0.4591	0.4599	0.4608	0.4616	0.4625	0.4633
1.8	0.4641	0.4649	0.4656	0.4664	0.4671	0.4678	0.4686	0.4693	0.4699	0.4706
1.9	0.4713	0.4719	0.4726	0.4732	0.4738	0.4744	0.4750	0.4756	0.4761	0.4767
2.0	0.4772	0.4778	0.4783	0.4788	0.4793	0.4798	0.4803	0.4808	0.4812	0.4817
2.1	0.4821	0.4826	0.4830	0.4834	0.4838	0.4842	0.4846	0.4850	0.4854	0.4857
2.2	0.4861	0.4864	0.4868	0.4871	0.4875	0.4878	0.4881	0.4884	0.4887	0.4890
2.3	0.4893	0.4896	0.4898	0.4901	0.4904	0.4906	0.4909	0.4911	0.4913	0.4916
2.4	0.4918	0.4920	0.4922	0.4925	0.4927	0.4929	0.4931	0.4932	0.4934	0.4936
2.5	0.4938	0.4940	0.4941	0.4943	0.4945	0.4946	0.4948	0.4949	0.4951	0.4952
2.6	0.4953	0.4955	0.4956	0.4957	0.4959	0.4960	0.4961	0.4962	0.4963	0.4964
2.7	0.4965	0.4966	0.4967	0.4968	0.4969	0.4970	0.4971	0.4972	0.4973	0.4974
2.8	0.4974	0.4975	0.4976	0.4977	0.4977	0.4978	0.4979	0.4979	0.4980	0.4981
2.9	0.4981	0.4982	0.4982	0.4983	0.4984	0.4984	0.4985	0.4985	0.4986	0.4986
3.0	0.4987	0.4987	0.4987	0.4988	0.4988	0.4989	0.4989	0.4989	0.4990	0.4990
3.1	0.4990	0.4991	0.4991	0.4991	0.4992	0.4992	0.4992	0.4992	0.4993	0.4993
3.2	0.4993	0.4993	0.4994	0.4994	0.4994	0.4994	0.4994	0.4995	0.4995	0.4995
3.3	0.4995	0.4995	0.4995	0.4996	0.4996	0.4996	0.4996	0.4996	0.4996	0.4997
3.4	0.4997	0.4997	0.4997	0.4997	0.4997	0.4997	0.4997	0.4997	0.4997	0.4998
3.5	0.4998	0.4998	0.4998	0.4998	0.4998	0.4998	0.4998	0.4998	0.4998	0.4998

→ 6.4 Exercises

Let X be a normal random variable with mean $\mu = 12$ and standard deviation $\sigma = 3$. Find each probability as an area under the normal curve.

1. $P(9 \leq X \leq 15)$

2. $P(3 \leq X \leq 21)$

3. $P(12 \leq X \leq 15)$

Find the z-score corresponding to each x-value.

4. $x = 6$ with $\mu = 4$ and $\sigma = 2$

5. $x = 15$ with $\mu = 20$ and $\sigma = 5$

Use the normal approximation to the binomial random variable X with $n = 20$ and $p = 0.7$ to find each probability.

6. $P(X = 10)$

7. $P(15 \leq X \leq 20)$

Answer each question using an appropriate normal probability.

8. **Weights** Weights of men are approximately normally distributed with mean 163 pounds and standard deviation 28 pounds. If the minimum and maximum weights for a man to be a volunteer firefighter in Martinsville are 128 and 254, what proportion of men meet this qualification?

9. **Airline Overbooking** To avoid empty seats, airlines generally sell more tickets than there are seats. Suppose the number of ticketed passengers who show up for a flight with 100 seats is a normal random variable with mean 97 and standard deviation 6. What is the probability that at least one person will have to be "bumped" from the flight?

10. **Smoking** In a large-scale study of male smokers 35–45 years old, the number of cigarettes smoked daily was approximately normally distributed with mean 28 and standard deviation 10. What proportion of the smokers smoked between 30 and 40 (two packs!) each day?

Markov Chains

7.1 States and Transitions

When you check the latest news, you may have noticed that much of the information is presented as the *change* since yesterday: The stock market is up, the Dodgers won again, and the weather is moderating. Sometimes the change is given in terms of probabilities ("There is a 70% probability that the rain will end tomorrow"). In this chapter we will explore the behavior of repeated changes that depend on probabilities.

A calculator with matrix arithmetic is appropriate for this chapter.

Transition Diagrams and Matrices

The Russell 2000 Index is an average of the prices of 2000 stocks traded on the New York Stock Exchange. Each trading day this index *gains*, *loses*, or remains *unchanged* compared with the previous day, and we can denote these three possible *states* by the letters G, L, and U. Watching the market for several successive trading days might result in a sequence of states such as

$$U, G, G, L, G, L, L, L, U, G, \ldots$$

Each *transition* from one day's state to the next begins from one of the three states G, L, and U and ends in one of them, so there are $3 \cdot 3 = 9$ possible transitions. From past stock market records, it is possible to estimate the probabilities of these transitions, and we can specify them by a *state-transition diagram* or, equivalently, by a *transition matrix*.

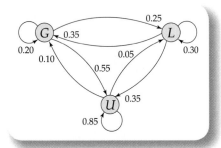

Next state

$$\begin{array}{c} \text{Current state} \end{array} \begin{array}{c} G \\ L \\ U \end{array} \begin{pmatrix} 0.20 & 0.25 & 0.55 \\ 0.35 & 0.30 & 0.35 \\ 0.10 & 0.05 & 0.85 \end{pmatrix}$$

State-Transition Diagram   Transition Matrix

In the state-transition diagram, the number near the arrowhead gives the probability of the transition from the starting state to the ending state. For example, the transition from state G to state L (which we denote as $G \rightarrow L$) has probability 0.25, while the transition $U \rightarrow U$ (that is, the market will remain unchanged) has probability 0.85. In the transition matrix, the number in a particular row and column gives the probability of the transition *from the row state to the column state*. We always list the row states from top to bottom in the same order as the column states from left to right. The matrix shows that the transition $U \rightarrow L$ has probability 0.05 and that the transition $L \rightarrow L$ (the market will continue to lose) has probability 0.30.

finite

→ **Practice Problem 1**

What is the probability that the market will continue to gain?

➤ **Solution on page 257**

Next state

$$\begin{array}{c} \text{C} \\ \text{u} \\ \text{r} \\ \text{r} \\ \text{e} \\ \text{n} \\ \text{t} \end{array} \begin{array}{c} \\ \text{s} \\ \text{t} \\ \text{a} \\ \text{t} \\ \text{e} \end{array} \begin{array}{c} \quad G \quad L \quad U \\ G \begin{pmatrix} 0.20 & 0.25 & 0.55 \\ 0.35 & 0.30 & 0.35 \\ 0.10 & 0.05 & 0.85 \end{pmatrix} \\ L \\ U \end{array}$$

Notice that the sum of the entries in each row is 1, since any given state must be followed by one of the states G, L, or U. Such a matrix of probabilities is called a *transition matrix*.

Transition Matrix

A transition matrix is a square matrix such that the entries are nonnegative and the sum of each row is 1.

→ **Practice Problem 2**

Which of the following are transition matrices?

a. $\begin{pmatrix} 0.3 & 0.7 \\ 0.4 & 0.6 \end{pmatrix}$ b. $\begin{pmatrix} 0.8 & 0.1 \\ 0.6 & 0.4 \end{pmatrix}$ c. $\begin{pmatrix} 0.5 & 0.5 \\ 1.1 & -0.1 \end{pmatrix}$

➤ **Solution on page 257**

Markov Chains

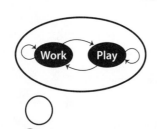

Any collection of states (such as G, L, and U) together with the probabilities of passing from any state to any state (including from a state to itself) is called a *Markov chain*, after the Russian mathematician A. A. Markov (1856–1922) who first studied them.

Markov Chain

A *Markov chain* is a collection of *states* $S_1, S_2, \ldots, S_n$ together with a *transition matrix* T whose entry in row i and column j is the probability of the transition $S_i \rightarrow S_j$.

© SPL/Photo Researchers, Inc.

A transition matrix alone may be considered a Markov chain by calling the states $S_1, S_2, \ldots, S_n$, corresponding to the rows (from top to bottom) and the columns (from left to right). Notice that the transition probabilities depend only on the present state and not on any earlier state. For this reason, a Markov chain is sometimes said to have *no memory*.

Types of Transition Matrices

What can we say about the effect of several successive transitions? Transition matrices may represent or contain several of the three basic types of transitions: *oscillating* (switching back and forth), *mixing* (moving among all possibilities), or *absorbing* (never leaving a state once it is reached).

Oscillating   Mixing   Absorbing

➡ **EXAMPLE 1**   **Types of Transitions**

Characterize the transition corresponding to each transition matrix as "oscillating," "mixing," or "absorbing."

a. $\begin{pmatrix} 0 & 1 \\ 1 & 0 \end{pmatrix}$ **b.** $\begin{pmatrix} \frac{1}{3} & \frac{1}{3} & \frac{1}{3} \\ \frac{1}{3} & \frac{1}{3} & \frac{1}{3} \\ \frac{1}{3} & \frac{1}{3} & \frac{1}{3} \end{pmatrix}$ **c.** $\begin{pmatrix} 1 & 0 & 0 \\ \frac{1}{2} & \frac{1}{2} & 0 \\ \frac{1}{2} & 0 & \frac{1}{2} \end{pmatrix}$

Solution

We make a state-transition diagram for each transition matrix.

$\begin{pmatrix} 0 & 1 \\ 1 & 0 \end{pmatrix}$

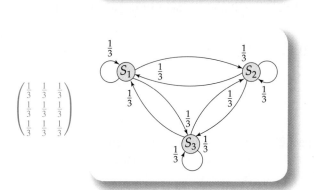

a. Since states S_1 and S_2 alternate, this is an *oscillating* transition. Repeating it several times would just flip back and forth between the two states.

$\begin{pmatrix} \frac{1}{3} & \frac{1}{3} & \frac{1}{3} \\ \frac{1}{3} & \frac{1}{3} & \frac{1}{3} \\ \frac{1}{3} & \frac{1}{3} & \frac{1}{3} \end{pmatrix}$

b. Since each state may lead to itself or to any of the other states, this is a *mixing* transition. Repeating it several times would give some combination of the states, and any mixture is possible.

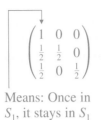

Means: Once in
S_1, it stays in S_1

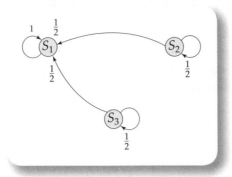

c. Since each of states S_2 and S_3 can reach state S_1 but then can never leave, this is an *absorbing* transition. Repeating it several times might lead to several occurrences of S_2 or S_3, but once the chain reached state S_1, it would remain there forever.

A state that can never be left once it is entered is called an *absorbing state*.

→ **Practice Problem 3**

Describe the action of the transition matrix $\begin{pmatrix} 1 & 0 & 0 \\ 0 & 1 & 0 \\ 0 & 0 & 1 \end{pmatrix}$ and draw its transition diagram.

➤ Solution on page 257

State Distribution Vectors

Given the transition matrix for a Markov chain together with information about its current state, what can we say about the chain after a transition? Consider an example.

→ **EXAMPLE 2   Calculating the Result of a Transition**

We return to the stock market situation with states G, L, and U (for *gaining*, *losing*, and *unchanged*) and the transition matrix shown on the left. If we have a portfolio of 100 stocks of which 50 gained, 20 lost, and 30 remained unchanged yesterday, how many may we expect to gain, lose, and remain unchanged in today's market?

Next state

$$
\begin{array}{c} \text{Current state} \end{array}
\begin{array}{c} \\ G \\ L \\ U \end{array}
\begin{array}{ccc} G & L & U \\ \begin{pmatrix} 0.20 & 0.25 & 0.55 \\ 0.35 & 0.30 & 0.35 \\ 0.10 & 0.05 & 0.85 \end{pmatrix} \end{array}
$$

Solution

We are given the numbers of stocks in the states G, L, and U, and the first column of the transition matrix gives the probabilities of transitions from these states into the state G, so multiplying and adding will give the number of gaining stocks expected today:

$$
(50\ 20\ 30) \begin{pmatrix} 0.20 \\ 0.35 \\ 0.10 \end{pmatrix} = 20
$$

$$
\underbrace{50 \cdot 0.20}_{\substack{\text{Gained } P(G \to G) \\ (G) \\ \text{yesterday}}} + \underbrace{20 \cdot 0.35}_{\substack{\text{Lost } P(L \to G) \\ (L) \\ \text{yesterday}}} + \underbrace{30 \cdot 0.10}_{\substack{\text{Unchanged } P(U \to G) \\ (U) \\ \text{yesterday}}} = 20
$$

Expected value is the sum of all the $n \cdot p$

Thus we can expect 20 of the stocks to gain today. Similarly, multiplying the given numbers by the probabilities in the second column (the probabilities of transitions into the state L) gives today's expected number of losers:

$$(50 \ 20 \ 30) \begin{pmatrix} L \\ 0.25 \\ 0.30 \\ 0.05 \end{pmatrix} = 20 \qquad \underbrace{50 \cdot 0.25}_{\substack{\text{Gained} \\ \text{yesterday}}} + \underbrace{20 \cdot 0.30}_{\substack{\text{Lost} \\ \text{yesterday}}} + \underbrace{30 \cdot 0.05}_{\substack{\text{Unchanged} \\ \text{yesterday}}} = 20 \qquad \begin{array}{l} \text{Expected} \\ \text{number of} \\ \text{losing stocks} \end{array}$$

Gained $P(G \to L)$ Lost $P(L \to L)$ Unchanged $P(U \to L)$

The remaining 60 stocks should remain unchanged today, and multiplying the given numbers by the *third* column of the transition matrix (the probabilities of transitions into the state U) gives exactly 60, as you should check.

Notice that these calculations are just the matrix product using the usual row and column multiplication described on page 117:

$$\underbrace{(50 \ \ 20 \ \ 30)}_{\text{Portfolio yesterday}} \cdot \underbrace{\begin{pmatrix} 0.20 & 0.25 & 0.55 \\ 0.35 & 0.30 & 0.35 \\ 0.10 & 0.05 & 0.85 \end{pmatrix}}_{\text{Transition matrix}} = \underbrace{(20 \ \ 20 \ \ 60)}_{\text{Portfolio today}}$$

Rephrasing our calculation in Example 2 in terms of probabilities for the 100 stocks in the portfolio, yesterday there were 50% gaining, 20% losing, and 30% unchanged, while after the transition to today, they became 20% gaining, 20% losing, and 60% unchanged. A row matrix consisting of such probabilities is called a *state distribution vector*, and multiplying it by the transition matrix gives the state distribution vector after one transition.

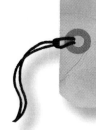

State Distribution Vector

A *state distribution vector* is a row matrix $D = (d_1 \ d_2 \dots d_n)$ of nonnegative numbers whose sum is 1. If D represents the current probability distribution for a Markov chain with transition matrix T, then the state distribution vector one step later is $D \cdot T$.

The *k*th State Distribution Vector

What can we say about the future of a Markov chain given some information about the present? If we have an *initial* state distribution vector D_0 and transition matrix T, then we have just seen that the state distribution vector after the first transition is

$$D_1 = D_0 \cdot T \qquad \qquad T \text{ gives one transition}$$

After another transition we obtain

$$D_2 = (D_0 \cdot T) \cdot T = D_0 \cdot T^2 \qquad \qquad \begin{array}{l}\text{Multiplying by } T \text{ again} \\ \text{gives two transitions}\end{array}$$

Multiplying again by T (that is, $D_0 \cdot T^3$) would give the third state distribution vector, and so on.

kth State Distribution Vector

$$D_k = D_0 \cdot T^k$$

Multiplying by T^k gives the kth state distribution vector

To put this another way, T^k gives the probabilities for k successive transitions, so the entry in row i and column j of T^k gives the probability of going from state S_i to state S_j in exactly k transitions.

→ **EXAMPLE 3** **Calculating State Distribution Vectors**

$$T = \begin{pmatrix} 1 & 0 & 0 \\ \frac{1}{2} & \frac{1}{2} & 0 \\ \frac{1}{2} & 0 & \frac{1}{2} \end{pmatrix}$$

For the Markov chain with transition matrix on the left and initial state distribution vector $D_0 = \left(\frac{1}{5} \quad \frac{2}{5} \quad \frac{2}{5} \right)$, calculate the next two state distribution vectors D_1 and D_2.

Solution

We obtain the next state distribution vector by multiplying the previous one by the transition matrix:

$$D_1 = D_0 \cdot T = \left(\frac{1}{5} \quad \frac{2}{5} \quad \frac{2}{5} \right) \cdot \begin{pmatrix} 1 & 0 & 0 \\ \frac{1}{2} & \frac{1}{2} & 0 \\ \frac{1}{2} & 0 & \frac{1}{2} \end{pmatrix}$$

$$= \left(\frac{1}{5} \cdot 1 + \frac{2}{5} \cdot \frac{1}{2} + \frac{2}{5} \cdot \frac{1}{2} \quad 0 + \frac{2}{5} \cdot \frac{1}{2} + 0 \quad 0 + 0 + \frac{2}{5} \cdot \frac{1}{2} \right) \qquad \text{Row} \times \text{columns}$$

$$= \left(\frac{3}{5} \quad \frac{1}{5} \quad \frac{1}{5} \right)$$

and

$$D_2 = D_1 \cdot T = \left(\frac{3}{5} \quad \frac{1}{5} \quad \frac{1}{5} \right) \cdot \begin{pmatrix} 1 & 0 & 0 \\ \frac{1}{2} & \frac{1}{2} & 0 \\ \frac{1}{2} & 0 & \frac{1}{2} \end{pmatrix} = \left(\frac{4}{5} \quad \frac{1}{10} \quad \frac{1}{10} \right) \qquad \text{Omitting the details}$$

$D_0 = \left(\frac{1}{5} \quad \frac{2}{5} \quad \frac{2}{5} \right)$

$D_2 = \left(\frac{4}{5} \quad \frac{1}{10} \quad \frac{1}{10} \right)$

Comparing D_0 with D_2, we see that the first entry has increased from $\frac{1}{5}$ to $\frac{4}{5}$. That is, in only two transitions, the probability of being in state S_1 has increased from 20% to 80%. Since we saw in Example 1c on page 254 that the state S_1 is absorbing for this transition matrix, the probability of being in S_1 should increase with each transition.

Duration in a Given State

In general, how long can we expect a Markov chain to remain in its current state before moving to another state? The expected number of steps is given by the following formula.

Expected Duration in a Given State

For a Markov chain in a state S_i, let p be the element in row i and column i of the transition matrix. If $p = 1$, then S_i is an absorbing state, so the Markov chain will remain in S_i forever. If $p < 1$, then the expected number of times that the chain will be in that state before moving to another state is

$$E = \frac{1}{1-p}$$

→ **EXAMPLE 4** **Finding an Expected Time**

$$\begin{array}{c} \\ H \\ M \\ I \end{array} \begin{array}{ccc} H & M & I \\ \begin{pmatrix} 0.75 & 0.20 & 0.05 \\ 0.20 & 0.70 & 0.10 \\ 0.25 & 0.20 & 0.55 \end{pmatrix} \end{array}$$

The EasyDotCom internet service provider classifies its residential customers by the number of connection hours used per day: H for high usage of more than 20 hours, M for moderate usage of between 5 and 20 hours, and I for infrequent usage of less than 5 hours. The company has found that the usage behavior of their subscribers can be modeled as a Markov chain with the transition matrix shown on the left.

If the Davidson household is presently a high user, how many days can it be expected to be a high user before changing to some other level of internet use?

Solution

Since the probability that state H stays state H for one transition of this Markov chain is 0.75 (from the upper left entry in the transition matrix), the number of days (including today) that the Davidsons are expected to be heavy users before changing to some other level is

$$E = \frac{1}{1 - 0.75} = \frac{1}{0.25} = 4 \text{ days} \qquad E = \frac{1}{1-p} \quad \text{with} \quad p = 0.75$$

→ **Solutions to Practice Problems**

1. The probability of the transition $G \to G$ is 0.20.

2. Only matrix (a). Matrix (b) fails because the first row does not sum to 1, and matrix (c) fails because one of the entries is negative.

3. The identity matrix sends each state to itself, so every state is absorbing.

→ 7.1 Exercises

Construct a transition matrix for the state-transition diagram and identify the transition as "oscillating," "mixing," or "absorbing."

1.

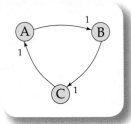

2.

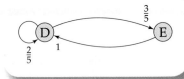

3.

Construct a state-transition diagram for the transition matrix and identify the transition as "oscillating," "mixing," or "absorbing."

4. $\begin{pmatrix} 0 & 1 & 0 \\ 0.2 & 0 & 0.8 \\ 0 & 1 & 0 \end{pmatrix}$ **5.** $\begin{pmatrix} 0.10 & 0.90 \\ 0.35 & 0.65 \end{pmatrix}$

6. $\begin{pmatrix} 1 & 0 & 0 \\ 0.1 & 0.2 & 0.7 \\ 0.1 & 0.2 & 0.7 \end{pmatrix}$

For each Markov chain transition matrix T and initial state distribution vector D_0, calculate the next two state distribution vectors D_1 and D_2.

7. $T = \begin{pmatrix} \frac{1}{4} & \frac{3}{4} \\ \frac{1}{2} & \frac{1}{2} \end{pmatrix}$, $D_0 = \begin{pmatrix} \frac{3}{5} & \frac{2}{5} \end{pmatrix}$

8. $T = \begin{pmatrix} 0 & 0.2 & 0.8 \\ 0.6 & 0.4 & 0 \\ 0.4 & 0 & 0.6 \end{pmatrix}$, $D_0 = (0.35\ 0.45\ 0.20)$

For each Markov chain transition matrix T and present state, find the expected number of times the chain will be in that state before moving to some other state.

9. $T = \begin{pmatrix} 0.6 & 0.2 & 0.2 \\ 0.1 & 0.8 & 0.1 \\ 0.6 & 0.2 & 0.2 \end{pmatrix}$,   presently in state S_2

10. $T = \begin{pmatrix} 0.8 & 0.1 & 0.1 & 0 \\ 0.2 & 0.5 & 0.1 & 0.2 \\ 0 & 0 & 1 & 0 \\ 0 & 0.3 & 0.1 & 0.6 \end{pmatrix}$,   presently in state S_4

Represent each situation as a Markov chain by constructing a state-transition diagram and a transition matrix. Be sure to state your final answer in terms of the original question.

11. *Breakfast Habits* A survey of weekly breakfast eating habits found that although many people eat cereal, boredom and other reasons cause 5% to switch to something else (eggs, muffins, and so on) each week, while among those who are not eating cereal for breakfast, 15% will switch to cereal each week. If 70% of the residents of Cincinnati are eating cereal for breakfast this week, what percentage will be eating cereal next week? in two weeks?

12. *Portfolio Management* An investment banker estimates the financial stability of midsized manufacturing companies as "secure," "doubtful," and "at risk." He has noticed that of the "secure" companies he follows, each year 5% decline to "doubtful" and the rest remain as they are; 10% of the "doubtful" companies improve to "secure," 5% decline to "at risk," and the rest remain as they are; and 5% of the "at risk" companies become bankrupt and never recover, 10% improve to "doubtful," and the rest remain as they are. If his current portfolio of investments is 80% "secure," 15% "doubtful," and 5% "at risk," what percentage will be "secure" in 2 years? (Assume that the present trends continue and no changes are made to the portfolio.) In the long run, how many of the companies in the portfolio will become bankrupt?

ATTENTION
NEED MORE PRACTICE? FIND MORE HERE:
CENGAGEBRAIN.COM

Regular Markov Chains

GoodMood Photo/ Shutterstock

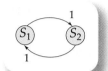

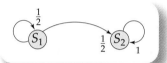

People who take their tea or coffee with cream and sugar know that it makes no difference which is added first nor on which side of the cup it is done—a few stirs mixes everything uniformly. In this section we will study Markov chains that exhibit this sort of mixing—that is, ones that move among all their states towards some sort of equilibrium.

Not all Markov chains exhibit this kind of mixing. The first chain on the left (from Example 1a on page 253) is an *oscillating* Markov chain, which bounces back and forth and never settles down to an equilibrium behavior. The second is an *absorbing* Markov chain, which eventually gets stuck in a single state rather than mixing around in all of them. In this section we begin by developing a condition on the transition matrix that guarantees mixing by avoiding oscillation and absorption, and then we show that the eventual equilibrium behavior of such a chain is independent of how it began.

Regular Markov Chains

Consider a Markov chain for which, in some number of moves, it is possible to go from any state to any state. Such a chain cannot rigidly oscillate nor be absorbing. We saw on page 256 that powers of the transition matrix give the probabilities of transitions in several moves, so some power of the transition matrix has all positive entries. Such chains are called *regular*.

> ### Regular Markov Chains
> A transition matrix is *regular* if some power of it has only positive entries. A Markov chain is regular if its transition matrix is regular.

We may determine whether a Markov chain is regular either from its transition matrix or its state-transition diagram.

→ EXAMPLE 1 Regular Transition Matrices

Which of the following transition matrices are regular?

a. $A = \begin{pmatrix} \frac{1}{2} & \frac{1}{2} \\ 1 & 0 \end{pmatrix}$ **b.** $B = \begin{pmatrix} \frac{1}{2} & \frac{1}{2} \\ 0 & 1 \end{pmatrix}$ **c.** $C = \begin{pmatrix} 1 & 0 \\ 0 & 1 \end{pmatrix}$

Solution
We will first find the answer by calculating powers of the matrix and then check it from the state-transition diagram.

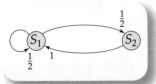

a. Since all entries of A^2 (calculated below) are positive, A is regular:

$$A^2 = \begin{pmatrix} \frac{1}{2} & \frac{1}{2} \\ 1 & 0 \end{pmatrix} \cdot \begin{pmatrix} \frac{1}{2} & \frac{1}{2} \\ 1 & 0 \end{pmatrix} = \begin{pmatrix} \frac{3}{4} & \frac{1}{4} \\ \frac{1}{2} & \frac{1}{2} \end{pmatrix}$$

The corresponding state-transition diagram shows that this is a mixing transition: Any state can be reached from every state in two steps. (Note that $S_2 \rightarrow S_2$ takes two steps.)

b. Since the lower left entry in B^k remains zero with each multiplication, B is not regular:

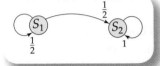

$$B = \begin{pmatrix} \frac{1}{2} & \frac{1}{2} \\ 0 & 1 \end{pmatrix}, \quad B^2 = \begin{pmatrix} \frac{1}{4} & \frac{3}{4} \\ 0 & 1 \end{pmatrix}, \quad B^3 = \begin{pmatrix} \frac{1}{8} & \frac{7}{8} \\ 0 & 1 \end{pmatrix}, \dots$$

Do you see why this entry will be 0 for *any* power of B?

The corresponding state-transition diagram shows that S_2 is an absorbing state for this transition, so S_1 is not reachable from S_2 in any number of steps.

c. Since every power of the identity matrix is still the identity matrix, $C = I$ is not regular:

$$C = \begin{pmatrix} 1 & 0 \\ 0 & 1 \end{pmatrix}, \quad C^2 = \begin{pmatrix} 1 & 0 \\ 0 & 1 \end{pmatrix}, \quad C^3 = \begin{pmatrix} 1 & 0 \\ 0 & 1 \end{pmatrix}, \dots$$

The state-transition diagram for C shows that both S_1 and S_2 are absorbing states.

Given a particular Markov chain, we can use the methods of the previous section to calculate the long-term distribution to see whether it is affected by the initial distribution.

➡ EXAMPLE 2   Calculating a Long-Term Distribution

A rental truck company has branches in Springfield, Tulsa, and Little Rock, and each rents trucks by the week. The trucks may be returned to any of the branches. The return location probabilities are shown in the following state-transition diagram. Find the distribution of the company's 180 trucks among the three cities after one year if the trucks are initially distributed as follows:

a. 36 in Springfield and 72 each in Tulsa and in Little Rock;

b. 90 in Springfield, 30 in Tulsa, and 60 in Little Rock;

c. all 180 in Tulsa.

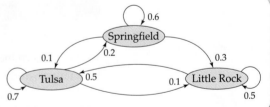

Solution

We will describe what happens in terms of the *proportion* of trucks in each city, with S_1, S_2, and S_3 standing for Springfield, Tulsa, and Little Rock, respectively. From the state-transition diagram, the transition matrix is

$$T = \begin{array}{c} \\ \\ \\ \\ \end{array} \begin{matrix} \text{Springfield} & \text{Tulsa} & \text{Little Rock} \\ \begin{pmatrix} 0.6 & 0.1 & 0.3 \\ 0.2 & 0.7 & 0.1 \\ 0 & 0.5 & 0.5 \end{pmatrix} & & \begin{matrix} \text{Springfield} \\ \text{Tulsa} \\ \text{Little Rock} \end{matrix} \end{matrix}$$

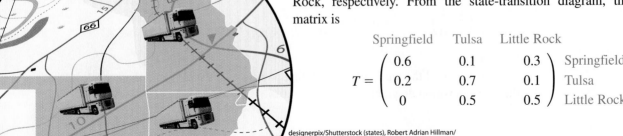

Since T is regular (the state-transition diagram shows that it is possible to go from any state to any state in two steps), the distribution of trucks among the rental offices over successive weeks should show that they "mix around" in some predictable fashion rather than "bunching up" in one city. For each initial state distribution D_0, the distribution after one year is the result of 52 weekly transitions:

$$D_{52} = D_0 \cdot T^{52}$$

a. For the numbers of trucks given in part (a), the initial state distribution vector is

$$D_0 = \begin{pmatrix} \frac{36}{180} & \frac{72}{180} & \frac{72}{180} \end{pmatrix} = \begin{pmatrix} \frac{1}{5} & \frac{2}{5} & \frac{2}{5} \end{pmatrix}$$

Each initial number divided by 180 trucks

Then, after 52 weekly transitions, the distribution will be

$$D_{52} = \begin{pmatrix} \frac{1}{5} & \frac{2}{5} & \frac{2}{5} \end{pmatrix} \cdot \begin{pmatrix} 0.6 & 0.1 & 0.3 \\ 0.2 & 0.7 & 0.1 \\ 0 & 0.5 & 0.5 \end{pmatrix}^{52} = \begin{pmatrix} \frac{1}{4} & \frac{1}{2} & \frac{1}{4} \end{pmatrix}$$

Using a calculator

After one year there will be $\frac{1}{4} \cdot 180 = 45$ trucks in Springfield, $\frac{1}{2} \cdot 180 = 90$ trucks in Tulsa, and $\frac{1}{4} \cdot 180 = 45$ trucks in Little Rock.

b. The initial numbers of trucks in part (b) give

$$D_0 = \begin{pmatrix} \frac{90}{180} & \frac{30}{180} & \frac{60}{180} \end{pmatrix} = \begin{pmatrix} \frac{1}{2} & \frac{1}{6} & \frac{1}{3} \end{pmatrix}$$

so

$$D_{52} = \begin{pmatrix} \frac{1}{2} & \frac{1}{6} & \frac{1}{3} \end{pmatrix} \cdot \begin{pmatrix} 0.6 & 0.1 & 0.3 \\ 0.2 & 0.7 & 0.1 \\ 0 & 0.5 & 0.5 \end{pmatrix}^{52} = \begin{pmatrix} \frac{1}{4} & \frac{1}{2} & \frac{1}{4} \end{pmatrix}$$

Using a calculator

This is the same distribution found in part (a), so again there will be 45 trucks in Springfield, 90 trucks in Tulsa, and 45 trucks in Little Rock.

c. For the initial numbers in part (c), we have

$$D_0 = \begin{pmatrix} 0 & \frac{180}{180} & 0 \end{pmatrix} = \begin{pmatrix} 0 & 1 & 0 \end{pmatrix}$$

so

$$D_{52} = \begin{pmatrix} 0 & 1 & 0 \end{pmatrix} \cdot \begin{pmatrix} 0.6 & 0.1 & 0.3 \\ 0.2 & 0.7 & 0.1 \\ 0 & 0.5 & 0.5 \end{pmatrix}^{52} = \begin{pmatrix} \frac{1}{4} & \frac{1}{2} & \frac{1}{4} \end{pmatrix}$$

Using a calculator

Again we get the same distribution as in parts (a) and (b), so again there will be 45 trucks in Springfield, 90 trucks in Tulsa, and 45 trucks in Little Rock.

As we expected, the long-term distribution of the trucks did not depend on the initial distribution. Moreover, further mixing will not change the result:

$$\begin{pmatrix} \frac{1}{4} & \frac{1}{2} & \frac{1}{4} \end{pmatrix} \cdot \begin{pmatrix} 0.6 & 0.1 & 0.3 \\ 0.2 & 0.7 & 0.1 \\ 0 & 0.5 & 0.5 \end{pmatrix} = \begin{pmatrix} \frac{1}{4} & \frac{1}{2} & \frac{1}{4} \end{pmatrix}$$

This distribution remains unchanged by another transition

A distribution such as this that remains unchanged by a transition is called a *steady-state distribution*.

Steady-State Distribution

A state distribution vector D is a *steady-state distribution* for the Markov chain with transition matrix T if $\;\; D \cdot T = D.$

> **Practice Problem 1**
>
> Is $\;\; D = \left(\tfrac{1}{2} \;\; \tfrac{1}{2} \right)$ a steady-state distribution for the Markov chain with transition
>
> matrix $\;\; T = \begin{pmatrix} \tfrac{1}{4} & \tfrac{3}{4} \\ \tfrac{3}{4} & \tfrac{1}{4} \end{pmatrix}$? ➤ **Solution on page 264**

The Fundamental Theorem of Regular Markov Chains

For the truck rental company in Example 2 we obtained the same long-run distribution regardless of the initial distribution. Can we obtain this long-run distribution directly from the transition matrix? The answer is "yes," and in two different ways.

For a regular Markov chain, calculating high powers T^k of the transition matrix T will result in rows that "settle down" to this steady-state distribution. To see this with the matrix from Example 2,

$$T = \begin{pmatrix} 0.6 & 0.1 & 0.3 \\ 0.2 & 0.7 & 0.1 \\ 0 & 0.5 & 0.5 \end{pmatrix}$$

gives

$$T^{52} = \begin{pmatrix} 0.25 & 0.5 & 0.25 \\ 0.25 & 0.5 & 0.25 \\ 0.25 & 0.5 & 0.25 \end{pmatrix}$$

(rounded).

This result is so important that it is called the Fundamental Theorem of Regular Markov Chains.

Fundamental Theorem of Regular Markov Chains

A regular Markov chain with transition matrix T has exactly one steady-state distribution D solving $\;\; D \cdot T = D.$ Higher powers of the transition matrix T approximate arbitrarily closely a matrix whose rows each equal D.

How to Solve $D \cdot T = D$

How can we find the steady-state distribution D that solves $D \cdot T = D$ without having to calculate arbitrarily high powers of the transition matrix? We begin by writing the equation with zero on the right:

$$D \cdot T - D = 0 \qquad \text{From} \quad D \cdot T = D$$

$$D \cdot (T - I) = 0 \qquad \text{Factoring (since} \quad D \cdot I = D)$$

$$(T - I)^t \cdot D^t = 0 \qquad \begin{array}{l} \text{Taking transposes since} \\ (A \cdot B)^t = B^t \cdot A^t \end{array}$$

$$(T^t - I) \cdot D^t = 0 \qquad \begin{array}{l} \text{Since} \quad (A - B)^t = A^t - B^t \\ \text{and} \quad I^t = I \end{array}$$

Since D is a state distribution vector, the sum of the entries of D must be 1, which we can write as the matrix equation

$$(1 \cdots 1) \cdot D^t = 1 \qquad \text{Entries of } D \text{ add up to 1}$$

Combining these last two matrix equations into one large matrix equation, we need to solve

$$\begin{pmatrix} T^t - I \\ 1 \cdots 1 \end{pmatrix} \cdot D^t = \begin{pmatrix} 0 \\ 1 \end{pmatrix} \qquad \begin{pmatrix} 0 \\ 1 \end{pmatrix} \text{ means } \begin{pmatrix} 0 \\ \vdots \\ 0 \\ 1 \end{pmatrix}$$

We may now solve for D^t by row-reducing the augmented matrix $\left(\begin{array}{c|c} T^t - I & 0 \\ \hline 1 \cdots 1 & 1 \end{array} \right)$ to obtain $\left(\begin{array}{c|c} I & D^t \\ \hline 0 & 0 \end{array} \right)$, just as we did in Section 3 of Chapter 3.

Steady-State Distribution for a Regular Markov Chain

The steady-state distribution D for a *regular* Markov chain with transition matrix T may be found by row-reducing the augmented matrix

$$\left(\begin{array}{c|c} T^t - I & 0 \\ \hline 1 \cdots 1 & 1 \end{array} \right) \quad \text{to obtain} \quad \left(\begin{array}{c|c} I & D^t \\ \hline 0 & 0 \end{array} \right)$$

and then transposing all but the bottom entry in the last column.

→ **EXAMPLE 3** Finding a Steady-State Distribution

Find the steady-state distribution for the transition matrix from the truck rental situation described in Example 2.

$$T = \begin{pmatrix} 0.6 & 0.1 & 0.3 \\ 0.2 & 0.7 & 0.1 \\ 0 & 0.5 & 0.5 \end{pmatrix}$$

Solution

Following the procedure described on the previous page, we transpose T by changing rows into columns and then calculate $T^t - I$:

$$T^t - I = \begin{pmatrix} 0.6 & 0.2 & 0 \\ 0.1 & 0.7 & 0.5 \\ 0.3 & 0.1 & 0.5 \end{pmatrix} - \begin{pmatrix} 1 & 0 & 0 \\ 0 & 1 & 0 \\ 0 & 0 & 1 \end{pmatrix} = \begin{pmatrix} -0.4 & 0.2 & 0 \\ 0.1 & -0.3 & 0.5 \\ 0.3 & 0.1 & -0.5 \end{pmatrix}$$

We need to solve the matrix equation

$$\begin{pmatrix} -0.4 & 0.2 & 0 \\ 0.1 & -0.3 & 0.5 \\ 0.3 & 0.1 & -0.5 \\ 1 & 1 & 1 \end{pmatrix} \begin{pmatrix} d_1 \\ d_2 \\ d_3 \end{pmatrix} = \begin{pmatrix} 0 \\ 0 \\ 0 \\ 1 \end{pmatrix} \qquad \begin{pmatrix} T^t - I \\ 1 \cdots 1 \end{pmatrix} \cdot D^t = \begin{pmatrix} 0 \\ 1 \end{pmatrix}$$

We row-reduce the augmented matrix

$$\left(\begin{array}{ccc|c} -0.4 & 0.2 & 0 & 0 \\ 0.1 & -0.3 & 0.5 & 0 \\ 0.3 & 0.1 & -0.5 & 0 \\ \hline 1 & 1 & 1 & 1 \end{array} \right) \text{ to obtain } \left(\begin{array}{ccc|c} 1 & 0 & 0 & \frac{1}{4} \\ 0 & 1 & 0 & \frac{1}{2} \\ 0 & 0 & 1 & \frac{1}{4} \\ 0 & 0 & 0 & 0 \end{array} \right) \qquad \begin{array}{c} \left(\begin{array}{c|c} T^t - I & 0 \\ 1 \cdots 1 & 1 \end{array} \right) \\ \text{reduces to} \\ \left(\begin{array}{c|c} I & D^t \\ 0 & 0 \end{array} \right) \end{array}$$

(omitting the details). The last column (omitting the bottom zero) is the transpose D^t of the steady-state distribution

$$D = \begin{pmatrix} \frac{1}{4} & \frac{1}{2} & \frac{1}{4} \end{pmatrix}$$

This answer agrees with the fractions that we found in Example 2 on page 261.

➤ Practice Problem 2

Use row-reduction to find the steady-state distribution of the Markov chain with transition matrix $T = \begin{pmatrix} 0 & 1 \\ \frac{1}{2} & \frac{1}{2} \end{pmatrix}$. ➤ **Solution below and on next page**

➤ Solutions to Practice Problems

1. $\begin{pmatrix} \frac{1}{2} & \frac{1}{2} \end{pmatrix} \cdot \begin{pmatrix} \frac{1}{4} & \frac{3}{4} \\ \frac{3}{4} & \frac{1}{4} \end{pmatrix} = \begin{pmatrix} \frac{1}{2} \cdot \frac{1}{4} + \frac{1}{2} \cdot \frac{3}{4} & \frac{1}{2} \cdot \frac{3}{4} + \frac{1}{2} \cdot \frac{1}{4} \end{pmatrix} = \begin{pmatrix} \frac{4}{8} & \frac{4}{8} \end{pmatrix} = \begin{pmatrix} \frac{1}{2} & \frac{1}{2} \end{pmatrix}$

Therefore, $\begin{pmatrix} \frac{1}{2} & \frac{1}{2} \end{pmatrix}$ *is* a steady-state distribution for the transition matrix.

2. $T^t - I = \begin{pmatrix} 0 & 1 \\ \frac{1}{2} & \frac{1}{2} \end{pmatrix}^t - \begin{pmatrix} 1 & 0 \\ 0 & 1 \end{pmatrix} = \begin{pmatrix} 0 & \frac{1}{2} \\ 1 & \frac{1}{2} \end{pmatrix} - \begin{pmatrix} 1 & 0 \\ 0 & 1 \end{pmatrix} = \begin{pmatrix} -1 & \frac{1}{2} \\ 1 & -\frac{1}{2} \end{pmatrix}$

so $\left(\begin{array}{c|c} T^t - I & 0 \\ 1 \cdots 1 & 1 \end{array} \right) = \left(\begin{array}{cc|c} -1 & \frac{1}{2} & 0 \\ 1 & -\frac{1}{2} & 0 \\ \hline 1 & 1 & 1 \end{array} \right)$.

One of the many ways to row-reduce this matrix (all leading to the same result) is

$$
\begin{matrix} R_2 + R_1 \rightarrow \\ R_3 + R_1 \rightarrow \end{matrix}
\left(\begin{array}{cc|c} -1 & \frac{1}{2} & 0 \\ 0 & 0 & 0 \\ 0 & \frac{3}{2} & 1 \end{array} \right)
\Rightarrow
\begin{matrix} -R_1 \rightarrow \\ R_3 \rightarrow \\ R_2 \rightarrow \end{matrix}
\left(\begin{array}{cc|c} 1 & -\frac{1}{2} & 0 \\ 0 & \frac{3}{2} & 1 \\ 0 & 0 & 0 \end{array} \right)
\Rightarrow \frac{2}{3}R_2 \rightarrow
\left(\begin{array}{cc|c} 1 & -\frac{1}{2} & 0 \\ 0 & 1 & \frac{2}{3} \\ 0 & 0 & 0 \end{array} \right)
\Rightarrow
\begin{matrix} R_1 + \frac{1}{2}R_2 \rightarrow \end{matrix}
\left(\begin{array}{cc|c} 1 & 0 & \frac{1}{3} \\ 0 & 1 & \frac{2}{3} \\ 0 & 0 & 0 \end{array} \right)
$$

The steady-distribution is $D = \left(\frac{1}{3} \quad \frac{2}{3} \right)$.

7.2 Exercises

Identify each transition matrix as "regular" or "not regular."

1. $\begin{pmatrix} 0.75 & 0.25 \\ 1 & 0 \end{pmatrix}$

2. $\begin{pmatrix} 0.75 & 0.25 \\ 0 & 1 \end{pmatrix}$

For each transition matrix T:

a. Find the steady-state distribution.

b. Calculate T^k for $k = 5, 10, 20,$ and 50 to verify that the rows of T^k approach the steady-state distribution.

3. $\begin{pmatrix} 0.4 & 0.6 \\ 0.6 & 0.4 \end{pmatrix}$

4. $\begin{pmatrix} 0.1 & 0.8 & 0.1 \\ 0.2 & 0.7 & 0.1 \\ 0.4 & 0.5 & 0.1 \end{pmatrix}$

Represent each situation as a Markov chain by constructing a state-transition diagram and the corresponding transition matrix. Find the steady-state distribution and interpret it in terms of the original situation. Be sure to state your final answer in terms of the original question.

5. *Voting Patterns* The students in the political science summer program at Edson State College are studying voting patterns in Marston County. Half of the students reviewed voter records at the County Clerk's Office and found that a person who voted in an election has an 80% chance of voting in the next election, while someone who did not vote in an election has a 30% chance of voting in the next election. The other half of the students conducted surveys door to door and at shopping centers and found that voter perceptions were somewhat different: 90% of those who claimed to have voted in the last election said they would vote in the next, while 40% of those who said they hadn't voted in the last election said they would vote in the next. If these findings are valid for predicting long-term trends,

which survey is consistent with the national average of about 61% of eligible voters actually voting?

6. *Mass Transit* Commuters in the Pittsburgh metropolitan area travel by one of three ways: drive alone, join a car pool, or take the bus. Each month, of those who drive by themselves, 20% join a car pool, 30% switch to the bus, and the rest continue driving alone; of those who are in a car pool, 30% switch to driving alone, 20% switch to the bus, and the rest stay in a car pool; and of those who take the bus, 20% switch to driving alone, 30% join a car pool, and the rest continue taking the bus. After many months, how many of the three million commuters will be driving alone?

7. *Market Share* The Peerless Products Corporation has decided to market a new toothpaste, DentiMint, designed to compete successfully with the market leaders: ConfiDent, SensaFresh, and MaxiWhite. Test marketing results from several cities indicate that each week, of those who used ConfiDent the previous week, 40% will buy it again, 20% will switch to SensaFresh, 30% will switch to MaxiWhite, and 10% will switch to DentiMint. Of those who used SensaFresh the previous week, 40% will buy it again, 20% will switch to ConfiDent, 10% will switch to MaxiWhite, and 30% will switch to DentiMint. Of those who used MaxiWhite the previous week, 40% will buy it again, 30% will switch to ConfiDent, 10% will switch to SensaFresh, and 20% will switch to DentiMint. And of those who used DentiMint the previous week, 40% will buy it again, 10% will switch to ConfiDent, 10% will switch to SensaFresh, and 40% will switch to MaxiWhite. If these buying patterns continue, what will the long-term market share be for DentiMint toothpaste?

8. Car Insurance The records of an insurance broker in Boston classify the auto policy holders as preferred, satisfactory, poor, or in the assigned-risk pool. Each year 20% of the preferred policies are downgraded to satisfactory; 30% of the satisfactory policies are upgraded to preferred and another 20% are downgraded to poor; 60% of the poor policies are upgraded to satisfactory and another 30% are placed in the assigned-risk pool; and just 20% of those in the assigned-risk pool are moved up to the poor classification. Assuming that these trends continue for many years, what percentage of the broker's auto policies are rated satisfactory or better?

7.3 Absorbing Markov Chains

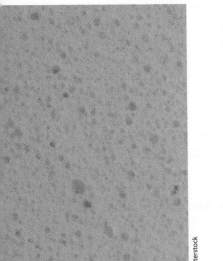

A state is *absorbing* if, once you enter it, you cannot leave it. If a Markov chain has several absorbing states, will it always end up in one of them? If so, which one and how soon? Unlike regular Markov chains, where the initial distribution has no influence on the long-term outcome, with absorbing states the initial distribution *does* affect when and where the eventual absorption occurs. As in the regular case, we will see that much useful information is given by high powers of the transition matrix.

Absorbing Markov Chains

For an *absorbing state*, the transition probabilities are 1 from that state to itself and 0 to every other state. A Markov chain is *absorbing* if it has at least one absorbing state *and* if from every nonabsorbing state it is possible to reach an absorbing state (in some number of steps).

→ EXAMPLE 1 **Finding Whether a Markov Chain Is Absorbing**

Determine whether each Markov chain is absorbing.

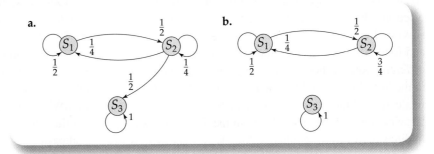

Solution

a. This *is* an absorbing Markov chain since it has an absorbing state, S_3, which can be reached from S_1 in two steps and from S_2 in one step.

b. This is *not* an absorbing Markov chain. Although S_3 is an absorbing state, it cannot be reached from either S_1 or S_2 (the nonabsorbing states).

➡ EXAMPLE 2 Wetlands Pollution

The Southern Electric Company has filed an application with the Environmental Protection Agency to store heavy-metal waste (mostly cadmium and mercury) in a clay-lined storage pool at its manufacturing plant in Mayfield. An independent engineering consultant has certified that only 1% of the waste will leach out of the pool into the groundwater each year. But a hydrologist working with the local chapter of People for a Cleaner Planet has pointed out that each year 2% of the contaminants in the groundwater reach the Marlin Memorial Wetlands Conservation Area, where they remain indefinitely. Represent this situation by a transition diagram and matrix. Show that the wetlands are an absorbing state for the pollution.

Solution

The waste can be in one of three locations: at the plant (state P), in the groundwater (state G), or in the wetlands (state W). The given information is shown in the following state-transition diagram:

$$T = \begin{array}{c} \\ \\ P \\ G \\ W \end{array} \begin{array}{c} \text{Next State} \\ \begin{array}{ccc} P & G & W \end{array} \\ \begin{pmatrix} 0.99 & 0.01 & 0 \\ 0 & 0.98 & 0.02 \\ 0 & 0 & 1 \end{pmatrix} \end{array}$$

Since these changes occur each year, we have a Markov chain with transition matrix T. Because once the contamination reaches the Wetlands, it does not leave, W is an absorbing state; the other states P and G are nonabsorbing. The state-transition diagram shows that it is possible for some of the waste to move from the plant storage pool to the groundwater and on to the wetlands, so this *is* an absorbing Markov chain.

$$T^2 = \begin{pmatrix} 0.9801 & 0.0197 & 0.0002 \\ 0 & 0.9604 & 0.0396 \\ 0 & 0 & 1 \end{pmatrix}$$

We could also use the transition matrix to show that from each nonabsorbing state it is possible to reach the absorbing state. Squaring the transition matrix T gives the matrix on the left, and since the last column (corresponding to the wetlands) is all positive, it is possible to reach the wetlands from any other state in two steps.

How much of the heavy-metal waste will reach the Wetlands, and how soon is it expected to get there? Will *all* the contamination

ultimately reach the wetlands? In Example 4 on pages 270–271 we will answer these important questions.

Standard Form

The transition matrix of an absorbing Markov chain is in *standard form* if it is written so that the absorbing states are listed *before* the nonabsorbing states. Since the absorbing states themselves may be in any order (as may the nonabsorbing states), the standard form is not necessarily unique.

Standard Form of an Absorbing Transition Matrix

An absorbing transition matrix is in *standard form* if the absorbing states appear before the nonabsorbing states, in which case it takes the form

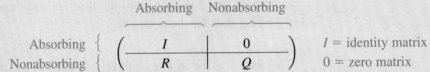

where R is the matrix of transition probabilities from the nonabsorbing states to the absorbing states and Q is the matrix of transition probabilities from the nonabsorbing states to the nonabsorbing states.

Any absorbing transition matrix can be rewritten in standard form by reordering the rows so that the absorbing states are above the nonabsorbing states and then reordering the columns to have the same order as the rows.

➡ EXAMPLE 3 Rewriting a Transition Matrix in Standard Form

Rewrite the following absorbing transition matrix in standard form and identify the matrices R and Q:

$$\begin{pmatrix} 0.80 & 0.03 & 0.10 & 0.02 & 0.05 \\ 0 & 1 & 0 & 0 & 0 \\ 0.10 & 0.07 & 0.20 & 0.03 & 0.60 \\ 0 & 0 & 0 & 1 & 0 \\ 0.05 & 0.01 & 0.10 & 0.04 & 0.80 \end{pmatrix}$$

Solution
We first identify the absorbing states. Since the row of an absorbing state consists of all zeros except one 1 in the *same column as the row*, the second and fourth rows represent absorbing states, which we will call A_1 and A_2. The other rows represent the nonabsorbing states, which we will call N_1, N_2 and N_3:

$$\begin{array}{c} \\ \text{Absorbing} \longrightarrow \\ \text{rows (1's on} \\ \text{the main} \\ \text{diagonal)} \longrightarrow \\ \\ \end{array} \begin{array}{c} N_1 \\ A_1 \\ N_2 \\ A_2 \\ N_3 \end{array} \left(\begin{array}{ccccc} 0.80 & 0.03 & 0.10 & 0.02 & 0.05 \\ 0 & 1 & 0 & 0 & 0 \\ 0.10 & 0.07 & 0.20 & 0.03 & 0.60 \\ 0 & 0 & 0 & 1 & 0 \\ 0.05 & 0.01 & 0.10 & 0.04 & 0.80 \end{array} \right)$$

Putting the rows in the order A_1, A_2, N_1, N_2, N_3, we have

$$\begin{array}{c} \\ A_1 \\ A_2 \\ N_1 \\ N_2 \\ N_3 \end{array} \begin{array}{ccccc} N_1 & A_1 & N_2 & A_2 & N_3 \\ \end{array} \left(\begin{array}{ccccc} 0 & 1 & 0 & 0 & 0 \\ 0 & 0 & 0 & 1 & 0 \\ 0.80 & 0.03 & 0.10 & 0.02 & 0.05 \\ 0.10 & 0.07 & 0.20 & 0.03 & 0.60 \\ 0.05 & 0.01 & 0.10 & 0.04 & 0.80 \end{array} \right)$$

First fix the rows

and then putting the columns in the same order gives

$$\begin{array}{c} \\ A_1 \\ A_2 \\ N_1 \\ N_2 \\ N_3 \end{array} \begin{array}{ccccc} A_1 & A_2 & N_1 & N_2 & N_3 \\ \end{array} \left(\begin{array}{cc|ccc} 1 & 0 & 0 & 0 & 0 \\ 0 & 1 & 0 & 0 & 0 \\ \hline 0.03 & 0.02 & 0.80 & 0.10 & 0.05 \\ 0.07 & 0.03 & 0.10 & 0.20 & 0.60 \\ 0.01 & 0.04 & 0.05 & 0.10 & 0.80 \end{array} \right)$$

$$\underbrace{}_{R} \quad \underbrace{}_{Q}$$

Then fix the columns

where R is the matrix of transition probabilities from the nonabsorbing states to the absorbing states and Q is the matrix of transition probabilities from the nonabsorbing states to the nonabsorbing states:

$$R = \begin{array}{c} N_1 \\ N_2 \\ N_3 \end{array} \begin{array}{cc} A_1 & A_2 \\ \end{array} \left(\begin{array}{cc} 0.03 & 0.02 \\ 0.07 & 0.03 \\ 0.01 & 0.04 \end{array} \right) \qquad Q = \begin{array}{c} N_1 \\ N_2 \\ N_3 \end{array} \begin{array}{ccc} N_1 & N_2 & N_3 \\ \end{array} \left(\begin{array}{ccc} 0.80 & 0.10 & 0.05 \\ 0.10 & 0.20 & 0.60 \\ 0.05 & 0.10 & 0.80 \end{array} \right)$$

> **Practice Problem 1**
>
> Rewrite the transition matrix $\begin{array}{c} P \\ G \\ W \end{array} \begin{array}{ccc} P & G & W \\ \end{array} \left(\begin{array}{ccc} 0.99 & 0.01 & 0 \\ 0 & 0.98 & 0.02 \\ 0 & 0 & 1 \end{array} \right)$ from Example 2 in standard form.
>
> ➤ Solution on page 273

Transition Times and Absorption Probabilities

Just as with regular Markov chains, the long-term behavior of an absorbing chain can be seen from high powers of the transition matrix. However, even more information about the ultimate behavior of the chain is provided by the following theorem about the limiting form of these powers.

Expected Transition Times and Long-Term Absorption Probabilities

For an absorbing transition matrix in standard form

$$T = \left(\begin{array}{c|c} I & 0 \\ \hline R & Q \end{array}\right)$$

Absorbing states before nonabsorbing states

high powers of T approach a matrix of the form

$$T^* = \left(\begin{array}{c|c} I & 0 \\ \hline F \cdot R & 0 \end{array}\right)$$

T^* gives the long-term transition probabilities

where the matrix F is given by $F = (I - Q)^{-1}$ and is called the *fundamental matrix* of the chain. It provides the following information:

1. The entry in row i and column j of F gives the expected number of times that the chain, if it begins in state N_i, will be in state N_j before being absorbed.
2. The sum of the entries in row i of F gives the expected number of times that the chain, if it begins in state N_i, will be in the nonabsorbing states before being absorbed.
3. The entry in row i and column j of the matrix $F \cdot R$ gives the probability that the chain, if it begins in state N_i, will be absorbed into the state A_j.

➡ **EXAMPLE 4** Finding Expected Time Until Absorption

Find the expected number of years until the heavy-metal waste in the plant (state P) will reach the Wetlands Conservation Area (state W) for the Markov chain described in Example 2.

Solution
Writing the transition matrix from Example 2 in standard form as

$$T = \begin{array}{c} \\ W \\ G \\ P \end{array} \begin{array}{c} \begin{array}{ccc} W & G & P \end{array} \\ \left(\begin{array}{c|cc} 1 & 0 & 0 \\ \hline 0.02 & 0.98 & 0 \\ 0 & 0.01 & 0.99 \end{array}\right) \end{array}$$

Listing first the absorbing state W (for "wetlands")

we have

$$Q = \begin{array}{c} G \\ P \end{array} \begin{array}{c} \begin{array}{cc} G & P \end{array} \\ \left(\begin{array}{cc} 0.98 & 0 \\ 0.01 & 0.99 \end{array}\right) \end{array}$$

From $T = \left(\begin{array}{c|c} I & 0 \\ \hline R & Q \end{array}\right)$

and

$$I - Q = \begin{pmatrix} 1 & 0 \\ 0 & 1 \end{pmatrix} - \begin{pmatrix} 0.98 & 0 \\ 0.01 & 0.99 \end{pmatrix} = \begin{pmatrix} 0.02 & 0 \\ -0.01 & 0.01 \end{pmatrix}$$

We calculate the fundamental matrix $F = (I - Q)^{-1}$ by row-reducing $(A \mid I)$ to obtain $(I \mid A^{-1})$, as we did on page 123.

$$\begin{pmatrix} 0.02 & 0 & \big| & 1 & 0 \\ -0.01 & 0.01 & \big| & 0 & 1 \end{pmatrix} \qquad (A \mid I)$$

$$R_2 + \tfrac{1}{2}R_1 \rightarrow \begin{pmatrix} 0.02 & 0 & \big| & 1 & 0 \\ 0 & 0.01 & \big| & \tfrac{1}{2} & 1 \end{pmatrix}$$

$$\begin{matrix} 50R_1 \rightarrow \\ 100R_2 \rightarrow \end{matrix} \begin{pmatrix} 1 & 0 & \big| & 50 & 0 \\ 0 & 1 & \big| & \underbrace{50 \quad 100}_{F} \end{pmatrix} \qquad (I \mid A^{-1})$$

The fundamental matrix F is the matrix on the right of the bar in the row-reduced matrix. From part (1) of the box on the previous page, its entries have the following interpretations.

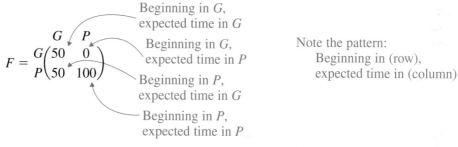

$$F = \begin{matrix} & G & P \\ G & \big(50 & 0 \\ P & \big(50 & 100 \big) \end{matrix}$$

Beginning in G, expected time in G
Beginning in G, expected time in P
Beginning in P, expected time in G
Beginning in P, expected time in P

Note the pattern:
Beginning in (row), expected time in (column)

From part (2) of the box on the previous page, the expected number of years until the heavy-metal waste in the plant (state P) is absorbed into the wetlands (the absorbing state W) is the sum of the numbers in row P: $50 + 100 = 150$ years.

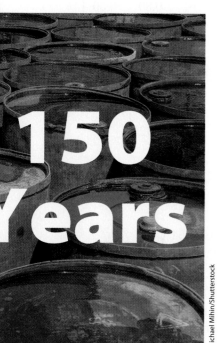

→ **Practice Problem 2**

How long is it expected that the heavy-metal waste will spend in the plant storage pool (state P) before absorption? How long in the groundwater (state G)?

➤ Solution on page 273

→ **EXAMPLE 5** Finding an Absorption Probability

For the Markov chain with transition matrix as given in Example 3 (see pages 268–269), find the probability of absorption into state A_2 given that it started in state N_3.

Solution

By rewriting the transition matrix in standard form, Example 3 showed that

$$
R = \begin{array}{c} \\ N_1 \\ N_2 \\ N_3 \end{array} \overset{\begin{array}{cc} A_1 & A_2 \end{array}}{\begin{pmatrix} 0.03 & 0.02 \\ 0.07 & 0.03 \\ 0.01 & 0.04 \end{pmatrix}} \quad \text{and} \quad Q = \begin{array}{c} \\ N_1 \\ N_2 \\ N_3 \end{array} \overset{\begin{array}{ccc} N_1 & N_2 & N_3 \end{array}}{\begin{pmatrix} 0.80 & 0.10 & 0.05 \\ 0.10 & 0.20 & 0.60 \\ 0.05 & 0.10 & 0.80 \end{pmatrix}}
$$

Thus

$$
I - Q = \begin{pmatrix} 1 & 0 & 0 \\ 0 & 1 & 0 \\ 0 & 0 & 1 \end{pmatrix} - \begin{pmatrix} 0.80 & 0.10 & 0.05 \\ 0.10 & 0.20 & 0.60 \\ 0.05 & 0.10 & 0.80 \end{pmatrix} = \begin{pmatrix} 0.20 & -0.10 & -0.05 \\ -0.10 & 0.80 & -0.60 \\ -0.05 & -0.10 & 0.20 \end{pmatrix}
$$

and its inverse can be found as usual by row reduction:

$$
\underbrace{\left(\begin{array}{ccc|ccc} 0.20 & -0.10 & -0.05 & 1 & 0 & 0 \\ -0.10 & 0.80 & -0.60 & 0 & 1 & 0 \\ -0.05 & -0.10 & 0.20 & 0 & 0 & 1 \end{array} \right)}_{I\,-\,Q} \xrightarrow{\text{row reduces to}} \left(\begin{array}{ccc|ccc} 1 & 0 & 0 & 8 & 2 & 8 \\ 0 & 1 & 0 & 4 & 3 & 10 \\ 0 & 0 & 1 & 4 & 2 & 12 \end{array} \right)
$$

omitting the details

(where the right block is F)

Now, $F \cdot R$ becomes

$$
\begin{pmatrix} 8 & 2 & 8 \\ 4 & 3 & 10 \\ 4 & 2 & 12 \end{pmatrix} \cdot \begin{pmatrix} 0.03 & 0.02 \\ 0.07 & 0.03 \\ 0.01 & 0.04 \end{pmatrix} = \begin{pmatrix} 0.46 & 0.54 \\ 0.43 & 0.57 \\ 0.38 & 0.62 \end{pmatrix}
$$

By part (3) of the result on page 270, the probability that the chain, beginning in state N_3, will be absorbed into state A_2 is the entry in row 3 and column 2 of $F \cdot R$, which is 0.62.

➡ Practice Problem 3

From the matrix $F \cdot R$ in the preceding Example, what is the probability that the chain, if it begins in state N_1, will be absorbed into state A_2?

➤ Solution on next page

Both Example 5 and Practice Problem 3 found probabilities of being absorbed into state A_2, but beginning in different states. The fact that we obtained two different probabilities (0.62 and 0.54) shows that for an absorbing Markov chain, the initial state *does* affect the ultimate behavior. This is quite different from a *regular* Markov chain, where the initial state had no effect on the ultimate distribution.

1. Both $P\begin{array}{c}W\\P\\G\end{array}\begin{pmatrix}\begin{array}{ccc}W&P&G\\1&0&0\\0&0.99&0.01\\0.02&0&0.98\end{array}\end{pmatrix}$ and $\begin{array}{c}W\\G\\P\end{array}\begin{pmatrix}\begin{array}{ccc}W&G&P\\1&0&0\\0.02&0.98&0\\0&0.01&0.99\end{array}\end{pmatrix}$ are standard forms of the transition matrix.

2. The expected times are 100 years in the plant storage pool and then 50 years in the groundwater.

3. 0.54 (from the entry in row 1 and column 2 of $F \cdot R$)

➡️ 7.3 Exercises

For each state-transition diagram, identify the absorbing and the nonabsorbing states. For the given starting state, find the smallest number of steps required to reach an absorbing state and the probability of that sequence of steps.

1. Starting in state S_1

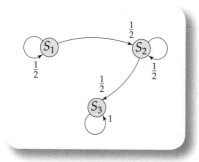

2. Starting in state S_4

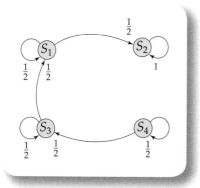

For each absorbing transition matrix T:

a. Draw a state-transition diagram, identify the absorbing and the nonabsorbing states, and find the smallest integer k such that after k transitions every nonabsorbing state can reach an absorbing state.

b. Verify your value for k by finding T^k and checking that every nonabsorbing state row has a positive probability in at least one absorbing state column.

3. $\begin{pmatrix} 1 & 0 & 0 \\ 0.20 & 0.55 & 0.25 \\ 0 & 0.25 & 0.75 \end{pmatrix}$

4. $\begin{pmatrix} 0 & 1 & 0 & 0 \\ 0 & 1 & 0 & 0 \\ 0.2 & 0 & 0.6 & 0.2 \\ 0 & 0 & 0.2 & 0.8 \end{pmatrix}$

Rewrite each absorbing transition matrix in standard form and identify the matrices R and Q.

5. $\begin{pmatrix} 0.88 & 0.02 & 0.10 \\ 0 & 1 & 0 \\ 0.04 & 0.16 & 0.80 \end{pmatrix}$

6. $\begin{pmatrix} 0.80 & 0 & 0.20 & 0 & 0 \\ 0 & 1 & 0 & 0 & 0 \\ 0 & 0.20 & 0.50 & 0.30 & 0 \\ 0 & 0 & 0 & 1 & 0 \\ 0.20 & 0 & 0.05 & 0 & 0.75 \end{pmatrix}$

Find the fundamental matrix for each absorbing transition matrix in standard form.

7. $\begin{pmatrix} 1 & 0 & 0 \\ \hline 0.4 & 0.4 & 0.2 \\ 0.2 & 0.2 & 0.6 \end{pmatrix}$

8. $\begin{pmatrix} 1 & 0 & 0 & 0 \\ 0 & 1 & 0 & 0 \\ \hline 0.10 & 0.10 & 0.20 & 0.60 \\ 0.05 & 0.15 & 0.20 & 0.60 \end{pmatrix}$

For Exercises 9 and 10, use the results of Exercises 7 and 8 as follows: For each given initial state, find the expected number of times in the given state before absorption and find the probability of absorption in the given final state using the indicated transition matrix.

	Initial State	Expected Number of Times in State	Probability of Absorption into State	Use the Transition Mattrix from Exercise
9.	N_1	N_1	A_1	7
10.	N_2	N_1	A_2	8

Represent each situation as an absorbing Markov chain by constructing a state-transition diagram and a transition matrix in standard form. Use the fundamental matrix to solve each problem. Be sure to state your final answers in terms of the original questions.

11. **Geriatric Care** Each year at the Shady Oaks Assisted Living Facility, 10% of the independent residents are reclassified as requiring assistance, 2% die, and 8% are transferred to a nursing home, while 10% of those needing assistance die and 15% are transferred to a nursing home. What is the probability that a resident requiring assistance will ultimately be transferred to a nursing home?

12. **Term Insurance** The term life insurance records of the All-County Insurance Company show that every five years those policies that were renewed are renewed again with probability 0.60, renewed with increased coverage with probability 0.20, closed with a death benefit payment with probability 0.07, and discontinued with probability 0.13, while those that had been renewed with increased coverage are renewed again with probability 0.30, renewed with increased coverage with probability 0.60, closed with a death benefit payment with probability 0.08, and discontinued with probability 0.02.

   a. What is the expected number of years that a policy that has just been renewed will be in force?
   b. What is the probability that a policy that is renewed with increased coverage will be closed with a death benefit payment?

STUDY YOUR WAY

At no additional cost, you have access to online learning resources that include **tutorial videos, printable flashcards, quizzes,** and more!

Watch videos that offer step-by-step conceptual explanations and guidance for each chapter in the text.

Along with the printable flashcards and other online resources, you will have a multitude of ways to check your comprehension of key mathematical concepts.

You can find these resources at **login.cengagebrain.com.**

Game Theory

8.1 Two-Person Games and Saddle Points

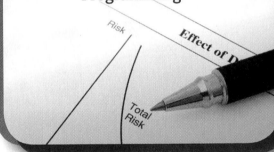

When faced with choices, we naturally wish for a clear way to decide what to do. Game theory provides a framework for analyzing situations in which opposing players make choices that affect the benefits received by each. The "games" that we consider include many of practical importance in business, economics, and political affairs. We restrict our attention to competitions involving just two players where one's gain is the other's loss. Such situations are called *two-person, zero-sum games* because whatever one player gains the other loses. Throughout this chapter, whenever we speak of a "game" we shall mean such a situation.

Payoff Matrix

A *game* consists of a list of choices for the first player and a list of choices for the second player, along with specified "payoffs" for the players that are determined by the choices they make. Since in a zero-sum game one player's gain is the other player's loss, we need only list

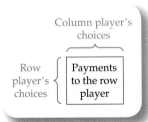

the payoff to one player, the other *losing* that amount. We list the payoffs in a grid called the *payoff matrix*, with one player's possible choices corresponding to the *rows* of the matrix, the other player's choices corresponding to the *columns* of the matrix, and the entries in the matrix are *amounts paid to the row player by the column player.* A negative number means a loss for the row player and so a gain for the column player.

→ **EXAMPLE 1** **Business Competition**

Two competing coffee houses, Rachel's House of Java and Carla's Colombian Brew, are planning to enter a tri-city market by each locating a single new store in one of the towns of

Lawrence (population 50,000), Moltonsville (population 90,000), and Northridge (population 70,000). If both stores are built in the same town, they will split the entire market equally. If not, the only store in town will get all that town's market as well as half that in any town without a store. What are the possible choices for Rachel and for Carla? If the payoff is the difference between their markets (in thousands), find the payoff matrix.

Solution

Rachel's choices are L (for choosing Lawrence), M (for Moltonsville), and N (for Northridge), and similarly for Carla, so there are $3 \cdot 3 = 9$ different situations to consider. We list Rachel's choices as rows and Carla's as columns.

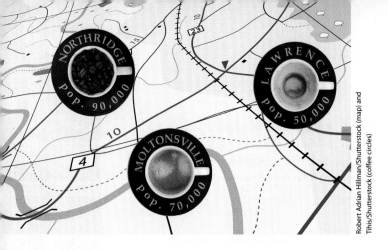

Should they both choose to locate in the same town, neither will have an advantage, giving zero payoffs for those choices:

	Carla		
	L	M	N
Rachel — L	0		
Rachel — M		0	
Rachel — N			0

Should Rachel choose L and Carla choose M, Rachel gets

$$50 \;+\; 0 \;+\; \tfrac{1}{2}\cdot 70 = 85$$

All of Lawrence   None of Moltonsville   Half of Northridge

In thousands of potential customers

while Carla gets

$$0 \;+\; 90 \;+\; \tfrac{1}{2}\cdot 70 = 125$$

None of Lawrence   All of Moltonsville   Half of Northridge

This is an advantage of $125 - 85 = 40$ thousand for Carla over Rachel, and we indicate it as -40 in the table because it is a *loss* for Rachel:

	Carla		
	L	M	N
Rachel — L	0	−40	
Rachel — M		0	
Rachel — N			0

Making similar calculations for the remaining five possibilities gives the payoff matrix for this game:

$$\begin{array}{c} \\ L \\ M \\ N \end{array}\begin{array}{c} \begin{array}{ccc} L & M & N \end{array} \\ \begin{pmatrix} 0 & -40 & -20 \\ 40 & 0 & 20 \\ 20 & -20 & 0 \end{pmatrix} \end{array}$$

A different ordering of the choices would give a different ordering of the rows and columns

➡ Practice Problem 1

Verify the bottom left-hand value in the above payoff matrix by calculating Rachel's advantage over Carla when Rachel chooses N and Carla chooses L.

➤ Solution on page 280

Any game can be expressed as a matrix, and any matrix defines a game. Mathematically, a game is completely defined by its payoff matrix, calling the players R (choosing the row) and C (choosing the column), with no other explanation necessary. When setting up the matrix, it makes no difference which player chooses rows and which chooses columns as long as the entries are the payoffs to R, so a positive entry is a gain for R (and a loss for C) while a negative entry is a loss for R (with the corresponding gain for C).

Optimal Strategy

Remember: The smallest number in {2, −3, −5} is −5.

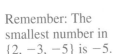

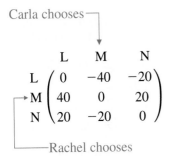

The goal of game theory is to find the *optimal strategy* for each player in a game. Remember that R wants the *highest* number in the matrix, since R receives that amount, while C wants the *smallest* number, since C loses that amount. *Solving* a game means finding an optimal strategy for each player.

Returning to the situation of Example 1, from Rachel's viewpoint row 2 is a good choice because the worst she could get would be zero. (There are better numbers, 40 and 20, but if Rachel repeatedly chose row 2 in hopes of getting them, Carla would choose column 2 to give Rachel only the zero.) From Carla's viewpoint, column 2 is appealing because the worst she could do is zero. (Carla would rather have the −40 and −20, since these mean positive payments to her, but if she chose column 2 repeatedly, Rachel would choose row 2 to give her only the zero.) Rachel's best strategy is row 2 and Carla's is column 2, both settling on the zero, corresponding to both locating in Moltonsville. These choices put them in equilibrium with each other: Neither can gain by changing if the other stays with the same choice.

Notice that the number at the intersection of the chosen row and column (here zero, but it won't always be zero) is both the *minimum number in its row* and also the *maximum number in its column*. Such a number is called a *saddle point* of the game.

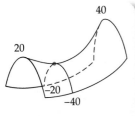

A saddle point is the lowest point along one curve and the highest along another

Saddle Point and Value of a Game

A *saddle point* of a game is an entry in the payoff matrix that is both the *minimum value in its row* and the *maximum value in its column*. If a game has a saddle point, it represents the *optimal strategy* for both players. The saddle point gives the *value* of the game.

A game with a saddle point is said to be *strictly determined*, and each player's selection of its corresponding row or column is called a *pure* optimal strategy. The *value* of the game (the saddle point) is a payoff to the row player, so a *positive* value means that the game is favorable to the row player, a *negative* value means that the game is favorable to the column player, and a *zero* value means that it is a "fair" game, favorable to neither player. For the game in Example 1, the saddle point is the 0 in the Moltonsville row and column, so the pure optimal strategies for both Rachel and Carla are to locate their new stores in Moltonsville and split the market equally, for a game value of 0.

Finding Saddle Points

A game may have any number of saddle points—none, one, or many—but if it has many, they will all have the same value. There is a simple way to find the saddle points of a matrix. A saddle point is the minimum value in its row, so we first circle ○ the smallest entry in each row; a saddle point is also the maximum in its column, so we box □ the largest entry in each column. If an entry is *both circled and boxed*, it is a saddle point. The circled (or boxed) entries of G are called *the security values* of the rows (or the

VR Photos/Shutterstock (saddle) and Vicente Barcelo Varona/Shutterstock (magnifying glass)

columns) since they represent the guaranteed return from choosing that row (or that column).

→ EXAMPLE 2 Finding Saddle Points

Find the saddle points of the game $\begin{pmatrix} 2 & 2 & 2 & 5 & 4 \\ 0 & 2 & 1 & 3 & 3 \\ 1 & 2 & -1 & 4 & 2 \\ 2 & 4 & 2 & 3 & 5 \end{pmatrix}$.

Solution

First we circle the smallest value in each row.

$$\begin{pmatrix} ② & ② & ② & 5 & 4 \\ ⓪ & 2 & 1 & 3 & 3 \\ 1 & 2 & ⊝1 & 4 & 2 \\ ② & 4 & ② & 3 & 5 \end{pmatrix}$$

Then we box the largest value in each column.

$$\begin{pmatrix} \boxed{②} & ② & \boxed{②} & \boxed{5} & 4 \\ ⓪ & 2 & 1 & 3 & 3 \\ 1 & 2 & ⊝1 & 4 & 2 \\ \boxed{②} & \boxed{4} & \boxed{②} & 3 & \boxed{5} \end{pmatrix}$$

The saddle points are the circled *and* boxed numbers.

$$\begin{pmatrix} \boxed{②} & ② & \boxed{②} & \boxed{5} & 4 \\ ⓪ & 2 & 1 & 3 & 3 \\ 1 & 2 & ⊝1 & 4 & 2 \\ \boxed{②} & \boxed{4} & \boxed{②} & 3 & \boxed{5} \end{pmatrix}$$

This game has four saddle points. The value of the game is 2, and each player has two optimal strategies corresponding to choosing the row or column of a saddle point: R chooses row 1 or 4, and C chooses column 1 or 3.

→ Practice Problem 2

Find the saddle points of the game $\begin{pmatrix} 8 & 2 & 5 \\ 5 & 4 & 7 \\ 2 & 3 & 6 \\ 6 & 4 & 5 \end{pmatrix}$.

What are the pure optimal strategies?

➤ **Solution on next page**

→ Solutions to Practice Problems

1. If Rachel (R) chooses Northridge (population 70,000) and Carla (C) chooses Lawrence (population 50,000), then

R gets: $\underbrace{0}_{\substack{\text{None of} \\ \text{Lawrence}}} + \underbrace{\tfrac{1}{2} \cdot 90}_{\substack{\text{Half of} \\ \text{Moltonsville}}} + \underbrace{70}_{\substack{\text{All of} \\ \text{Northridge}}} = 115$ In thousands

C gets: $\underbrace{50}_{\substack{\text{All of} \\ \text{Lawrence}}} + \underbrace{\tfrac{1}{2} \cdot 90}_{\substack{\text{Half of} \\ \text{Moltonsville}}} + \underbrace{0}_{\substack{\text{None of} \\ \text{Northridge}}} = 95$

$$\begin{array}{c} & \begin{matrix} L & \quad M & \quad N \end{matrix} \\ \begin{matrix} L \\ M \\ N \end{matrix} & \begin{pmatrix} 0 & -40 & -20 \\ 40 & 0 & 20 \\ 20 & -20 & 0 \end{pmatrix} \end{array}$$

Therefore, R gets $115 - 95 = 20$ thousand more than C, verifying the positive 20 in row N, column L of the payoff matrix.

2. We circle the smallest entry in each row.

Then we box the largest entry in each column.

$$\begin{pmatrix} 8 & ② & 5 \\ 5 & ④ & 7 \\ ② & 3 & 6 \\ 6 & ④ & 5 \end{pmatrix}$$

$$\begin{pmatrix} \boxed{8} & ② & 5 \\ 5 & \boxed{④} & \boxed{7} \\ ② & 3 & 6 \\ 6 & \boxed{④} & 5 \end{pmatrix}$$ Saddle points

There are two saddle points. R has two optimal strategies: Choose row 2 or row 4. C has one optimal strategy: Choose column 2.

→ 8.1 Exercises

For each situation, identify the two players and their possible choices, and construct a payoff matrix for their conflict.

1. *TV Scheduling* In an attempt to gain more viewers, Channel 86 and Channel 7 are each trying to decide whether to schedule a quiz show or a reality series in their 8:00 prime time slot. Market research indicates that if Channel 86 chooses a quiz show, it will gain 5% of the market if Channel 7 runs a quiz show and lose 8% if Channel 7 runs a reality series, while if Channel 86 chooses a reality series, it will gain 10% if Channel 7 runs a quiz show and lose 10% if Channel 7 runs a reality series. [*Hint*: Use Q and R for quiz show and reality series.]

2. *Market Share* Andersonville has two gas stations, Ralph's Qwik-Serv and Charlie's Gas-n-Go. Both Ralph and Charlie are considering raising prices by 1¢, staying with their current prices, or lowering prices by 1¢. If they both make the same choice, there will be no change in their market shares, but if they make different choices, the one with the lower price will gain 5% of the market for each penny difference in their prices.

For each game, identify the saddle point and determine the corresponding optimal strategy for each player.

3. $\begin{pmatrix} 3 & 7 \\ 2 & 5 \end{pmatrix}$

4. $\begin{pmatrix} 3 & 1 & 2 \\ 5 & -1 & -2 \end{pmatrix}$

5. $\begin{pmatrix} 3 & 2 & 6 \\ -3 & -2 & 7 \\ 4 & 1 & -1 \end{pmatrix}$

6. $\begin{pmatrix} 24 & 8 & 7 & 1 \\ 21 & 15 & 16 & 26 \\ 11 & 13 & 18 & 10 \end{pmatrix}$

Determine the optimal strategy for each situation by representing it as a game and finding the saddle point. State your final answer in the terms of the original question.

7. *Market Share* In an ongoing price war between Burger Haven (locally owned) and MacArches (a chain), both restaurant managers plan to change the price of a hamburger by 10¢. If they both raise their prices, there will be no change in their market shares, but if they both lower their prices, the chain's national advertising will ensure that MacArches gains 6% of the market. Again because of advertising, if Burger Haven lowers their price and MacArches raises their price, Burger Haven will gain only 4% of the market, but if Burger Haven raises their price and MacArches lowers their price, MacArches will gain 8% of the market. What should the managers do?

8. *Political Campaigns* A Republican and a Democratic candidate are running for office in a heavily Republican county. A recent newspaper poll comparing their views on taxes and welfare reform showed that when compared on taxes or on welfare reform, the Republican leads by 15%. When comparing the Democrat on welfare reform to the Republican on taxes, the Republican leads by 20%. But when comparing the Democrat on taxes to the Republican on welfare reform, the Republican leads by only 10%. What should each discuss at their next debate?

We know that if a game has a saddle point, then the optimal strategies for the row and column players are to choose the row and column corresponding to that saddle point. But what if there is no saddle point? If the game is played several times and one player repeatedly makes a particular choice, the other will quickly notice and shift to the best response to the first player's choice. But then the first player will notice the consistency of the other player's choices and will react accordingly. Such continuing reacting and shifting cannot be the optimal solution because it does not represent an equilibrium between the desires of the opposing players. Thus, we must consider *random* changes between the possible choices, since only random actions are not susceptible to pattern detection.

Dmitri Shironosov/Shutterstock

Mixed Strategies and Expected Values

How do we change randomly among choices? For each possible choice we select a probability between 0 and 1 (inclusive) such that the sum of the probabilities is 1. (More later on how to choose the probabilities.) Such a collection of probabilities for the possible choices is called a *mixed strategy*. But if the row and column players each use mixed strategies, how can we find the value of the outcome? As we saw on page 216, the *long-term average winnings* from repeated plays of a game may be found by multiplying each possible outcome by its probability and adding the products. The result is called the *expected value* of the game.

→ **EXAMPLE 1**   **Finding the Expected Value of a Game**

For the game $\begin{pmatrix} -1 & 2 \\ 2 & -2 \end{pmatrix}$, suppose the row player R uses a mixed strategy with probabilities $\frac{2}{5}$ and $\frac{3}{5}$ (meaning choosing the first row with probability $\frac{2}{5}$ and the second row with probability $\frac{3}{5}$) while the column player C uses a mixed strategy with probabilities $\frac{1}{2}$ and $\frac{1}{2}$. What is the expected value of one play of the game?

Solution

If C chooses the first column, then R obtains the payoffs in the first column $\begin{pmatrix} -1 \\ 2 \end{pmatrix}$ with probabilities $\frac{2}{5}$ and $\frac{3}{5}$, for an expected value of

$$\tfrac{2}{5} \cdot (-1) + \tfrac{3}{5} \cdot 2 = \tfrac{-2 + 6}{5} = \tfrac{4}{5}$$

See page 215 to review how to calculate expected value

In matrix notation, this is just the product of a row of probabilities with a column of payoffs:

$$\left(\tfrac{2}{5} \quad \tfrac{3}{5}\right)\begin{pmatrix} -1 \\ 2 \end{pmatrix} = \tfrac{2}{5} \cdot (-1) + \tfrac{3}{5} \cdot 2 = \tfrac{4}{5}$$

See page 117 to review row times column multiplication

Carrying this out for *each* column:

$$\left(\tfrac{2}{5} \quad \tfrac{3}{5}\right)\begin{pmatrix} -1 & 2 \\ 2 & -2 \end{pmatrix} = \left(\tfrac{4}{5} \quad -\tfrac{2}{5}\right)$$

See page 118 to review matrix multiplication

The two resulting numbers are the expected values for the columns, which we then multiply by the probabilities of choosing those columns, $\tfrac{1}{2}$ each, and add the results. This calculation, however, amounts to multiplying the above row $\left(\tfrac{4}{5} \quad -\tfrac{2}{5}\right)$ by the column of probabilities $\begin{pmatrix} \tfrac{1}{2} \\ \tfrac{1}{2} \end{pmatrix}$, which is equivalent to multiplying *all three matrices together*:

$$\begin{pmatrix} \text{Expected} \\ \text{value} \end{pmatrix} = \left(\tfrac{2}{5} \quad \tfrac{3}{5}\right)\begin{pmatrix} -1 & 2 \\ 2 & -2 \end{pmatrix}\begin{pmatrix} \tfrac{1}{2} \\ \tfrac{1}{2} \end{pmatrix} = \tfrac{1}{5}$$

Expected value of one play of the game

In general, the row player's mixed strategy consists of a probability for each row and the column player's mixed strategy consists of a probability for each column, with the results most conveniently represented as matrices.

Mixed Strategies and Expected Value

For a game G with m rows and n columns:

A mixed strategy for the row player consists of a row matrix

$$r = (r_1 \ \ldots \ r_m)$$

Each number between 0 and 1 and adding to 1

A mixed strategy for the column player consists of a column matrix

$$c = \begin{pmatrix} c_1 \\ \vdots \\ c_n \end{pmatrix}$$

Each number between 0 and 1 and adding to 1

The expected value E for this mixed strategy is the matrix product

$$E = r \cdot G \cdot c$$

G is the payoff matrix

➡ Practice Problem 1

For the game in Example 1, G is $\begin{pmatrix} -1 & 2 \\ 2 & -2 \end{pmatrix}$. If the row player chooses mixed strategy $r = \left(\tfrac{1}{2} \quad \tfrac{1}{2}\right)$ and the column player chooses mixed strategy $c = \begin{pmatrix} \tfrac{4}{5} \\ \tfrac{1}{5} \end{pmatrix}$, what is the expected value of each play of the game?

➤ Solution on page 288

How do the *pure strategies* of the preceding section relate to the *mixed* strategies discussed here? Pure strategies are simply mixed strategies in which one of the choices has probability 1 and the others have probability 0 so that the same row and column (corresponding to the saddle point) are always chosen.

Given a mixed strategy, how in practice does a player choose moves according to the probabilities? If there are only two possible choices and each has probability $\frac{1}{2}$, then it is simple enough to just flip a coin and make the first choice on "heads" and the other choice on "tails." For other probabilities it is usually easiest to base your decision on *random numbers*, which can be found in tables or obtained from calculators and computers. To use random numbers between 0 and 1 to carry out the strategy $\begin{pmatrix} \frac{2}{5} & \frac{3}{5} \end{pmatrix}$ for choosing Row 1 or Row 2 in Example 1, divide the interval from 0 to 1 into parts whose lengths are these probabilities:

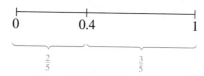

For each successive random number, choose Row 1 if the number falls in the *first* part and choose Row 2 if it falls in the *second*. If it is exactly one of the division points (in this case, just the 0.4), continue on to your next random number. Be sure to use a new random number each time you play the game. For mixed strategies with more choices, make a similar diagram with the interval from 0 to 1 divided into as many parts with the lengths matching the probabilities, and then choose according to the part in which the random number falls.

> **Practice Problem 2**
>
> For the mixed strategy from Example 1 discussed above, if the random number were 0.38491, would you select Row 1 or Row 2? What if the random number were 0.71108?
>
> ➤ Solution on page 288

Optimal Mixed Strategies for 2 × 2 Games

The row player wants to find the mixed strategy that *maximizes* the expected value per play (since the value goes to the row player), while the column player wants the mixed strategy that *minimizes* the expected value per play. How can we find such *optimal* mixed strategies for each player? In the next section, we will answer this question for *any* game, but in this section we will solve this problem for 2 × 2 games.

Let $\begin{pmatrix} a & b \\ c & d \end{pmatrix}$ be a 2 × 2 game without a saddle point. A mixed strategy $r = \begin{pmatrix} r_1 & r_2 \end{pmatrix}$ for R is really just $\begin{pmatrix} r_1 & 1 - r_1 \end{pmatrix}$ since $r_1 + r_2 = 1$. (For example, if one probability is $\frac{1}{3}$, the other must be $1 - \frac{1}{3} = \frac{2}{3}$.) If C chooses the first column, $\begin{pmatrix} a \\ c \end{pmatrix}$, then the expected value is

$$E_1 = \begin{pmatrix} r_1 & 1 - r_1 \end{pmatrix}\begin{pmatrix} a \\ c \end{pmatrix} = ar_1 + c(1 - r_1) \qquad \text{For} \quad 0 \le r_1 \le 1$$

Graphing values of $E_1 = ar_1 + c(1 - r_1)$ for values of r_1 between 0 and 1 gives a line between heights a and c as shown below.

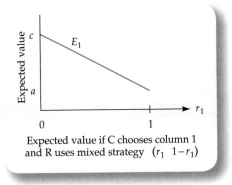

Expected value if C chooses column 1 and R uses mixed strategy $(r_1 \quad 1-r_1)$

Similarly, if C chooses the second column, the expected value is

$$E_2 = (r_1 \quad 1 - r_1)\binom{b}{d} = br_1 + d(1 - r_1)$$

and the graph becomes

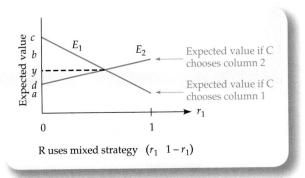

Expected value if C chooses column 2

Expected value if C chooses column 1

R uses mixed strategy $(r_1 \quad 1-r_1)$

In this graph, C chooses the line, but R chooses the *point* on the line (by choosing r_1), so if R chooses any point *other* than the intersection point, C will choose the lower line to decrease R's expected value. It is clear that R should choose the intersection point, in which case it does not matter which line C chooses. Solving for the intersection point of the lines:

$$ar_1 + c(1 - r_1) = br_1 + d(1 - r_1) \qquad \text{Setting} \quad E_1 = E_2$$

$$(a + d - b - c)r_1 = d - c \qquad \text{Collecting like terms}$$

$$r_1 = \frac{d - c}{(a + d) - (b + c)} \qquad \text{Dividing to find } r_1$$

Dividing by $(a + d) - (b + c)$ in the last step is permissible because if G does not have a saddle point, then $(a + d) - (b + c) \neq 0$. The other probability for R's optimal mixed strategy is found by subtracting r_1 from 1:

$$r_2 = 1 - r_1 = 1 - \frac{d - c}{(a + d) - (b + c)} = \frac{a - b}{(a + d) - (b + c)}$$

The value of the game when R uses the mixed strategy $(r_1 \quad r_2)$ is then the height of the intersection point in the preceding graph, which is

$$v = ar_1 + cr_2 = \frac{a(d-c) + c(a-b)}{(a+d) - (b+c)} = \frac{ad - bc}{(a+d) - (b+c)}$$

A similar calculation gives the probabilities c_1 and c_2 for C's optimal strategy, which are included in the summary in the box below.

Optimal Mixed Strategy for a 2 × 2 Game

A game $\begin{pmatrix} a & b \\ c & d \end{pmatrix}$ without a saddle point has optimal mixed strategies:

For row player R: $\begin{cases} r_1 = \dfrac{d-c}{(a+d) - (b+c)} \\ r_2 = 1 - r_1 \end{cases}$

For column player C: $\begin{cases} c_1 = \dfrac{d-b}{(a+d) - (b+c)} \\ c_2 = 1 - c_1 \end{cases}$

The value of the game is $\quad v = \dfrac{ad - bc}{(a+d) - (b+c)}$

→ **EXAMPLE 2**   **Calculating an Optimal Mixed Strategy**

Find the optimal mixed strategy and the value of the game $\begin{pmatrix} 7 & -2 \\ 3 & 6 \end{pmatrix}$.

Solution
Since this game does not have a saddle point, we may use the above formulas. For the row player we find probabilities

$\begin{cases} r_1 = \dfrac{6 - 3}{(7 + 6) - (-2 + 3)} = \dfrac{3}{12} = \dfrac{1}{4} \\ r_2 = 1 - \dfrac{1}{4} = \dfrac{3}{4} \end{cases}$ $\begin{cases} r_1 = \dfrac{d-c}{(a+d) - (b+c)} \\ r_2 = 1 - r_1 \end{cases}$
with $a = 7, \quad b = -2,$
$c = 3, \quad$ and $\quad d = 6$

For the column player we find

$\begin{cases} c_1 = \dfrac{6 - (-2)}{(7 + 6) - (-2 + 3)} = \dfrac{8}{12} = \dfrac{2}{3} \\ c_2 = 1 - \dfrac{2}{3} = \dfrac{1}{3} \end{cases}$ $\begin{cases} c_1 = \dfrac{d-b}{(a+d) - (b+c)} \\ c_2 = 1 - c_1 \end{cases}$
with $a = 7, \quad b = -2,$
$c = 3, \quad$ and $\quad d = 6$

The value of the game is

$$v = \frac{7 \cdot 6 - (-2) \cdot 3}{(7 + 6) - (-2 + 3)} = \frac{48}{12} = 4$$

$$v = \frac{ad - bc}{(a + d) - (b + c)}$$
with $a = 7$, $b = -2$, $c = 3$, and $d = 6$

Therefore, the row player should play rows 1 and 2 with probabilities $\frac{1}{4}$ and $\frac{3}{4}$ and the column player should play columns 1 and 2 with probabilities $\frac{2}{3}$ and $\frac{1}{3}$. The value of the game is 4. (That is, the game is favorable to the row player, who, in the long run, should expect a gain of about 4 per play.)

→ **Practice Problem 3**

Find the optimal mixed strategy and value of the game $\begin{pmatrix} 3 & 0 \\ -2 & 1 \end{pmatrix}$.

➤ **Solution on next page**

Photosani/Shutterstock

Other Interpretations of Mixed Strategies

While we developed mixed strategies to give the best probabilities for choices that must be made repeatedly in competitive situations, these strategies have other interpretations as well. Even if you are in the situation only once, there may be many pairs of people in similar situations, and the mixed strategies then give the *proportions* of people who should make each choice. In fact, it is not even necessary to have an actual opposing player. In many situations it is useful to imagine a *fictitious opponent*. For example, an investor in stocks and bonds might imagine the *economy* as the opponent, bringing about changes in economic conditions, and a strategy of $\begin{pmatrix} \frac{2}{3} & \frac{1}{3} \end{pmatrix}$ might suggest investing $\frac{2}{3}$ of one's capital in bonds and $\frac{1}{3}$ in stocks. Other examples are farmers with unpredictable weather and doctors with uncertain diagnoses. While these situations may not involve an actual "opponent" who is actively trying to minimize your gain, using the optimal strategy will guarantee a certain average value no matter what happens.

→ **EXAMPLE 3   Optimal Farm Policy**

After carefully studying the economy and the likelihood of price supports, a midwestern farmer has estimated his profits per acre for both wheat and soybeans under recession and expansion conditions, as shown in the following table.

		Economy	
		Recession	Expansion
Farmer's	Wheat	$10	$40
Profits	Soybeans	$30	$20

How should the farmer manage his 10,000 acres in the face of uncertain economic conditions?

Solution

Let us view the economy as a fictitious opponent. The game $\begin{pmatrix} 10 & 40 \\ 30 & 20 \end{pmatrix}$ does not

have a saddle point, and the farmer's optimal mixed strategy is

$$\begin{cases} r_1 = \dfrac{20 - 30}{(10 + 20) - (40 + 30)} = \dfrac{-10}{30 - 70} = \dfrac{1}{4} \\ r_2 = 1 - \dfrac{1}{4} = \dfrac{3}{4} \end{cases}$$

From the box on page 286

The farmer should devote $\frac{1}{4}$ of the farm to wheat (that is, 2500 acres) and $\frac{3}{4}$ to soybeans (that is, 7500 acres). With this mixed strategy, should the economy be in recession, the farmer's profits will be

$$\$10 \cdot 2500 + \$30 \cdot 7500 = \$250{,}000$$

while if it is in expansion, the farmer's profit will still be

$$\$40 \cdot 2500 + \$20 \cdot 7500 = \$250{,}000$$

The profit of $250,000 represents a security level for the farmer because, while he could do better if he knew the future state of the economy, there is no way he can receive less.

➡ Solutions to Practice Problems

1. $\begin{pmatrix} \frac{1}{2} & \frac{1}{2} \end{pmatrix} \underbrace{\begin{pmatrix} -1 & 2 \\ 2 & -2 \end{pmatrix} \begin{pmatrix} \frac{4}{5} \\ \frac{1}{5} \end{pmatrix}}_{\left(\frac{1}{2} \ 0 \right)} = \begin{pmatrix} \frac{1}{2} & 0 \end{pmatrix} \begin{pmatrix} \frac{4}{5} \\ \frac{1}{5} \end{pmatrix} = \begin{pmatrix} \frac{2}{5} \end{pmatrix}$

The expected value for each play is $\frac{2}{5}$.

2. Row 1 (since 0.38491 is between 0 and 0.4); Row 2 (since 0.71108 is between 0.4 and 1).

3. Since $G = \begin{pmatrix} 3 & 0 \\ -2 & 1 \end{pmatrix}$ does not have a saddle point, we substitute $a = 3$, $b = 0$, $c = -2$, and $d = 1$ into the formulas from the box on page 286.

$$(a + d) - (b + c) = (3 + 1) - (0 - 2) = 4 + 2 = 6$$

For the row player: $r_1 = \dfrac{1 - (-2)}{6} = \dfrac{3}{6} = \dfrac{1}{2}$ and $r_2 = 1 - \dfrac{1}{2} = \dfrac{1}{2}$

For the column player: $c_1 = \dfrac{1 - 0}{6} = \dfrac{1}{6}$ and $c_2 = 1 - \dfrac{1}{6} = \dfrac{5}{6}$

The value of the game is $\dfrac{3 \cdot 1 - 0 \cdot (-2)}{6} = \dfrac{1}{2}$

→ 8.2 Exercises

For each game and mixed strategies, find the expected value.

1. Let $G = \begin{pmatrix} 2 & -2 \\ 1 & 3 \end{pmatrix}$, $r = \begin{pmatrix} \frac{1}{2} & \frac{1}{2} \end{pmatrix}$ and $c = \begin{pmatrix} \frac{1}{2} \\ \frac{1}{2} \end{pmatrix}$

2. Let $G = \begin{pmatrix} 2 & -4 \\ -1 & 3 \end{pmatrix}$,

   $r = \begin{pmatrix} \frac{4}{5} & \frac{1}{5} \end{pmatrix}$ and $c = \begin{pmatrix} \frac{2}{5} \\ \frac{3}{5} \end{pmatrix}$

3. Let $G = \begin{pmatrix} 3 & 1 & 2 \\ -6 & 4 & -1 \end{pmatrix}$,

   $r = \begin{pmatrix} \frac{1}{2} & \frac{1}{2} \end{pmatrix}$ and $c = \begin{pmatrix} \frac{1}{3} \\ \frac{1}{3} \\ \frac{1}{3} \end{pmatrix}$

4. Let $G = \begin{pmatrix} 9 & 1 \\ -9 & 3 \\ -1 & 2 \end{pmatrix}$,

   $r = \begin{pmatrix} \frac{1}{2} & 0 & \frac{1}{2} \end{pmatrix}$ and $c = \begin{pmatrix} \frac{3}{5} \\ \frac{2}{5} \end{pmatrix}$

For each 2 × 2 game, find the optimal strategy for each player and the value of the game. Be sure to check for saddle points before using the formulas.

5. $\begin{pmatrix} 2 & -1 \\ 1 & 2 \end{pmatrix}$ 6. $\begin{pmatrix} 3 & 1 \\ -4 & 2 \end{pmatrix}$

7. $\begin{pmatrix} 3 & -3 \\ -1 & 3 \end{pmatrix}$

Represent each situation as a game and find the optimal strategy for each player. State your final answer in the terms of the original question.

8. *Political Campaigns* Political scientists distinguish between two kinds of issues in elections campaigns: "positional" issues (voters have sharply divided and incompatible views) and "valence" issues (voters agree on goals but are divided on the best ways to achieve them). In a race between the incumbent and the challenger for state governor, a marketing survey found that if the incumbent focused on positional issues, she trailed the challenger by 5% if he focused on the same kind of issues, but she led him by 5% if he concentrated on valence issues. If she focused on valence issues, she trailed him by 10% if he focused on the same kind of issues, but she led him by 20% if he concentrated on positional issues. In designing the incumbent's TV ads for the campaign, what proportion should focus on positional issues?

9. *Farm Management* A farmer grows apples on her 600-acre farm and must cope with occasional infestations of worms. If she refrains from using pesticides, she can get a premium for "organically grown" produce and her profits per acre increase by $600 if there is no infestation, but they decrease by $400 if there is. If she does use pesticides and there is an infestation, her crop is saved and the resulting apple shortage (since other farms are decimated) raises her profits by $500 per acre. Otherwise, her profits remain at their usual levels. How should she divide her farm into a "pesticide-free" zone and a "pesticide-use" zone? What will be her expected increase in profits per acre with this strategy?

10. *Quizzes in English Class* In a section of Freshman Composition at your college, the English professor gives a daily quiz on either vocabulary or writing, and allows his students to bring either a dictionary or a grammar textbook (but not both) to use during the quiz. Joe estimates that if he brings a dictionary, he can get a 100 on a vocabulary quiz but only an 80 on a paragraph revision, while with his grammar textbook, he can get a 90 on the revision but only a 70 on the vocabulary quiz. How should he decide each day what to bring and what grade can he be expected to earn?

8.3 Games and Linear Programming

In the previous section we found the optimal strategy for a 2×2 game. In this section we will see how to find the optimal strategy for *any* game. Rather than develop a collection of complicated formulas analogous to those on page 286, we will show that a game can be expressed as a pair of dual linear programming problems (see page 169), which can then be solved by the simplex method that we developed in Sections 3 and 4 of Chapter 4.

The idea of representing a game as a pair of dual linear programming problems follows from the observation that the profits P to be maximized and the costs C to be minimized always satisfy $P \le C$. These dual problems are a game in which one player tries to increase the value of P while the other tries to decrease the value of C. If the first succeeds in making P as large as possible while the second succeeds in making C as small as possible, they will have found the solution to both problems at feasible points giving $P = C$. The Duality Theorem (page 170) states that the solutions of both dual problems appear in the final simplex tableau that solves the maximum problem.

Games as Linear Programming Problems

$$\begin{pmatrix} 3 & 7 \\ 4 & 1 \end{pmatrix}$$

To see how to express a game as a linear programming problem, consider the game shown on the left. Given any strategy $(r_1 \quad r_2)$ for R, we may calculate the expected value depending on whether C chooses the first or second column:

If C chooses the first column   If C chooses the second column

$$E_1 = (r_1 \quad r_2)\begin{pmatrix} 3 \\ 4 \end{pmatrix} = 3r_1 + 4r_2 \qquad E_2 = (r_1 \quad r_2)\begin{pmatrix} 7 \\ 1 \end{pmatrix} = 7r_1 + 1r_2$$

Since C will choose the column to give R the *lower* of these expected gains, the value will be at or below *both* of these expected values, so R wants to solve the problem

$$\text{Maximize } v \qquad\qquad \text{R tries to maximize } v$$

$$\text{Subject to } \begin{cases} 3r_1 + 4r_2 \ge v \\ 7r_1 + 1r_2 \ge v \\ r_1 + r_2 = 1 \\ r_1 \ge 0 \quad \text{and} \quad r_2 \ge 0 \end{cases}$$

C keeps v at or below both expected values

r_1 and r_2 are probabilities

We now must make this into a *standard* linear programming problem (see page 155). Since every payoff for this game is positive, the value v is also positive, so we may divide the inequalities by v without changing their sense.

$$\begin{cases} 3r_1 + 4r_2 \ge v \\ \\ 7r_1 + 1r_2 \ge v \end{cases} \quad \text{becomes} \quad \begin{cases} 3\left(\dfrac{r_1}{v}\right) + 4\left(\dfrac{r_2}{v}\right) \ge 1 \\ \\ 7\left(\dfrac{r_1}{v}\right) + 1\left(\dfrac{r_2}{v}\right) \ge 1 \end{cases} \qquad \text{Dividing by } v$$

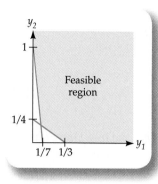

Using new variables $y_1 = r_1/v$ and $y_2 = r_2/v$ (which are nonnegative), these inequalities take the simpler form $3y_1 + 4y_2 \geq 1$ and $7y_1 + 1y_2 \geq 1$. Furthermore, the equality $r_1 + r_2 = 1$ divided by v becomes $y_1 + y_2 = \frac{1}{v}$. Since maximizing the positive quantity v is the same as minimizing $\frac{1}{v}$ (a bigger v means a smaller $\frac{1}{v}$), R wants to solve the standard linear programming problem

$$\text{Minimize} \quad y_1 + y_2$$

$$\text{Subject to} \quad \begin{cases} 3y_1 + 4y_2 \geq 1 \\ 7y_1 + 1y_2 \geq 1 \\ y_1 \geq 0 \text{ and } y_2 \geq 0 \end{cases}$$

Now consider the game from the point of view of the column player. For any strategy $\begin{pmatrix} c_1 \\ c_2 \end{pmatrix}$ for C, we can calculate the expected value depending on the row that R chooses:

$$\begin{pmatrix} 3 & 7 \\ 4 & 1 \end{pmatrix}$$

If R chooses the first row

$$(3 \quad 7)\begin{pmatrix} c_1 \\ c_2 \end{pmatrix} = 3c_1 + 7c_2$$

If R chooses the second row

$$(4 \quad 1)\begin{pmatrix} c_1 \\ c_2 \end{pmatrix} = 4c_1 + 1c_2$$

C seeks the *smallest* possible value, while R chooses the row to keep it at or above these expectations, so C wants to solve the problem

$$\text{Minimize} \quad v \qquad \text{C tries to minimize } v$$

$$\text{Subject to} \quad \begin{cases} 3c_1 + 7c_2 \leq v \\ 4c_1 + 1c_2 \leq v \\ c_1 + c_2 = 1 \\ c_1 \geq 0 \text{ and } c_2 \geq 0 \end{cases}$$

R keeps v at or above both expected values

c_1 and c_2 are probabilities

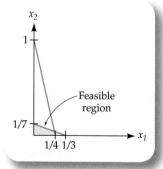

Using new variables $x_1 = c_1/v$ and $x_2 = c_2/v$, we can write this problem as a standard linear programming problem

$$\text{Maximize} \quad x_1 + x_2$$

$$\text{Subject to} \quad \begin{cases} 3x_1 + 7x_2 \leq 1 \\ 4x_1 + 1x_2 \leq 1 \\ x_1 \geq 0 \text{ and } x_2 \geq 0 \end{cases}$$

$$x_1 + x_2 = \frac{1}{v}$$

$$3\left(\frac{c_1}{v}\right) + 7\left(\frac{c_2}{v}\right) \leq \frac{v}{v}$$

$$4\left(\frac{c_1}{v}\right) + 1\left(\frac{c_2}{v}\right) \leq \frac{v}{v}$$

Notice that the problems for R and C are *dual* linear programming problems:

For C:	For R:
Maximize $x_1 + x_2$	Minimize $y_1 + y_2$
Subject to $\begin{cases} 3x_1 + 7x_2 \leq 1 \\ 4x_1 + 1x_2 \leq 1 \\ x_1 \geq 0 \text{ and } x_2 \geq 0 \end{cases}$	Subject to $\begin{cases} 3y_1 + 4y_2 \geq 1 \\ 7y_1 + 1y_2 \geq 1 \\ y_1 \geq 0 \text{ and } y_2 \geq 0 \end{cases}$

See page 169 to review dual linear programming problems

The maximum problem $\left(x_1 + x_2 = \frac{1}{v}\right)$ comes from the column player's attempt to minimize v, and the minimum problem $\left(y_1 + y_2 = \frac{1}{v}\right)$ results from the row player's desire to maximize the same v. Since the simplex method solves both problems at the same time, we may solve the original game by constructing the initial simplex tableau and pivoting to find the optimal values for the variables x_1, x_2, y_1, and y_2 and then recovering the probabilities r_1, r_2, c_1, and c_2 for the optimal strategies. The initial simplex tableau (see page 157) is

	x_1	x_2	s_1	s_2	
s_1	3	7	1	0	1
s_2	4	1	0	1	1
	-1	-1	0	0	0

	X	S	
S	A	I	b
	$-c^t$	0	0

with

$$A = \begin{pmatrix} 3 & 7 \\ 4 & 1 \end{pmatrix}, \quad b = \begin{pmatrix} 1 \\ 1 \end{pmatrix}, \quad c^t = (1 \; 1)$$

Pivoting twice (first at column 1 and row 2, and then at column 2 and row 1), we arrive at the final tableau, with the resulting x- and y-values as shown.

	x_1	x_2	s_1	s_2	
x_2	0	1	4/25	$-3/25$	1/25	$x_2 = 1/25$
x_1	1	0	$-1/25$	7/25	6/25	$x_1 = 6/25$
	0	0	3/25	4/25	7/25

$y_1 = 3/25, \quad y_2 = 4/25$ — Maximum is 7/25

Since the maximum value, 7/25, is the reciprocal $\frac{1}{v}$ of the value v of the game, we have $v = \frac{25}{7}$. We can recover the strategy from the xs and the ys by multiplying back by the value we have found for v:

$$r_1 = v \cdot y_1 = \frac{25}{7} \cdot \frac{3}{25} = \frac{3}{7}$$

$$r_2 = v \cdot y_2 = \frac{25}{7} \cdot \frac{4}{25} = \frac{4}{7}$$

and

$$c_1 = v \cdot x_1 = \frac{25}{7} \cdot \frac{6}{25} = \frac{6}{7}$$

$$c_2 = v \cdot x_2 = \frac{25}{7} \cdot \frac{1}{25} = \frac{1}{7}$$

The value $v = \frac{25}{7}$ and the optimal strategies $\left(\frac{3}{7} \; \frac{4}{7}\right)$ for R and $\left(\frac{6}{7} \; \frac{1}{7}\right)$ for C solve the game.

For our first example we chose a 2×2 game, which also can be solved by the formulas on page 286 (giving the same answer, as you may easily check). However, the linear programming method we developed here is not restricted to 2×2 games and may be used to solve a game of *any* size.

The general procedure for solving any game by expressing it as a linear programming problem is described in the following box. The first step ensures that the value of the game is positive so that division by v will not alter the sense of the inequalities.

The Simplex Method Solution of Any Game

To solve a game of any size:

1. If any entries in the payoff matrix G are less than 1, add a constant k to every entry to make every entry at least 1. We write $G + k$ for this new matrix.

2. Solve the linear programming problem on the left below by applying the simplex method (see page 164) to the initial tableau on the right:

$$\text{Maximize } \overbrace{(1 \ \ldots \ 1)}^{n \ 1\text{s}} X$$

$$\text{Subject to } \left\{ \begin{array}{l} (G + k)X \leq \begin{pmatrix} 1 \\ \vdots \\ 1 \end{pmatrix} \\ X \geq 0 \end{array} \right\} m \ 1\text{s}$$

	X	S	
S	$G + k$	I	$\left.\begin{array}{c} 1 \\ \vdots \\ 1 \end{array}\right\} m \text{ 1s}$
	$\underbrace{-1 \ldots -1}_{n-1\text{s}}$	$\underbrace{0 \ldots 0}_{m \text{ 0s}}$	0

3. In the final tableau (from step 2):

   a. The maximum value V of the objective function is in the bottom-right corner of the final tableau.

   b. The x-variables for the maximum problem take the following values: The basic variables take the values on the far right of the tableau, and the nonbasic variables take the value 0.

   c. The y-variables for the (dual) minimum problem take the values at the bottom of the tableau in the slack variable columns.

4. The value of the game is $v = \frac{1}{V} - k$. The optimal strategies are as follows:

   For the row player R:

$$r_1 = \frac{y_1}{V}, \quad r_2 = \frac{y_2}{V}, \quad \ldots, \quad r_m = \frac{y_m}{V}$$

      Using xs and ys from the final x- and y-values

   For the column player C:

$$c_1 = \frac{x_1}{V}, \quad c_2 = \frac{x_2}{V}, \quad \ldots, \quad c_n = \frac{x_n}{V}$$

The optimal strategy found by the simplex method may be either pure or mixed.

→ EXAMPLE 1 Solving a Game by the Simplex Method

Use the simplex method to find the optimal strategies for the game

$$G = \begin{pmatrix} 1 & -2 & 1 \\ 2 & -1 & -2 \\ -1 & 2 & 1 \end{pmatrix}.$$

Solution

Step 1: Since the smallest entry in the matrix is -2, adding 3 to every element of the matrix will make each at least 1:

$$\begin{pmatrix} 4 & 1 & 4 \\ 5 & 2 & 1 \\ 2 & 5 & 4 \end{pmatrix}$$

      $G + k$ with $k = 3$

Step 2: The initial simplex tableau is

	x_1	x_2	x_3	s_1	s_2	s_3	
s_1	4	1	4	1	0	0	1
s_2	⑤	2	1	0	1	0	1
s_3	2	5	4	0	0	1	1
	-1	-1	-1	0	0	0	0

See page 157 for more about the initial simplex tableau

The first pivot (on the 5 in column 1 and row 2) yields

	x_1	x_2	x_3	s_1	s_2	s_2	
s_1	0	$-3/5$	⓰⑥⁄⑤ 16/5	1	$-4/5$	0	1/5
x_1	1	2/5	1/5	0	1/5	0	1/5
s_3	0	21/5	18/5	0	$-2/5$	1	3/5
	0	$-3/5$	$-4/5$	0	1/5	0	1/5

See pages 160-161 for more about the pivot operation

The next pivot (on the 16/5 in column 3 and row 1) yields

	x_1	x_2	x_3	s_1	s_2	s_2	
x_3	0	$-3/16$	1	5/16	$-1/4$	0	1/16
x_1	1	7/16	0	$-1/16$	1/4	0	3/16
s_3	0	⟨39/8⟩	0	$-9/8$	1/2	1	3/8
	0	$-3/4$	0	1/4	0	0	1/4

The next pivot (on the 39/8 in column 2 and row 3) gives the final tableau:

	x_1	x_2	x_3	s_1	s_2	s_3		
x_3	0	0	1	7/26	$-3/13$	1/26	1/13	$x_3 = \frac{1}{13}$
x_1	1	0	0	1/26	8/39	$-7/78$	2/13	$x_1 = \frac{2}{13}$
x_2	0	1	0	$-3/13$	4/39	8/39	1/13	$x_2 = \frac{1}{13}$
	0	0	0	1/13	1/13	2/13	4/13	

$$y_1 = \tfrac{1}{13}, \quad y_2 = \tfrac{1}{13}, \quad y_3 = \tfrac{2}{13} \qquad V = \tfrac{4}{13}$$

Step 3: $V = \frac{4}{13}$ with $x_1 = \frac{2}{13}, \quad x_2 = \frac{1}{13}, \quad x_3 = \frac{1}{13}$
and $y_1 = \frac{1}{13}, \quad y_2 = \frac{1}{13}, \quad y_3 = \frac{2}{13}.$

Step 4: The value of the game is

$$v = \frac{1}{4/13} - 3 = \frac{13}{4} - 3 = \frac{1}{4} \qquad\qquad v = \frac{1}{V} - k$$

The optimal strategy is

$$r_1 = \frac{1/13}{4/13} = \frac{1}{4}, \quad r_2 = \frac{1/13}{4/13} = \frac{1}{4}, \quad \text{and} \quad r_3 = \frac{2/13}{4/13} = \frac{1}{2} \qquad r_i = \frac{y_i}{V}$$

$$c_1 = \frac{2/13}{4/13} = \frac{1}{2}, \quad c_2 = \frac{1/13}{4/13} = \frac{1}{4}, \quad \text{and} \quad c_3 = \frac{1/13}{4/13} = \frac{1}{4} \qquad c_i = \frac{x_i}{V}$$

The row player should play the rows with probabilities $\frac{1}{4}, \frac{1}{4}, \frac{1}{2}$ and the column player should play the columns with probabilities $\frac{1}{2}, \frac{1}{4}, \frac{1}{4}$, for a game value of $\frac{1}{4}$ (favorable to the row player).

294   *Chapter 8: Game Theory*

→ **Practice Problem**

Use the simplex method to find the optimal strategy for the game $\begin{pmatrix} 3 & 2 & 1 \\ 4 & 5 & 6 \end{pmatrix}$.

➤ Solution on pages 296–297

Every Game Has a Solution

For any payoff matrix we may add a constant to make every entry at least 1. The linear programming problem (in step 2 of the box on pages 292–293) has a bounded region with the origin as a feasible point, so the simplex method *will* find an optimal strategy for the game. We therefore have the following important result.

Fundamental Theorem of Game Theory

Every two-person zero-sum game has an optimal strategy.

Agricultural?
Residential?
Commercial?
Industrial?

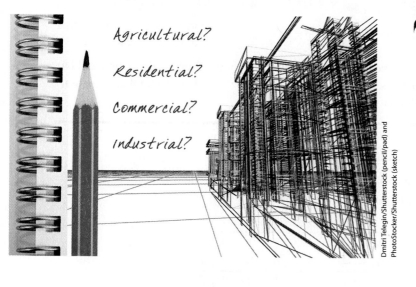

Dmitri Telegin/Shutterstock (pencil/pad) and PhotoStocker/Shutterstock (sketch)

→ **EXAMPLE 2** Real Estate Speculation

An investor's option to purchase 5000 acres of undeveloped land expires next Friday. The investor may have the land zoned for agricultural, residential, commercial, or industrial use. Rumors indicate that either or both an interstate highway and a nuclear power plant may be built near this property within the next few years. The investor estimates the returns (in percentages gained) for the various zoning possibilities as shown in the following table. Should the investor exercise the option to buy? If so, how would he prefer the land to be zoned?

Nearby Construction

		None	Highway	Power Plant	Both
	Agricultural	4%	8%	−8%	−4%
Zoning	Residential	−5%	12%	−6%	−9%
	Commercial	−3%	6%	−4%	−7%
	Industrial	−2%	5%	6%	8%

Solution

By imagining a fictitious opponent, the investor can find the best that can be achieved no matter what happens. Adding 10 to every entry in the table above, the initial simplex tableau for the corresponding game is

	x_1	x_2	x_3	x_4	s_1	s_2	s_3	s_4	
s_1	14	18	2	6	1	0	0	0	1
s_2	5	22	4	1	0	1	0	0	1
s_3	7	16	6	3	0	0	1	0	1
s_4	8	15	16	18	0	0	0	1	1
	-1	-1	-1	-1	0	0	0	0	0

Pivoting at column 1, row 1 and then at column 3, row 4, we reach the final tableau:

	x_1	x_2	x_3	x_4	s_1	s_2	s_3	s_4	
x_1	1	129/104	0	15/52	1/13	0	0	$-1/104$	7/104
s_2	0	1511/104	0	$-227/52$	$-3/13$	1	0	$-23/104$	57/104
s_3	0	563/104	0	$-255/52$	$-4/13$	0	1	$-35/104$	37/104
x_3	0	33/104	1	51/52	$-1/26$	0	0	7/104	3/104
	0	29/52	0	7/26	1/26	0	0	3/52	5/52

The maximum value is

$$V = \frac{5}{52} \quad \text{with} \quad \begin{cases} x_1 = \frac{7}{104} & y_1 = \frac{1}{26} \\ x_2 = 0 & y_2 = 0 \\ x_3 = \frac{3}{104} & y_3 = 0 \\ x_4 = 0 & y_4 = \frac{3}{52} \end{cases} \qquad \begin{array}{l} x_2 \text{ and } x_4 \text{ are} \\ \text{not in the basis} \end{array}$$

The value of the game is $v = \frac{1}{5/52} - 10 = \frac{2}{5}$. The optimal strategy for the investor is

$$r_1 = \tfrac{2}{5}, \quad r_2 = 0, \quad r_3 = 0, \quad r_4 = \tfrac{3}{5}$$

Dividing each y-value by 5/52 (or multiplying each by 52/5)

Since the value of this game is positive, the investor should exercise the option and buy the land. Of the 5000 acres, 2000 (that is, $\frac{2}{5}$) should be zoned agricultural and the other 3000 should be zoned industrial.

➡ Solution to Practice Problem

Since every entry of this 2×3 game is at least one, we do not need to add anything to the matrix. The initial simplex tableau is

	x_1	x_2	x_3	s_1	s_2	
s_1	3	2	1	1	0	1
s_2	④	5	6	0	1	1
	-1	-1	-1	0	0	0

Pivoting on the 4 in column 1 and row 2, gives the final tableau:

	x_1	x_2	x_3	s_1	s_2	
s_1	0	$-7/4$	$-7/2$	1	$-3/4$	1/4
x_1	1	5/4	3/2	0	1/4	1/4
	0	1/4	1/2	0	1/4	1/4

The maximum value of the objective function is

$$V = \frac{1}{4} \quad \text{with} \quad \begin{cases} x_1 = \frac{1}{4} \\ x_2 = 0 \\ x_3 = 0 \end{cases} \text{and} \quad \begin{cases} y_1 = 0 \\ y_2 = \frac{1}{4} \end{cases} \qquad \begin{array}{l} x_2 \text{ and } x_3 \text{ are} \\ \text{not in the basis} \end{array}$$

The value of the game is $v = \frac{1}{1/4} - 0 = 4$, and the optimal strategy is

$$\begin{cases} r_1 = \frac{0}{1/4} = 0 \\ r_2 = \frac{1/4}{1/4} = 1 \end{cases} \quad \text{and} \quad \begin{cases} c_1 = \frac{1/4}{1/4} = 1 \\ c_2 = \frac{0}{1/4} = 0 \\ c_3 = \frac{0}{1/4} = 0 \end{cases}$$

Notice that this optimal strategy is the pure strategy that selects the saddle point in row 2 and column 1:

$$\begin{pmatrix} 3 & 2 & \textcircled{1} \\ \boxed{4} & \boxed{5} & \boxed{6} \end{pmatrix}$$

→ 8.3 Exercises

Use the simplex method to find the solution of each game.

1. $\begin{pmatrix} 3 & 1 \\ 2 & 4 \end{pmatrix}$

2. $\begin{pmatrix} -3 & 6 \\ 2 & -4 \end{pmatrix}$

3. $\begin{pmatrix} 1 & -4 & -3 \\ -1 & 5 & 3 \end{pmatrix}$

4. $\begin{pmatrix} -1 & 1 & 2 \\ 0 & 1 & -1 \\ 2 & 0 & 3 \end{pmatrix}$

5. $\begin{pmatrix} -3 & 3 & 1 & 3 \\ 2 & 1 & 2 & -2 \\ -1 & 2 & 2 & 2 \end{pmatrix}$

Determine an optimal strategy for each situation by representing it as a game and using the simplex method. State your final answer in the terms of the original question.

6. **Portfolio Management** An investment counselor recommends three mutual funds (in precious metals, municipal bonds, and technology stocks) and estimates their potential returns under bull, stagnant, and bear market conditions, as shown in the table. How much of a $50,000 retirement portfolio should be invested in each fund, and what is the expected minimum return?

Market Conditions

		Bull	Stagnant	Bear
	Precious Metals	1%	1%	4%
Investment Returns	Municipal Bonds	2%	4%	2%
	Technology Stocks	6%	2%	3%

7. **Political Campaigns** Midway through the fall campaign, the mayor's election manager conducted focus groups to estimate the challenger's perceived lead on a variety of paired issues. The results in percentage lead are shown in the table. On what issues may the challenger be expected to concentrate during the rest of the campaign?

		Challenger			
		Lower Taxes	Downtown Mall	More Police	Better Schools
	Recycling	−2	3	2	−1
Mayor	New Stadium	−1	−2	0	0
	Mass Transit	−1	1	−2	0

8. **Guess Which Coin** Adana has a dime, a quarter, and a half-dollar in her pocket. She picks one coin and hides it in her fist. If Daryl can guess which coin it is, Adana gives it to him, but if Daryl is wrong, he pays Adana the value of his guess. Is this game fair? How often should Adana hide the quarter in her hand?

LEARN YOUR WAY

With **FINITE**, you have a multitude of study aids at your fingertips.

The Student Solutions Manual contains worked-out solutions to every odd-numbered exercise, to further reinforce your understanding of mathematical concepts in **FINITE**.

In addition to the Student Solutions Manual, Cengage Learning's **CourseMate** offers exercises and questions that correspond to every section and chapter in the text for extra practice.

For access to these study aids, sign in at **login.cengagebrain.com**.

F

Face value, 72
Factorials, 186
Fair coin, 194
Fair die, 194
Fair game, 216
False negative test results, 212
False positive test results, 212
Favorable game, 216
Feasible region, 136, 138, 155
    vertex, 139
Feasible system, 138
Fictitious opponent, 287
Final tableau, 163
Five-point summary, 233
Fractional exponents, 17–19
Free variable, 108
Function, 23
    absolute value, 36
    composite, *see* Composite function
    domain, 23
    exponential, *see* Exponential function
    graph of, 23
    inverse, 53, 55
    linear, 25
    logarithmic, *see* Logarithm, Natural logarithm
    piecewise linear, 35–37
    polynomial, *see* Polynomial
    quadratic, 26–32
    range, 23
    rational, *see* Rational function
    vertical line test for, 24
    zero of, 28
Fundamental matrix, absorbing Markov
        chain, 270
Fundamental theorem of
    game theory, 295
    linear programming, 147
    regular Markov chains, 262
Future value, 62, 68

G

Game, 276
    dual linear programming
        problems, 290
    expected value, 283
    fair, 216
    favorable, 216
    fictitious opponent, 287
    fundamental theorem, 295
    optimal strategy, 279, 286, 293
    simplex method solution, 292–293
    strictly determined, 279
    two-person, zero-sum, 276
    unfair, 216
    value, 279, 286, 293
    without saddle point, 286
GAUSS, CARL F., 106, 238
Gauss-Jordan method, 106
General linear equation, 11
Generalized multiplication principle
        for counting, 183
Geometric series, 75
Graph, 23
    shift of, 40
    x-intercept, 28
Growth times, 69–71

H

Histogram, 224
Hollow dot (○), 4
Horizontal line, 10, 11
Horizontal shift, 40

I

Identity matrix, 114, 119
Inconsistent linear equations, 90, 93, 98,
        107, 112
Independent events, 206, 207
Independent variable, 23, 109
Inequality, 2, 136
    boundary of, 136
    double, 4
    feasible region, 136, 138
    feasible system, 138
    graphing, 138
    infeasible system, 138
    multiplying by a negative, 4, 174
    nonnegativity, 141
    vertex, 139

prepcard CHAPTER 1

Learning Objectives

1.1 Real Numbers, Inequalities, and Lines
- Translate an interval into set notation and graph it on the real line. (*Practice Test Exercise 1.*)

$$[a, b] \quad (a, b) \quad [a, b) \quad (a, b]$$
$$(-\infty, b] \quad (-\infty, b) \quad [a, \infty) \quad (a, \infty) \quad (-\infty, \infty)$$

- Express given information in interval form.
- Find an equation for a line that satisfies certain conditions. (*Practice Test Exercises 2 and 3.*)

$$m = \frac{y_2 - y_1}{x_2 - x_1} \quad y = mx + b \quad y - y_1 = m(x - x_1)$$
$$x = a \quad y = b \quad ax + by = c$$

- Use straight-line depreciation to find the value of an asset. (*Practice Test Exercise 4.*)

Key Concepts · Objective 1.1

Interval, 4–6
Cartesian plane, 6
Slope, 6–7
y-intercept, 8
Slope-intercept form, 8
Point-slope form, 9
Horizontal line, 10
Vertical line, 10
General linear equation, 11
Straight-line depreciation, 13

1.2 Exponents
- Evaluate negative and fractional exponents without a calculator. (*Practice Test Exercise 5.*)

$$x^0 = 1 \quad x^{-n} = \frac{1}{x^n} \quad x^{m/n} = \sqrt[n]{x^m} = \left(\sqrt[n]{x}\right)^m$$

- Evaluate an exponential expression using a calculator.

Key Concepts · Objective 1.2

Exponent, 14–19

1.3 Functions
- Evaluate and find the domain of a function. (*Practice Test Exercise 6.*)

  A function f is a rule that assigns to each number x in a set (the domain) a (single) number $f(x)$. The range is the set of all resulting values $f(x)$.

- Use the vertical line test to see if a graph defines a function. (*Practice Test Exercise 7.*)
- Graph a linear function. (*Practice Test Exercise 8.*)

$$f(x) = mx + b$$

- Graph a quadratic function. (*Practice Test Exercise 9.*)

$$f(x) = ax^2 + bx + c$$

- Solve a quadratic equation by factoring and by the Quadratic Formula.

Vertex	x-intercepts
$x = \dfrac{-b}{2a}$	$x = \dfrac{-b \pm \sqrt{b^2 - 4ac}}{2a}$

- Construct a linear function from a word problem or from real-life data and then use the function in an application. (*Practice Test Exercise 10.*)
- For given cost and revenue functions, calculate the break-even points and maximum profit. (*Practice Test Exercise 11.*)

Key Concepts · Objective 1.3

Domain, 22–23
Vertical line test, 24
Linear function, 25
Marginal cost, 26
Quadratic function, 26
Parabola, 26
Vertex of a parabola, 27
x-intercept, 28
Quadratic formula, 28
Break-even point, 28, 30
Profit, Revenue, and Cost, 29

1.4 More About Functions
- Evaluate and find the domain of a more complicated function.
- Solve a polynomial equation by factoring. (*Practice Test Exercise 12.*)
- Graph a piecewise linear function. (*Practice Test Exercise 13.*)
- Given two functions, find their composition. (*Practice Test Exercise 14.*)

$$f(g(x)) \quad g(f(x))$$

- Solve an applied problem involving the composition of functions.
- Draw a "shifted" graph. (*Practice Test Exercise 15.*)

Key Concepts · Objective 1.4

Polynomial, 33
Rational function, 34
Piecewise linear function, 35–37
Absolute value function, 36
Composite functions, 37–38
Shifts of graphs, 39–40

1.5 Exponential Functions

- Sketch the graph of an exponential function.
- Find the value of money invested at compound interest. (*Practice Test Exercise 16.*)

$$P\left(1 + \frac{r}{m}\right)^{mt}$$

- Depreciate an asset by a fixed percentage per year. (*Practice Test Exercise 17.*)

    (Above formula with $m = 1$ and a *negative* value for r)

- Find the value of a deposit invested with continuous compounding. (*Practice Test Exercise 18.*)

$$Pe^{rt}$$

Key Concepts · Objective 1.5

Exponential function, 42–43
Compound interest, 43
Depreciation by fixed
    percentage, 44
e, 45–47
Continuous compounding, 47
Exponential growth, 48

1.6 Logarithmic Functions

- Evaluate common and natural logarithms *without* using a calculator. (*Practice Test Exercise 19.*)

$$\log 1 = 0 \qquad \log 10 = 1 \qquad \log 10^x = x \qquad 10^{\log x} = x$$

$$\log(M \cdot N) = \log M + \log N$$

$$\log\left(\frac{M}{N}\right) = \log M - \log N$$

$$\log\left(\frac{1}{N}\right) = -\log N \qquad \log(M^P) = P \cdot \log M$$

$$\ln 1 = 0 \qquad \ln e = 1 \qquad \ln e^x = x \qquad e^{\ln x} = x$$

$$\ln(M \cdot N) = \ln M + \ln N$$

$$\ln\left(\frac{M}{N}\right) = \ln M - \ln N$$

$$\ln\left(\frac{1}{N}\right) = -\ln N \qquad \ln(M^P) = P \cdot \ln M$$

Key Concepts · Objective 1.6

Logarithm, 50
Common logarithm, 50, 53
Properties of common
    logarithms, 51–53
Natural logarithm, 54, 56
Properties of natural
    logarithms, 55
Carbon-14 dating, 57

- Use the properties of natural logarithms to simplify a function. (*Practice Test Exercise 20.*)
- Determine when an asset depreciates to a fraction of its value. (*Practice Test Exercise 21.*)
- Estimate the age of a fossil. (*Practice Test Exercise 22.*)

Instructional Hints

- Notice the difference between *geometric* objects (points, curves, etc.) and *analytic* objects (numbers, functions, etc.), and the connections between them. Descartes first made this connection: By drawing axes, he saw that points could be specified by numerical coordinates and so *curves* could be specified by *equations* governing their coordinates. This idea connected geometry to algebra, previously distinct subjects. You should be able to express geometric objects analytically and vice versa. For example, given a *graph* of a line, you should be able to find an *equation* for it, and given a quadratic *function*, you should be able to *graph* it.

- Interest rates are always *annual* (unless clearly stated otherwise), but the *compounding* may be done more frequently (in the formula, *m* times a year).
- When do you use the formula $P(1 + r/m)^{mt}$ and when do you use Pe^{rt}? Use Pe^{rt} if the word "continuous" occurs and use $P(1 + r/m)^{mt}$ if it does not.
- The formula Pe^{rt} has no *m* because there is no period to wait for the interest to be compounded—interest is added to the account *continuously*.
- For depreciation, use $P(1 + r/m)^{mt}$ with a *negative* value for *r*. For example, depreciating by 15% annually would mean $m = 1$ and $r = -0.15$.

Go to login.cengage.com to access additional resources.

Learning Objectives

2.1 Simple Interest

- Find the interest due on a simple interest loan. *(Practice Test Exercise 1.)*

$$I = Prt$$

- Find the total amount due on a simple interest loan. *(Practice Test Exercise 2.)*

$$A = P(1 + rt)$$

- Solve a simple interest situation for the interest, the interest rate, the principal, or the term. *(Practice Test Exercises 3 and 4.)*

- Find a simple interest future or present value.

$$A = P(1 + rt) \qquad P = \frac{A}{1 + rt}$$

- Find the effective simple interest rate of a discounted loan. *(Practice Test Exercise 5.)*

$$r_s = \frac{r}{1 - rt}$$

- Solve an applied simple interest problem.

Key Concepts · Objective 2.1

Simple interest, 60
Principal, 60
Term, 60
Rate of interest, 60
Total amount due, 62
Future value, 62
Present value, 63
Discounted loan, 64–65
Effective rate, 65

2.2 Compound Interest

- Determine the amount due on a compound interest loan or the value of an invested sum. *(Practice Test Exercise 6.)*

$$A = P(1 + r/m)^{mt}$$

- Find a future or present value. *(Practice Test Exercise 7.)*

$$A = P(1 + r/m)^{mt} \qquad P = \frac{A}{(1 + r/m)^{mt}}$$

- Find the term needed for a given principal to grow to a future value. *(Practice Test Exercises 8 and 9.)*

- Use the "rule of 72" to estimate a doubling time. *(Practice Test Exercise 10.)*

$$\left(\begin{array}{c}\text{Doubling} \\ \text{time}\end{array}\right) \approx \frac{72}{r \times 100}$$

- Find the effective rate of a loan or investment. *(Practice Test Exercise 11.)*

$$r_e = (1 + r/m)^m - 1$$

Key Concepts · Objective 2.2

Compound interest, 66–68
Future value, 68
Present value, 68
Growth time, 69–71
Doubling time, 71
Rule of 72, 71
Effective rate, 71

2.3 Annuities

- Find the accumulated amount of an ordinary annuity. *(Practice Test Exercises 12 and 13.)*

$$A = P\frac{(1 + r/m)^{mt} - 1}{r/m}$$

- Calculate the regular payment to make into a sinking fund to accumulate a given amount. *(Practice Test Exercise 14.)*

$$P = A\frac{r/m}{(1 + r/m)^{mt} - 1}$$

- Find the number of periods to make payments into a sinking fund to accumulate a given amount. *(Practice Test Exercise 15.)*

$$mt = \frac{\log\left((A/P)(r/m) + 1\right)}{\log(1 + r/m)}$$

Key Concepts · Objective 2.3

Annuity, 74
Geometric series, 75
Accumulated amount, 76
Sinking fund, 77
Number of periods, 78

2.4 Amortization

- Find the present value of an ordinary annuity. *(Practice Test Exercise 16.)*

$$PV = P \frac{1 - (1 + r/m)^{-mt}}{r/m}$$

- Find the regular payment to amortize a debt. *(Practice Test Exercise 17.)*

$$P = D \frac{r/m}{1 - (1 + r/m)^{-mt}}$$

- Find the amount still owed after making amortization payments for part of the term. *(Practice Test Exercise 18.)*

Key Concepts · Objective 2.4

Present value of an annuity, 81
Amortization, 82–83
Unpaid balance, 83
Equity, 84

Instructional Hints

- Compound interest is repeated simple interest with the interest added to the principal in each successive period. The *future value* is the amount the principal will become after all the interest is included. Reversing the point of view, the *present value* is the principal needed now that will grow to the final amount. The three basic amount formulas are $P(1 + rt)$ for simple interest, $P(1 + r/m)^{mt}$ for compound interest, and $P((1 + r/m)^{mt} - 1)/(r/m)$ for an annuity (multiple payments). The other formulas all follow from these basic formulas.

- The rate is usually stated as a percent but is always used in decimal form for calculations: 6.7% in decimal form is 0.067.

- Round only your final answer when using your calculator. Don't use the decimal 0.33 for the fraction $\frac{1}{3}$.

- Don't confuse a *percent increase* with the *multiplier*. *Increasing* an amount by 25% means *multiplying* it by 1.25.

- Round *up* to find the whole number of compoundings needed to reach a future value because a smaller number won't reach the stated goal.

- The "rule of 72" is only an approximation, but it is a helpful check that can catch "button-pushing" errors when using your calculator.

Go to login.cengage.com to access additional resources.

Learning Objectives

3.1 Systems of Two Linear Equations in Two Variables
- Represent a pair of statements as a system of two linear equations in two variables.

$$\begin{cases} ax + by = h \\ cx + dy = k \end{cases}$$

- Solve a system of two linear equations in two variables by graphing. (*Practice Test Exercise 1.*)

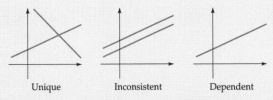

Unique	Inconsistent	Dependent

- Solve a system of two linear equations in two variables by the elimination method. (*Practice Test Exercises 2.*)
- Solve an applied problem by using a system of two equations in two variables and the elimination method.

Key Concepts · Objective 3.1

System of equations, 86
Unique solution, 90
Dependent linear equations, 90
Inconsistent linear
  equations, 90
Parameterized solution, 91
Equivalent systems of
  equations, 91
Elimination method, 92–93

3.2 Matrices and Linear Equations in Two Variables
- Find the dimension of a matrix and identify a particular element using double subscript notation. (*Practice Test Exercise 3.*)

$$A = \begin{pmatrix} a_{1,1} & \cdots & a_{1,n} \\ \vdots & & \\ a_{m,1} & \cdots & a_{m,n} \end{pmatrix}$$

- Carry out a given row operation on an augmented matrix.

$$\begin{cases} ax + by = h \\ cx + dy = k \end{cases} \longleftrightarrow \left(\begin{array}{cc|c} a & b & h \\ c & d & k \end{array} \right)$$

- Solve a system of two linear equations in two variables by row-reducing an augmented matrix. (*Practice Test Exercise 4.*)
- Solve an applied problem by using a system of two linear equations in two variables and row-reducing an augmented matrix. (*Practice Test Exercise 5.*)

Key Concepts · Objective 3.2

Matrix, 95
Element, 95
Row matrix, 95
Column matrix, 95
Square matrix, 95
Dimension, 95
Augmented matrix, 95–96
Coefficient matrix, 95
Constant term matrix, 95
Row operations, 96–97
Equivalent matrices, 96
Solving equations by row
  reduction, 98

3.3 Systems of Linear Equations and the Gauss–Jordan Method
- Interpret a row-reduced matrix as the solution of a system of equations. If the solution is not unique, identify the system as "inconsistent" or "dependent."

$$\begin{array}{cccc|c} 0 & 0 & \cdots & 0 & 1 \end{array} \quad \text{means inconsistent.}$$
$$\begin{array}{cccc|c} 0 & 0 & \cdots & 0 & 0 \end{array} \quad \text{means dependent.}$$

- Use the Gauss–Jordan method to find the equivalent row-reduced matrix.
- Solve a system of equations by the Gauss–Jordan method. If the solution is not unique, identify the system as "inconsistent" or "dependent." (*Practice Test Exercise 6.*)
- Formulate an application as a system of linear equations. Solve the system by the Gauss–Jordan method. State the final answer in terms of the original question. (*Practice Test Exercise 7.*)

Key Concepts · Objective 3.3

Row-reduced form, 104–105
Leftmost 1, 105
Gauss-Jordan method, 106
Determined variable, 108
Free variable, 108

3.4 Matrix Arithmetic
- Find a matrix product or determine that the product is not defined.

$$\begin{array}{ccccc} A & \cdot & B & = & C \\ m \times p & & p \times n & & m \times n \end{array}$$

- Find the value of a matrix expression involving scalar multiplication, matrix addition, subtraction, transposition, or multiplication. (*Practice Test Exercise 8.*)

$$A \cdot I = I \cdot A = A \qquad A + 0 = 0 + A = A$$

- Rewrite a system of linear equations as a matrix equation $AX = B$. (*Practice Test Exercise 9.*)
- Formulate an application in matrix form and indicate the meaning of each row and column. Find the requested quantity using matrix arithmetic. (*Practice Test Exercise 10.*)

Key Concepts · Objective 3.4

Equal matrices, 114
Transpose, 114
Identity matrix, 114, 119
Scalar multiplication, 115
Matrix addition, 115
Matrix subtraction, 115
Matrix multiplication, 117–118

3.5 Inverse Matrices and Systems of Linear Equations

- Find the product of a pair of matrices to identify the pair as "a matrix and its inverse" or "not a matrix and its inverse."

$$AA^{-1} = A^{-1}A = I$$

- Find the inverse of a matrix or identify it as singular. (*Practice Test Exercise 11.*)

$$(A \,|\, I) \rightarrow (I \,|\, A^{-1})$$

- Rewrite a system of equations as a matrix equation $AX = B$ and use the inverse of A to find the solution.

$$X = A^{-1}B$$

- Formulate an application in terms of systems of linear equations. Solve the collection of systems of equations by finding the inverse of a common coefficient matrix and using matrix multiplications of this inverse times the various constant term matrices. State the final answer in terms of the original question. (*Practice Test Exercise 12.*)

$$X = A^{-1}B_1, \dots, X = A^{-1}B_n$$

Key Concepts · Objective 3.5

Inverse matrix, 122–123
Invertible matrix, 122
Singular matrix, 122
Solving $AX = B$ using A^{-1}, 124–125

3.6 Introduction to Modeling: Leontief Models

- Find the technology matrix from an economy diagram. (*Practice Test Exercise 13.*)
- Draw an economy diagram from a technology matrix.
- Find the excess production of an economy with a given technology matrix and economic activity level.

$$Y = (I - A)X$$

- Find, for an economy with a given technology matrix, the economic activity level necessary to generate a specified excess production. (*Practice Test Exercise 14.*)

$$X = (I - A)^{-1}Y$$

- Represent an application as a Leontief "open" input–output model by constructing an economy diagram and the corresponding technology matrix. Find the required excess production or level of economic activity and state the final answer in terms of the original question. (*Practice Test Exercise 15.*)

Key Concepts · Objective 3.6

Leontief model, 129–130
Input-output diagram, 130
Technology matrix, 131–132
Excess production, 131–132
Total production, 131–132

Instructional Hints

- Row operations on matrices are a generalization of the elimination method of finding the intersection of two lines, extending the technique to problems with many equations in many variables. Matrix arithmetic is similar to real number operations except that matrix multiplication is not commutative and many matrices do not have inverses. However, if a square matrix A *does* have an inverse, then the equation $AX = B$ can be solved as $X = A^{-1}B$, just as a real number linear equation can be solved by dividing. Matrices are used to represent and solve large and complicated problems in a wide variety of applications.
- Although not every system of linear equations has a solution, they can all be identified as "having a unique solution," "inconsistent," or "dependent" by row-reducing the corresponding augmented matrix.
- When setting up a word problem, look first for the questions "how many" or "how much" to help identify the variables. Be sure that finding values for your variables will answer the question stated in the problem. Use the rest of the given information to build equations describing facts about your variables.
- Many row-reduction problems require many steps to solve, so don't give up; keep improving the matrix until it meets all the requirements to be row-reduced. Make sure that each row operation you choose will move you toward your goal without undoing the parts you already have gotten the way you need.

Go to login.cengage.com to access additional resources.

Learning Objectives

4.1 Linear Inequalities

- Graph a linear inequality $ax + by \leq c$ by drawing the boundary line $ax + by = c$ from its intercepts, using a test point to choose the correct side, and shading in the feasible region. (*Practice Test Exercise 1.*)

- Graph a system of linear inequalities $\begin{cases} ax + by \leq c \\ \vdots \end{cases}$ by drawing the boundary lines, finding the correct sides of the boundaries, and shading in the feasible region. List the vertices or "corners" of the region and identify the region as bounded or unbounded. (*Practice Test Exercise 2.*)

- Formulate an applied situation as a system of linear inequalities and sketch the feasible region with the vertices. (*Practice Test Exercise 3.*)

Key Concepts · Objective 4.1

Inequality, 136
Boundary of inequality, 136
Feasible region, 136
Vertex, 139
Bounded region, 140
Unbounded region, 140
Nonnegativity constraints, 141

4.2 Two-Variable Linear Programming Problems

- Find the maximum or minimum of a linear function on a given region or explain why such a value does not exist. (*Practice Test Exercises 4 and 5.*)

- Solve a two-variable linear programming problem by sketching the feasible region, determining whether a solution exists, and then finding it by evaluating the objective function at the vertices. (*Practice Test Exercises 6 and 7.*)

- Solve an applied linear programming problem by identifying the variables, the objective function, and the constraints, and then using the vertices of the feasible region. (*Practice Test Exercise 8.*)

Key Concepts · Objective 4.2

Linear programming problem, 146
Objective function, 146
Constraint, 146
Fundamental theorem of linear programming, 146–147
Existence of solutions, 147, 150

4.3 The Simplex Method for Standard Maximum Problems

- Check that a maximum linear programming problem is a standard problem and construct the initial simplex tableau.

$$\text{Maximize} \quad P = c^t X$$
$$\text{Subject to} \quad \begin{cases} AX \leq b \quad (b \geq 0) \\ X \geq 0 \end{cases}$$

	X	S	
S	A	I	b
P	$-c^t$	0	0

- Find the pivot element in a tableau and carry out one complete pivot operation. If there is no pivot element, explain what this means about the original problem.

- Solve a standard maximum problem by the simplex method (construct the initial simplex tableau, pivot until the tableau does not have a pivot element, and interpret this final tableau). (*Practice Test Exercise 9.*)

- Solve an applied linear programming problem by the simplex method. (*Practice Test Exercise 10.*)

Key Concepts · Objective 4.3

Standard maximum problem, 155–156
Slack variable, 156
Initial tableau, 156–157
Basic variable, 158–159
Nonbasic variable, 158–159
Pivot column, 159
Pivot row, 160
Pivot element, 160
Pivot operation, 160–161
Final tableau, 163
Simplex method, 164

4.4 Standard Minimum Problems and Duality

- Check that a minimum linear programming problem is a standard problem and construct the dual maximum problem and the initial simplex tableau.

$$\text{Minimize} \quad C = b^t Y \qquad\qquad \text{Maximize} \quad P = c^t X$$
$$\text{Subject to} \quad \begin{cases} A^t Y \geq c \\ Y \geq 0 \end{cases} \qquad \text{Subject to} \quad \begin{cases} AX \leq b \\ X \geq 0 \end{cases}$$
$$\text{where} \quad b \geq 0 \qquad\qquad\qquad \text{where} \quad b \geq 0$$

Key Concepts · Objective 4.4

Standard minimum problem, 168
Dual problems, 169, 173
Duality theorem, 170

- Solve a standard minimum problem by finding the dual maximum problem and using the simplex method. *(Practice Test Exercise 11.)*
- Solve an applied minimum problem by finding the dual maximum problem and using the simplex method. *(Practice Test Exercise 12.)*

Instructional Hints

- A linear programming problem asks for the maximum or minimum of a linear objective function subject to constraints in the form of linear inequalities. If there is a solution to the problem, it occurs at a *vertex* of the feasible region determined by the constraints. A problem with two variables may be solved graphically, but one with more than two variables must be solved algebraically using the simplex method, using duality if it is a standard minimum problem.

- While a problem with a *bounded* region always has a solution, a problem with an *unbounded* region may or may not have a solution (if it *does,* then the solution occurs at a vertex).

- When setting up a word problem, look first for the question "how many" or "how much" to help identify the variables. Be sure that finding values for your variables will answer the question stated in the problem. Use the rest of the given information to find the objective function and constraints. Include nonnegativity conditions for your variables as appropriate (for example, a farmer *cannot* raise a negative number of cows).

- A simplex tableau is *feasible* if it has no negative numbers in the rightmost *column* and is *optimal* if it has no negative numbers in the bottom *row.* If a simplex tableau is both feasible and optimal, it displays the solution of the problem.

- The pivot operation is not completed until *all* the rows of the tableau have been recalculated.

- If you are solving a standard maximum problem by the simplex method and you have a negative number in the rightmost column after pivoting, you've made an error in your choice of the pivot element or in your arithmetic.

- Many simplex tableaux require several pivots to solve, so don't give up: Keep pivoting until the tableau does not have a pivot element.

- In a *maximum* problem, the simplex method begins at the origin, which is feasible, and pivots to successively better vertices, making P larger while keeping the right column nonnegative. That is, the simplex method *keeps the tableau feasible while making it optimal.*

- In a *minimum* problem, the pivot operation begins at the origin, which is optimal (minimizing C to 0), but not feasible (not in the shaded region) and pivots to better vertices. That is, the simplex method *keeps the tableau optimal while making it feasible.*

- For a proof of the Duality Theorem (as stated on page 170), see "Derivations, Justifications, and Proofs" on this book's website, accessible at login.cengagebrain.com.

Learning Objectives

5.1 Sets, Counting, and Venn Diagrams

- Read and interpret a Venn diagram. (*Practice Test Exercise 1.*)
- Use the complementary or the addition principles of counting or Venn diagrams to solve an applied problem. (*Practice Test Exercise 2.*)

$$n(A^c) = n(U) - n(A)$$

$$n(A \cup B) = n(A) + n(B) - n(A \cap B)$$

- Use the multiplication principle of counting to solve an applied problem.

Key Concepts · Objective 5.1

Set, 178
Universal set, 178
Intersection, 179
Disjoint sets, 179
Union, 180
Complement, 180
Empty set, 180
Subset, 180
Complementary principle, 180
Addition principle, 180–181
Tree diagram, 182
Multiplication principle, 182–183
Number of subsets, 184

5.2 Permutations and Combinations

- Calculate numbers of permutations and combinations.

$$_nP_r = n \cdot (n - 1) \cdot \cdots \cdot (n - r + 1)$$

$$_nC_r = \frac{n(n - 1) \cdot \cdots \cdot (n - r + 1)}{r \cdot (r - 1) \cdot \cdots \cdot 1} = \frac{n!}{r!(n - r)!}$$

- Use the permutation and combination formulas (possibly using a calculator) to solve an applied problem. (*Practice Test Exercise 3.*)

Key Concepts · Objective 5.2

Factorials, 186
Permutations, 187–188
Combinations, 189–190

5.3 Probability Spaces

- Find an appropriate sample space for a random experiment. (*Practice Test Exercise 4.*)
- Describe an event in terms of the sample space.
- Assign probabilities to outcomes in a sample space and find probabilities of events.
- Use the techniques of this section to find a probability in an applied problem. (*Practice Test Exercises 5 through 9.*)

Key Concepts · Objective 5.3

Sample space, 192
Events, 193
Equally-likely outcomes, 194, 196
Probability of event, 195
Complementary probability, 196
Probability space, 197
Addition rule, 197
Disjoint events, 198–199
Mutually exclusive events, 198–199

5.4 Conditional Probability and Independence

- Find a conditional probability using either the definition or a restricted sample space. (*Practice Test Exercise 10.*)

$$P(A \text{ given } B) = \frac{P(A \cap B)}{P(B)}$$

- Use conditional probability to solve an applied problem.
- Use a tree diagram to solve an applied problem.
- Determine whether two events are independent or dependent. (*Practice Test Exercise 11.*)

$$P(A \cap B) = P(A) \cdot P(B)$$

- Use independence to find a probability in an applied problem. (*Practice Test Exercise 12.*)

Key Concepts · Objective 5.4

Conditional probability, 201
Product rule, 203
Independent events, 206, 207
Dependent events, 206

5.5 Bayes' Formula

- Use Bayes' formula to solve an applied problem. (*Practice Test Exercise 13.*)

Key Concepts · Objective 5.5

Bayes' formula, 209–210

5.6 Random Variables and Distributions

- Determine the outcomes that correspond to the values of a random variable.
- Find and graph the probability distribution of a random variable, and find the mean. *(Practice Test Exercises 14 through 16.)*
- For a binomial random variable, find and graph the probability distribution, and find its mean.

$$P(X = k) = {}_nC_k p^k (1 - p)^{n-k}$$

$$E(X) = n \cdot p$$

- Use the binomial probability distribution to solve an applied problem. *(Practice Test Exercises 17 and 18.)*

Key Concepts · Objective 5.6

Random variable, 213–214
Probability distribution, 214
Expected value, 215
Mean, 215
Binomial distribution, 217–218
Binomial random variable, 217–218

Instructional Hints

- Much of probability depends on counting, and there are several principles to simplify counting large numbers of objects. Roughly speaking, the *complementary* principle says that you can count the *opposite* set and then subtract, the *addition* principle says that you can add numbers from two sets but you must then subtract what was double-counted, and the *multiplication* principle says that the number of two-part choices is the product of the number of first-part choices times the number of second-part choices.

- *Permutations* are arrangements where a different order means a different object (such as letters in a word or rankings of people). *Combinations* are arrangements where reordering does *not* make a new object (such as hands of cards or committees of people).

- A set of *n* objects has 2^n subsets (counting the "empty" subset and the set itself).

- Probabilities are assigned to possible outcomes of a random experiment so that each probability is between 0 and 1 (inclusive) and they add to 1. Equally likely outcomes should be assigned equal probabilities. Probabilities of more complicated events are found by adding up the probabilities of the outcomes in the event.

- Conditional probability is the relative probability of the intersection of the events compared to the probability of the *given* event. Conditional probabilities are sometimes more easily found from the restricted sample space than from the definition.

- A complicated probability can sometimes be found by a tree diagram, branching on some event on which others depend.

- Independent events "have nothing to do with each other," and the probability of both occurring is the *product* of their probabilities. This is not the same as *disjoint* events, where one *precludes* the other and which therefore *cannot* be independent.

- Bayes' formula is useful for finding events of the form S_1 *given A* in terms of events of the form *A given* S_i where $S_1, \ldots, S_n$ are disjoint events whose union is the entire sample space.

- The mean of a random variable gives a *representative* or *typical* value.

- Bernoulli trials are repeated independent experiments with only two outcomes, *success* and *failure*. The number of successes in several Bernoulli trials is a *binomial random variable*.

- For a proof of the formula for the mean of the binomial distribution (as stated on page 218), see "Derivations, Justifications, and Proofs" on this book's website, accessible at login.cengagebrain.com.

Go to login.cengage.com to access additional resources.

Learning Objectives

6.1 Random Samples and Data Organization
- Construct a bar chart or a histogram for a given data set. *(Practice Test Exercises 1 and 2.)*

Key Concepts • Objective 6.1

Statistical population, 222
Random sample, 222
Bar chart, 223
Histogram, 224
Class width, 224

6.2 Measures of Central Tendency
- Find the mode, the median, and the mean of a data set. *(Practice Test Exercise 3.)*

  Mode: most frequent

  Median: middle

  Mean: $\bar{x} = \dfrac{x_1 + \cdots + x_n}{n}$

Key Concepts • Objective 6.2

Mode, 227
Median, 228
Mean, 228–229
Comparing mode, median, and mean, 229–230

6.3 Measures of Variation
- Find the range of a data set.
- Find the five-point summary of a data set and draw the box-and-whisker plot.
- Find the sample standard deviation of a data set.

$$s = \sqrt{\dfrac{(x_1 - \bar{x})^2 + \cdots + (x_n - \bar{x})^2}{n - 1}}$$

- Analyze the data in an application by finding the range, the five-point summary, and the sample standard deviation and then drawing the box-and-whisker plot. *(Practice Test Exercise 4.)*

Key Concepts • Objective 6.3

Range, 233
Quartile, 233
Box-and-whisker plot, 233–234
Sample standard deviation, 236

6.4 Normal Distributions and Binomial Approximation
- Find the probability that the normal random variable with mean μ and standard deviation σ is between two given values. *(Practice Test Exercise 5.)*
- Find the z-score of an x-value of the normal random variable with mean μ and standard deviation σ. *(Practice Test Exercise 6.)*

$$z = \dfrac{x - \mu}{\sigma}$$

- Use the normal approximation to the binomial random variable X to find the probability that X takes given values. *(Practice Test Exercise 7.)*
- Solve an applied problem involving probabilities using the normal distribution. *(Practice Test Exercise 8.)*

Key Concepts • Objective 6.4

Continuous random variable, 238
Normal distribution, 238–239
Standard normal distribution, 240–241
z-score, 240–241
Normal approximation to the binomial, 243–244
Continuity correction, 244
Normal probability table, 248

- Properties of "many" can be inferred from just "some" only if the "some" are a random sample from the population. Data may be organized into bar charts or histograms. Measures of central tendency (mode, median, and mean) summarize all the data as one typical value, while measures of variation (range, box-and-whisker plot, and sample standard deviation) show how closely the values cluster about the center. The bell-shaped curve of the normal distribution applies to many situations, and z-scores give the number of standard deviations the values are from the mean. The binomial distribution is approximated by the normal distribution, provided both np and $n(1-p)$ are sufficiently large.

- A bar chart may have its bars separated or touching. Histograms have bars that touch each other on both sides.

- To find the mode or the median of data values, it is helpful to begin by sorting them in order.

- The mean is sensitive to changes in a few extreme values, while the mode and median are not.

- Box-and-whisker plots graphically show how the data are distributed among the four quartiles.

- Areas under the normal distribution curve are probabilities for the normal random variable.

- When using the normal distribution to approximate the binomial distribution, be sure to increase the largest value by $\frac{1}{2}$ and to decrease the smallest by $\frac{1}{2}$ (the continuity correction) to cover all the area represented by the boxes making up the binomial distribution.

- Although not used in this book, there is an "alternative formula" for the sample standard deviation (page 236) that is developed in the "Derivations, Justifications, and Proofs" on this book's website, accessible at login.cengagebrain.com.

Go to login.cengage.com to access additional resources.

Learning Objectives

7.1 States and Transitions

- Construct a transition matrix from a state-transition diagram and identify the transition as "oscillating," "mixing," or "absorbing." (*Practice Test Exercise 1.*)
- Construct a state-transition diagram from a transition matrix and identify the transition as "oscillating," "mixing," or "absorbing." (*Practice Test Exercise 2.*)
- Calculate the kth state distribution vector for a Markov chain with transition matrix T and initial distribution D_0. (*Practice Test Exercise 3.*)

$$D_k = D_0 \cdot T^k$$

- Find the expected number of times a Markov chain will be in a given state before moving to some other state. (*Practice Test Exercise 4.*)

$$E = \frac{1}{1-p}$$

- Use a Markov chain to solve an applied problem. (*Practice Test Exercise 5.*)

Key Concept • Objective 7.1

State, 250
Transition, 250
State-transition diagram, 250
Transition matrix, 250, 252
Markov chain, 252
Oscillating transition, 253
Mixing transition, 253
Absorbing transition, 253
Absorbing state, 254
State-distribution vector,
  254–255
kth state distribution vector,
  255–256
Expected duration, 257

7.2 Regular Markov Chains

- Identify a transition matrix as "regular" or "not regular." (*Practice Test Exercise 6.*)
- Find the steady-state distribution of a regular Markov chain by row-reduction and by calculating powers of the transition matrix (using a calculator). (*Practice Test Exercise 7.*)

$$D \cdot T = D$$

$$\left(\begin{array}{c|c} T^t - I & 0 \\ \hline 1 \cdots 1 & 1 \end{array}\right) \longrightarrow \left(\begin{array}{c|c} I & D^t \\ \hline 0 & 0 \end{array}\right)$$

- Find the smallest positive number k of transitions needed to move from state S_i to state S_j in a regular Markov chain and then verify this value (using a calculator) by checking that T^k has a positive transition probability in row i and column j.
- Use a regular Markov chain to solve an applied problem. (*Practice Test Exercise 8.*)

Key Concept • Objective 7.2

Regular Markov chain, 259
Steady-state distribution,
  262, 263
Fundamental theorem, 262

7.3 Absorbing Markov Chains

- Identify the absorbing and nonabsorbing states in a state-transition diagram and find the smallest number of transitions needed to move from a given nonabsorbing state to an absorbing state. (*Practice Test Exercise 9.*)
- Identify absorbing and nonabsorbing states from a transition matrix and find the smallest number of transitions needed to ensure that every nonabsorbing state can reach an absorbing state.
- Rewrite an absorbing Markov chain transition matrix in standard form. (*Practice Test Exercise 10.*)

$$T = \left(\begin{array}{c|c} I & 0 \\ \hline R & Q \end{array}\right)$$

- Find the fundamental matrix of an absorbing Markov chain. (*Practice Test Exercise 11.*)

$$F = (I - Q)^{-1}$$

- Use the fundamental matrix to find the expected number of times in nonabsorbing states before absorption and the probability of absorption into a particular absorbing state. (*Practice Test Exercise 12.*)
- Use an absorbing Markov chain to solve an applied problem. (*Practice Test Exercise 13.*)

Key Concept • Objective 7.3

Absorbing Markov chain, 266
Standard form of absorbing
  transition matrix, 268
Fundamental matrix, 270
Expected transition times, 270
Long-term absorption
  probabilities, 270

- We discussed three general types of Markov chains: *oscillating*, *mixing*, and *absorbing*. Oscillating chains always move back and forth. Mixing (that is, regular) chains settle down to a steady-state that is independent of how they began. Absorbing chains end in absorption, with time until absorption and the final absorbing state depending on where they began.

- An entry in a particular row and column of a transition matrix gives the probability of moving *from* the *row* state *to* the *column* state.

- For a regular Markov chain, the steady-state distribution D can be found in two ways: by raising the transition matrix to a high power so the rows become D, or by row-reducing $\left(\begin{array}{c|c} T^t - I & 0 \\ \hline 1 \cdots 1 & 1 \end{array}\right)$ to obtain $\left(\begin{array}{c|c} I & D^t \\ \hline 0 & 0 \end{array}\right)$, where the superscript t means *transpose* (not a power).

- An *absorbing* Markov chain is not just a chain with an absorbing state. To be an absorbing Markov chain, it must be possible to move from each nonabsorbing state to some absorbing state (possibly in several moves).

- When writing an absorbing transition matrix in standard form, be sure to keep the state names with the correct rows and columns as you switch the absorbing rows to the top positions and then as you put the columns in the same order.

- For an absorbing Markov chain we could find the long-term behavior by raising the transition matrix to a high power, but we obtain much more information from the results in the box on page 270. These results depend on the *fundamental matrix* $F = (I - Q)^{-1}$. F gives expected times, and $F \cdot R$ gives probabilities, in each case from the row state to the column state.

- When doing an applied problem, always draw the state-transition diagram first (verifying that the probabilities of *leaving* each state add to 1) and then write the transition matrix (being sure that each *row* adds to 1).

- Watch out for careless errors like changing $\frac{1}{2}$ to 0.05 or 5% to 0.50.

- For a proof of the formula for the expected duration in a given state of a Markov chain (as stated on page 257), see "Derivations, Justifications, and Proofs" on this book's website, accessible at login.cengagebrain.com.

- For proofs of the formulas for the long-term behavior of an absorbing Markov chain (as stated on page 270), see "Derivations, Justifications, and Proofs" on this book's website, accessible at login.cengagebrain.com.

Learning Objectives

8.1 Two-Person Games and Saddle Points

- Construct a payoff matrix for a game by identifying the players, their possible choices, and the resulting gains and losses. (*Practice Test Exercise 1*)

- Find the saddle point and optimal strategy of a game. (*Practice Test Exercise 2*)

  ○ smallest in row   □ largest in column

- Determine an optimal strategy for a conflict situation by representing it as a game and then solving the game. (*Practice Test Exercise 3*)

Key Concepts • Objective 8.1

*Two-person, zero-sum
    game, 276
Payoff matrix, 276
Row player, 276
Column player, 276
Saddle point, 279
Optimal strategy, 279
Value, 279
Strictly determined game, 279*

8.2 Mixed Strategies

- Find the expected value of a mixed strategy for a game.

$$E = r \cdot G \cdot c$$

- Find an optimal strategy for a 2×2 game by identifying the saddle points or using the formulas for the optimal mixed strategy. (*Practice Test Exercise 4*)

$$G = \begin{pmatrix} a & b \\ c & d \end{pmatrix}$$

$$r_1 = \frac{d - c}{(a + d) - (b + c)}, \quad r_2 = 1 - r_1$$

$$c_1 = \frac{d - b}{(a + d) - (b + c)}, \quad c_2 = 1 - c_2$$

- Determine an optimal strategy for a conflict situation by representing it as a game and then solving the game. (*Practice Test Exercise 5*)

Key Concepts • Objective 8.2

*Mixed strategy, 282–283
Expected value, 282–283
Optimal mixed strategy, 286
Fictitious opponent, 287*

8.3 Games and Linear Programming

- Find an optimal strategy for a game using the simplex method. (*Practice Test Exercise 6*)

	X	S	
S	$G + k$	I	1
	-1	0	0

$\implies$

	X	S	
	$\vdots$	$\vdots$	$\vdots$
	$\cdots$	Y^*	V

$\left.\begin{matrix} \\ \\ \end{matrix}\right\}$

$c_i = \dfrac{x_i}{V}$

$r_i = \dfrac{y_i}{V}$

Key Concepts • Objective 8.3

*Simplex method solution,
    292–293*

- Determine an optimal strategy for a conflict situation by representing it as a game and then using the simplex method. (*Practice Test Exercise 7*)

Instructional Hints

- In a *two-person, zero-sum* game, whatever one player gains the other loses. In the payoff matrix the entries represent gains for the row player (and so losses for the column player). Therefore, negative numbers represent losses for the row player (and gains for the column player).

- Always first check a game for saddle points. A saddle point is a row minimum and a column maximum, and represents a *pure optimal solution* for a game. For a game with no saddle points, use the formulas on page 286 for 2×2 games and use the linear programming method on pages 292–293 for larger games.

- For a game you do not need to have an actual opponent actively trying to minimize your gain. You may have a game against *uncertainty* (such as weather or economic conditions) as long as you can estimate probabilities for the various random occurrences.

- For a proof that if a 2×2 game $G = \begin{pmatrix} a & b \\ c & d \end{pmatrix}$ does not have a saddle point, then $(a + d) - (b + c) \neq 0$ (as stated on page 285), see "Derivations, Justifications, and Proofs" on this book's website, accessible at login.cengagebrain.com.

- For a proof that $P \leq C$ for dual linear programming problems (as stated on page 290), see "Derivations, Justifications, and Proofs" on this book's website, accessible at login.cengagebrain.com.

Go to login.cengage.com to access additional resources.

Graphing Calculator Terminology

The Graphing Calculator Explorations on your tech cards have been kept as generic as possible for use with any of the popular graphing calculators. We assume that you or your instructor is familiar with the sequence of button pushes necessary to accomplish various basic operations on your calculator. Certain standard calculator terms are capitalized on these cards and are described below. Your calculator may use slightly different terminology.

The viewing window or graphing **WINDOW** is the part of the Cartesian plane shown in the display screen of your graphing calculator. **XMIN** and **XMAX** are the smallest and largest x-values shown, and **YMIN** and **YMAX** are the smallest and largest y-values shown. These can be set by using the **WINDOW** or **RANGE** command and are changed automatically by using any of the **ZOOM** operations. **XSCALE** and **YSCALE** define the distance between tick marks on the x- and y-axes.

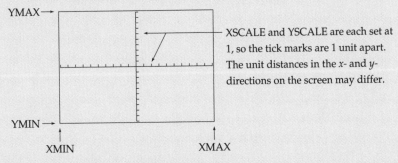

XSCALE and YSCALE are each set at 1, so the tick marks are 1 unit apart. The unit distances in the x- and y-directions on the screen may differ.

Viewing Window [−10, 10] by [−10, 10]

The viewing window is always [XMIN, XMAX] by [YMIN, YMAX]. We will set XSCALE and YSCALE so that there are a reasonable number of tick marks (generally 2 to 20) on each axis. Either or both of the x- and y-axes may not be visible when the viewing window does not include the origin.

Pixel, an abbreviation for *pic*ture *el*ement, refers to a tiny rectangle on the screen that can be darkened to represent a dot on a graph. Pixels are arranged in a rectangular array on the screen. In the above window, the axes and tick marks are formed by darkened pixels. The size of the screen and the number of pixels vary with different calculators.

TRACE allows you to move a flashing pixel (or *cursor*) along a curve in the viewing window, with the x- and y-coordinates shown at the bottom of the screen.

Useful Hint: To make the x-values in TRACE take simple values like 0.1, 0.2, and 0.3, choose XMIN and XMAX to be multiples of one less than the number of pixels across the screen. For example, on the *TI-83* or *TI-84*, which have 95 pixels across the screen, using an x-window like [−9.4, 9.4] or [−4.7, 4.7] or [−940, 940] will TRACE with simpler x-values than the standard windows used on these cards.

ZOOM IN allows you to magnify any part of the viewing window to see finer detail around a chosen point. **ZOOM OUT** does the opposite, like stepping back to see a larger portion of a picture but with less detail. These and any of several other **ZOOM** commands change the viewing window.

VALUE or **EVALUATE** finds the value of a previously entered expression at a specified x-value.

SOLVE or **ROOT** or **ZERO** finds the x-value that solves $f(x) = 0$, or equivalently, the x-intercepts of a curve $y = f(x)$. When applied to a difference $f(x) - g(x)$, it finds the x-value where the two curves meet (as also done by the **INTERSECT** command).

MAX and **MIN** find the highest and lowest points of a (previously entered) curve between specified x-values.

In **CONNECTED MODE** your calculator will darken pixels to connect calculated points on a graph in an attempt to show it as a continuous or "unbroken" curve. However, this may lead to "false lines" in a graph that should have breaks or "jumps." Such false lines can be eliminated by using **DOT MODE**.

The **TABLE** command on some calculators lists in table form the values of a function, just as you have probably done when graphing a curve. The x-values may be chosen by you or by the calculator.

The **Order of Operations** used by most calculators evaluates operations in the following order: first powers and roots, then operations like **LN** and **LOG**, then multiplications and divisions, and finally additions and subtractions, always working from left to right within each level. For example, $5{\wedge}2x$ means $(5{\wedge}2)x$ *not* $5{\wedge}(2x)$. Also, $1/x + 1$ means $(1/x) + 1$, *not* $1/(x + 1)$. See your calculator's instruction manual for further information. *Be careful*: Some calculators evaluate $1/2x$ as $(1/2)x$ and some as $1/(2x)$. When in doubt, use parentheses to clarify the expression.

Much more information can be found in the manual for your graphing calculator.

Graphing Calculator Explorations

SECTION 1.3

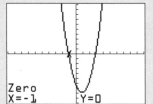

Zero
X=-1   Y=0

Solving a Quadratic Equation

Find the solutions to the equation in Example 6 on page 28 by graphing the function $f(x) = 2x^2 - 4x - 6$ and using ZERO or TRACE to find where the curve crosses the x-axis. Your answers should agree with those found in Example 6.

SECTION 1.4

Shifted Absolute Value Function

The absolute value function $y = |x|$ may be graphed on some graphing calculator as $y_1 = \text{ABS}(x)$.

a. Graph $y_1 = \text{ABS}(x - 2) - 6$ and observe that the absolute value function is shifted *right* 2 units and *down* 6 units. (The graph shown is drawn using ZOOM ZSquare.)

b. Predict the shift of $y_1 = \text{ABS}(x + 4) + 2$ and then verify your prediction by graphing the function on your calculator.

SECTION 1.5

Graphing Exponential Functions

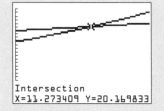

Intersection
X=11.273409 Y=20.169833

The most populous states are California and Texas, with New York third and Florida fourth but gaining. According to data from the Census Bureau, x years after 2000 the population of New York will be $19e^{0.0053x}$ and the population of Florida will be $15.9e^{0.0211x}$ (both in millions).

a. Graph these two functions on the window [0, 20] by [0, 25]. [Use the 2nd and In keys for entering e to powers.]

b. Use INTERSECT to find the x-value where the curves intersect.

c. From your answer to part (b), in which year is Florida projected to overtake New York as the third largest state? [*Hint: x* is years after 2000.]

SECTION 1.6

log(100)
                    2
log(1/10)
                   -1
log(10000)
                    4

Common Logarithms

On a graphing calculator, common logarithms are found using the LOG key. Verify that the logarithms we considered in Examples 1 and 2 and Practice Problem 1 on pages 50–51 can be found in this way.

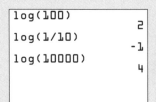

ln(e^(17))
                   17
e^(ln(29))
                   29

Natural Logarithms

a. Evaluate $\ln e^{17}$ to verify that the answer is 17. Change 17 to other numbers (positive, negative, or zero) to verify that $\ln e^x = x$ for any x.

b. Evaluate $e^{\ln 29}$ to verify that the answer is 29. Change the 29 to another positive number to verify that $e^{\ln x} = x$. What about negative numbers, or zero?

Graphing Calculator Explorations

SECTION 2.2

Compound Interest

To see that compound interest eventually surpasses simple interest (even with a higher rate and principal), compare $500 invested at 4% compounded annually with $2500 invested at 8% simple interest.

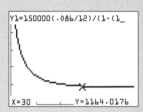

a. Graph the simple interest amount $A = P(1 + rt)$ as $y_1 = 2500(1 + .08x)$ on the window $[0, 150]$ by $[0, 35{,}000]$ so that y_1 is the amount of the investment after x years.

b. Graph the compound interest amount $A = P(1 + rt)^t$ as $y_2 = 500 (1 + .04)^x$ on the same window.

c. Use TRACE or INTERSECT to find when the compound interest amount equals the simple interest amount. What happens after this intersection point? Try your graphs with the larger windows $[0, 300]$ by $[0, 70{,}000]$ and by $[0, 500{,}000]$.

SECTION 2.4

Amortizing a Loan

When amortizing a loan, a longer term means smaller payments, but the total amount the borrower pays is much larger.

Y1=150000(.086/12)/(1-(1_

X=30 ⎿_____⏋ Y=1164.0176

a. To find the monthly payment y to amortize a debt over x years using the values $D = 150{,}000$, $r = .086$, and $m = 12$ similar to those in Practice Problem 2 on page 83, graph the curve $y_1 = 150000(.086/12)/(1 - (1 + .086/12)^{-12x})$ on the window $[0, 50]$ by $[0, 5000]$.

b. Notice that as the term increases the payment decreases, at first rapidly but later more slowly. Use TRACE or VALUE to find the payments for mortgage terms of 15, 20, 25, and 30 years. How do your answers compare to the answers to Example 2 on page 83 and Practice Problem 2, where the interest rate is lower?

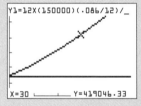

Y1=12X(150000)(.086/12)/_

X=30 ⎿_____⏋ Y=419046.33

c. To find the total amount the borrower pays, multiply the payment by m and t (here replaced by x). Graph the new curve $y_1 = 12x (150000(.086/12)/(1 - (1 + .086/12)^{-12x})$, along with the line $y_2 = 150000$ to represent the original debt, on the window $[0, 50]$ by $[0, 600{,}000]$ to find the total amount paid on a loan for x years.

d. Notice that as the term increases the total amount paid increases, well above the original debt of $150,000. Use TRACE or VALUE to find the total amount paid for mortgage terms of 15, 20, 25, and 30 years. Based on these two screens, does it make financial sense to extend the term of this loan beyond 30 years?

Constructing an Amortization Table

The program AMORTABL* constructs an *amortization table* that shows how each payment is allocated to paying the interest and reducing the outstanding debt. To explore the table for the situation used in *Amortizing a Loan* just above, run this program by selecting it from the program menu and proceed as follows:

```
DEBT (DOLLARS)
  150000
RATE (DECIMAL)
  .086
NUMBER OF YEARS
  30
COMPOUNDINGS/YR
  12
```

a. Enter 150000 for the debt, .086 for the interest rate, 30 for the number of years, and 12 for the number of compoundings per year (because the payments are monthly).

```
    VIEW TABLE
1:FROM START
2:JUMP TO YEAR
3:AT END
4:FINISHED
```

b. After the table of values has been calculated by finding the payment rounded to the nearest penny and then applying each payment to the rounded interest due on the debt for that period and reducing the debt by the excess, you may choose which part of the table you wish to see.

* Go to CourseMate for FINITE for graphing calculator programs, Excel spreadsheets, and technology guides. Access at login.cengagebrain.com.

X	Y₁	Y₂
0	**0**	**0**
1	1164	1075
2	1164	1074.4
3	1164	1073.7
4	1164	1073.1
5	1164	1072.4
6	1164	1071.8
X=0		

X	Y₃	Y₄
0	0	**150000**
1	89.02	149911
2	89.66	149821
3	90.3	149731
4	90.95	149640
5	91.6	149548
6	92.26	149456
Y₄=150000		

c. The amortization table contains five columns: X is the number of the payment (negative Xs or Xs larger than the number of payments display zeros in the other columns), Y_1 is the payment, Y_2 is the interest part of the payment, Y_3 is the remaining part used to reduce the debt, and Y_4 is the debt remaining after this reduction. Use the arrow keys to scroll right and left through these columns.

X	Y₁	Y₂
356	1164	40.8
357	1164	32.75
358	1164	24.64
359	1164	16.48
360	1160	8.25
361	**0**	**0**
362	419043	269043
X=361		

X	Y₃	Y₄
356	1123.2	4570
357	1131.3	3438.7
358	1139.4	2299.3
359	1147.5	1151.8
360	1151.8	0
361	0	**0**
362	150000	0
Y₄=0		

The second line after the end of the amortization (line 362) contains the total payments, interest, and debt reduction. Notice that the final payment is adjusted to correct for rounding errors and the debt is reduced to zero.

X	Y₃	Y₄
118	205.28	133573
119	206.75	133366
120	208.23	**133158**
121	209.72	132948
122	211.23	132737
123	212.74	132524
124	214.27	132310
Y₄=133157.56		

d. To use the table to find the remaining debt after a given number of years, "jump" to that position in the table and scroll to the Y_4 column. After 10 years of payments, the table shows $133,157.56 remaining.

Spreadsheet Exploration
SECTION 2.4
Amortization Table

The following spreadsheet* shows an "amortization table" for a debt of $100,000 to be paid off in five annual payments at 10% compounded yearly. Notice that the last payment is adjusted to *exactly* pay off the remaining debt and interest.

C2		=	=PMT(A5/(A11),A8*A11,A2)				
	A	B	C	D	E	F	G
1	Debt		Payment				
2	100000		(26,379.75)				
3							
4	Annual Rate						
5	10%		Year	Payment	Interest	Debt Reduction	Outstanding Debt
6			0.00	0.00	0.00	0.00	100,000.00
7	Term in Years		1.00	(26,379.75)	10,000.00	(16,379.75)	83,620.25
8	5		2.00	(26,379.75)	8,362.03	(18,017.72)	65,602.53
9			3.00	(26,379.75)	6,560.25	(19,819.50)	45,783.03
10	Compoundings per Year		4.00	(26,379.75)	4,578.30	(21,801.45)	23,981.58
11	1		5.00	(26,379.74)	2,398.16	(23,981.58)	0.00
12							
13			Totals:	(131,898.74)	31,898.74	(100,000.00)	

The payment amount in cell **C2** was found using the spreadsheet **PMT** function. The table shows how each payment is allocated to pay the interest and some of the debt, with the early payments paying more for interest than the later payments, which go almost entirely for debt.

---

* Go to CourseMate for FINITE for graphing calculator programs, Excel spreadsheets, and technology guides. Access at login.cengagebrain.com.

Graphing Calculator Explorations

SECTION 3.2

Row-Reducing Matrices

Some graphing calculators can row-reduce matrices. If your calculator has a RREF command (for "reduced row-echelon form"), you can easily check your row reduction. For the system of equations in Example 3 on page 98:

a. Enter $\begin{pmatrix} 1 & 3 & 15 \\ 2 & -5 & 8 \end{pmatrix}$ as matrix [A] using MATRX EDIT.

b. QUIT and select the RREF command from the MATRX MATH menu.

c. Apply RREF to the matrix [A].

Although this method serves as a useful check of your answer, do not rely on it completely because the calculator sometimes returns an answer with rounding errors. Furthermore, this calculator command may not work for matrices with more rows than columns. We will be interested in such matrices in the next section.

SECTION 3.3

Row Operations

The program ROWOPS* carries out the arithmetic for the type of row operation you select from a menu. To have your calculator perform the row operations used in Example 1 on pages 106–107, proceed as follows:

a. Enter the augmented matrix as matrix [A]. Because this matrix is too large to fit on the calculator screen, it will scroll from left to right and back again as you enter your numbers.

b. Run the program ROWOPS. It will display the current values in the matrix [A], and you can use the arrows to scroll the screen to see the rest of it. Press [ENTER] to select the type of row operation you want to use.

---

* Go to CourseMate for FINITE for graphing calculator programs, Excel spreadsheets, and technology guides. Access at login.cengagebrain.com.

```
      [1 2 4 7]]
       DIVIDE
  DIVIDE ROW 1
    BY 2
       [[1 2 1 1]
        [3 7 3 0]
        [1 2 4 7]]
```

c. Choose the type of row operation you want by using the arrows to move to its number and pressing ENTER, or just press the number. Enter the specific details for the operation and the program will carry out your request.

d. Press ENTER to select another row operation or to choose 7 and QUIT the program.

The program ROWOPS allows you to multiply rows by fractions and displays them in the usual 3/5 notation. The reduction of this matrix took six row operations to complete in Example 1. Can you find other ways to achieve the same result?

SECTION 3.5
Solving a System Using an Inverse Matrix

If A is invertible, you can solve $AX = B$ by entering the matrices A and B and then calculating $A^{-1}B$. To solve the problem from Example 3 on page 125, enter $\begin{pmatrix} 1 & 0 & 2 \\ 1 & 1 & 1 \\ 1 & 1 & 2 \end{pmatrix}$ in [A], enter $\begin{pmatrix} 22 \\ 11 \\ 20 \end{pmatrix}$ in [B], and calculate $[A]^{-1}[B]$:

```
[A]                   [B]                  [A]⁻¹[B]
     [[1 0 2]              [[22]                 [[4 ]
      [1 1 1]               [11]                  [-2]
      [1 1 2]]              [20]]                  [9 ]]
```

Now choose three other integers, enter them in [B], and find $[A]^{-1}[B]$. Do these values satisfy your new equations?

Spreadsheet Exploration
SECTION 3.5
Finding an Inverse

The following spreadsheet* uses the matrix inverse function **MINVERSE** to calculate inverses of

$$\begin{pmatrix} 1 & 0 & 2 \\ 1 & 1 & 1 \\ 1 & 1 & 2 \end{pmatrix} \quad \text{and} \quad \begin{pmatrix} 1 & -6 & 4 & 6 \\ 2 & 5 & 4 & 0 \\ 2 & 2 & 3 & 2 \\ 1 & 4 & 2 & -1 \end{pmatrix}.$$

	A	B	C	D	E	F	G	H	I
1		Matrix A					Inverse of A		
2									
3	1	0	2			1	2	-2	
4	1	1	1			-1	0	1	
5	1	1	2			0	-1	1	
6									
7	1	-6	4	6		-5	34	-8	-46
8	2	5	4	0		2	-15	4	20
9	2	2	3	2		-3.25E-17	2	-1	-2
10	1	4	2	-1		3	-22	6	29

However, the first answer is correct and the second is not. Spreadsheet programs sometimes produce rounding errors (such as the small number -3.25×10^{-17} instead of zero). The correct inverse of $\begin{pmatrix} 1 & -6 & 4 & 6 \\ 2 & 5 & 4 & 0 \\ 2 & 2 & 3 & 2 \\ 1 & 4 & 2 & -1 \end{pmatrix}$ is $\begin{pmatrix} -5 & 34 & -8 & -46 \\ 2 & -15 & 4 & 20 \\ 0 & 2 & -1 & -2 \\ 3 & -22 & 6 & 29 \end{pmatrix}$ as you can easily check.

Be careful! Because apparent rounding errors might also indicate that A is invertible when it is in fact singular, *always* verify that the spreadsheet (or computer or calculator) answer does indeed satisfy $A^{-1}A = I$.

---

* Go to CourseMate for FINITE for graphing calculator programs, Excel spreadsheets, and technology guides. Access at login.cengagebrain.com.

Graphing Calculator Explorations

SECTION 4.1

Boundary Line

You can view any nonvertical boundary line $ax + by = h$ by entering it as $y = (h - ax)/b$. To see the boundary line $2x + 3y = 12$ from Example 1 on pages 137–138, enter it as $y_1 = (12 - 2x)/3$ and graph it on the window $[-5, 10]$ by $[-5, 10]$. Some calculators will also "shade" above or below the line provided you change the \ marker to the left of the y_1.

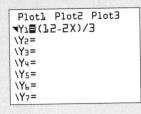
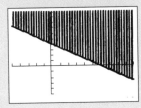

SECTION 4.3

Pivoting

The program PIVOT* carries out the pivot operation after you specify the pivot column and pivot row. To carry out the pivot operation with your calculator on the simplex tableau from Example 5 on page 162, proceed as follows:

a. Enter the simplex tableau as one large matrix [A] having 3 rows and 5 columns. Because this matrix is too large to fit on the calculator screen, it will scroll from left to right and back again as you enter your numbers.

b. Run the program PIVOT. It will display the current tableau in matrix [A] and you can use the arrows to scroll the screen to see the rest of it. Press ENTER to enter the **P**ivot **C**olumn and then again to enter the **P**ivot **R**ow of your pivot element.

c. After you ENTER the pivot row, the calculator will perform the pivot operation and display the new tableau. You can use the arrows to scroll the screen to see the rest of it. Press ENTER to exit the program.

If your problem requires several pivot operations, you can rerun the program with the new tableau by pressing ENTER again.

* Go to CourseMate for FINITE for graphing calculator programs, Excel spreadsheets, and technology guides. Access at login.cengagebrain.com.

Graphing Calculator Explorations
SECTION 5.2
Permutations

Values of $_nP_r$ are easy to find on some graphing calculators. The following displays show $_nP_r$ for $n = 26$ and several values of r, showing that $_nP_r$ gets large quickly as r increases.

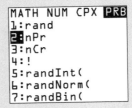

Combinations

Values of $_nC_r$ are easy to find if your calculator includes this command. The following screens show several values of $_nC_r$ for $n = 8$. Notice that $_nC_r$ gets larger and then smaller as r increases.

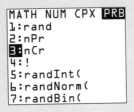

Spreadsheet Exploration

SECTION 5.6

Using a Random Number Generator

The spreadsheet* below uses a "random number generator" to simulate four coin tosses repeated 400 times. The numbers of heads on each of these four tosses are recorded in cells   A21:T40,   and the results are summarized in cells   G18:K18.   The graph shows that the results of this experiment approximate the probability distribution found in Example 6 on page 218.

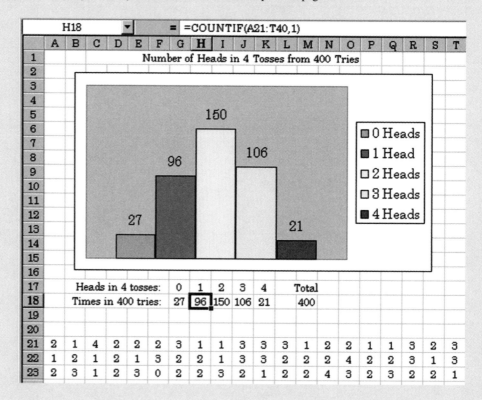

* Go to CourseMate for FINITE for graphing calculator programs, Excel spreadsheets, and technology guides. Access at login.cengagebrain.com.

Graphing Calculator Explorations

SECTION 6.1

Histograms

Histograms can be drawn on graphing calculators using the STAT PLOT command. To explore several histograms of the data from Example 2 on page 225, proceed as follows:

a. Enter the data values as a list and store it in the list L_1.

b. Turn off the axes in the window FORMAT menu, turn on STAT PLOT 1, select the histogram icon with Xlist L_1 and set the WINDOW parameters Xmin = 9.5, Xmax = 81.5, Xscl = 12 to match the start, finish, and class width we used. GRAPH and then TRACE to explore your histogram.

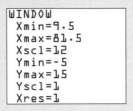

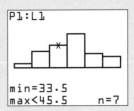

Of course, different choices of the class width, number of classes, and starting value will result in slightly different histograms. The following are several possibilities:

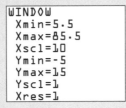

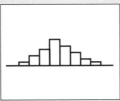

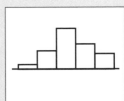

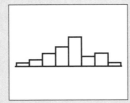

Notice that each histogram supports the conclusion that the website is most popular with those in the middle age range.

SECTION 6.3

Box-and-Whiskers Plots

Box-and-whisker plots can be drawn on graphing calculators by using the STAT PLOT command. To explore several box-and-whisker plots, including one of the data from Example 2 on page 234, proceed as follows.

a. Turn off the axes in the window FORMAT menu and set the WINDOW parameters to Xmin = 0 and Xmax = 100.

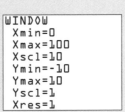

b. Enter the data values:

$\{20, 37, 65, 77, 78, 79, 81, 82, 83, 85, 87, 90\}$ as the list L_1,

$\{20, 31, 39, 41, 63, 80, 80, 80, 81, 87, 88, 90\}$ as the list L_2, and

$\{20, 55, 58, 60, 60, 61, 63, 64, 65, 79, 81, 90\}$ as the list L_3.

c. Turn on STAT PLOT 1 and select the box-and-whisker plot icon with *X*list L_1, and do the same for STAT PLOT 2 with *X*list L_2 and STAT PLOT 3 with *X*list L_3.

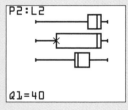

d. GRAPH and then TRACE to explore your box-and-whisker plots. The screen shows that the first quartile of the second list of values is 40. The first plot agrees with the one we drew in Example 2.

SECTION 6.4

Areas under Normal Distribution Curves

Areas under normal distribution curves can be found easily on a graphing calculator. To find the area under the normal distribution curve with mean $\mu = 20$ and standard deviation $\sigma = 5$ from 10 (which is $\mu - 2\sigma$) to 30 (which is $\mu + 2\sigma$), proceed as follows:

a. From the DISTRIBUTION menu, select the normalcdf command to find the cumulative values of the normal distribution (that is, the area under the curve) between two given values.

```
DISTR DRAW
1:normalpdf(
2:normalcdf(
3:invNorm(
4:tpdf(
5:tcdf(
6:X²pdf(
7↓X²cdf(
```

```
normalcdf(10,30,
20,5)
          .9544997876
```

b. Enter the (left) starting value and the (right) ending value as well as the mean and standard deviation. This area is the same as the probability that the normal random variable is between the two given values 10 and 30.

c. To *see* this area shaded under the normal curve, use the ShadeNorm command from the DISTRIBUTION DRAW menu with an appropriate window.

```
WINDOW
 Xmin=0
 Xmax=40
 Xscl=5
 Ymin=-.05
 Ymax=.1
 Yscl=1
 Xres=1
```

```
DISTR DRAW
1:ShadeNorm(
2:Shade_t(
3:ShadeX²(
4:ShadeF(
```

```
ShadeNorm(10,30,
20,5)
```

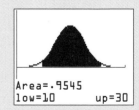

Graphing Calculator Explorations

SECTION 7.1

Transition Matrix

$$T = \begin{pmatrix} 0.20 & 0.25 & 0.55 \\ 0.35 & 0.30 & 0.35 \\ 0.10 & 0.05 & 0.85 \end{pmatrix}$$

To verify that the transition matrix from Example 2 on page 254 (shown on the left) transforms the state distribution vector $D = (0.50 \quad 0.20 \quad 0.30)$ into $(0.20 \quad 0.20 \quad 0.60)$, enter T and D into matrices [A] and [B] and find their product as follows.

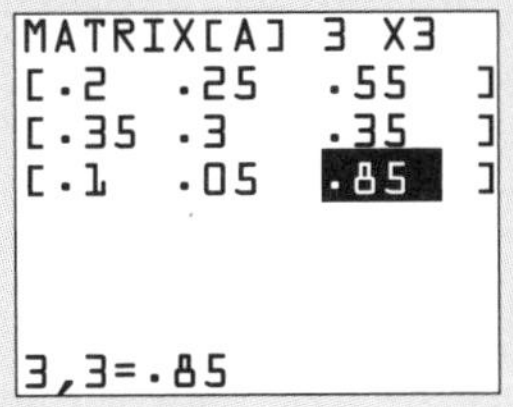

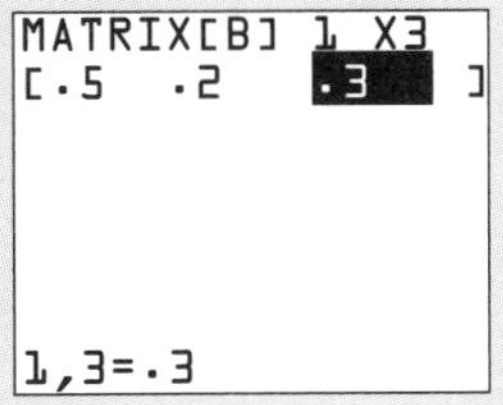

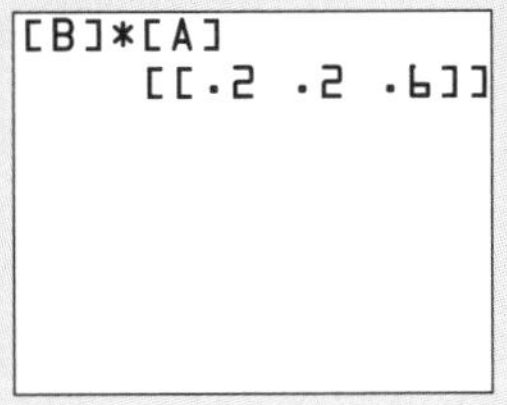

```
MATRIX[A] 3 X3          MATRIX[B] 1 X3          [B]*[A]
[.2   .25   .55  ]      [.5   .2   .3   ]          [[.2 .2 .6]]
[.35  .3    .35  ]
[.1   .05   .85  ]


3,3=.85                 1,3=.3
```

What happens if we repeat this calculation with the state distribution vector $D = (0.140 \quad 0.104 \quad 0.756)$?

State Distribution Vectors

For the absorbing Markov chain in Example 3 on page 256, we may easily check that the probability of being in state S_1 increases with each transition. Entering T and D_0 into matrices [A] and [B] and calculating $D_k = D_0 \cdot T^k$ for successively higher values of k gives

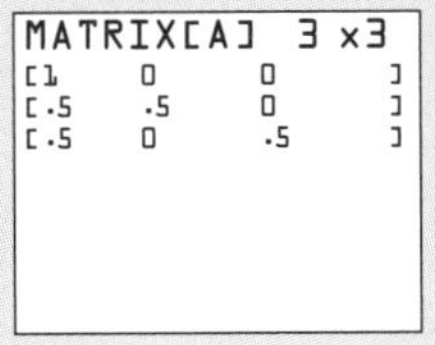

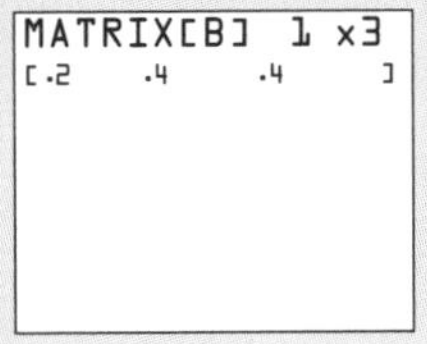

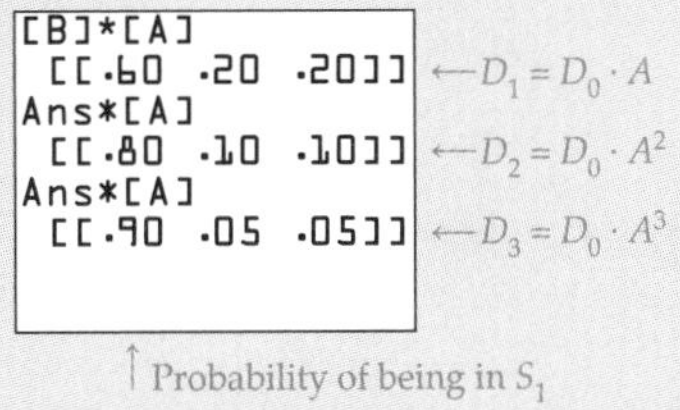

```
MATRIX[A] 3 x3          MATRIX[B] 1 x3          [B]*[A]
[1    0    0   ]        [.2   .4   .4  ]          [[.60 .20 .20]]    ←—D₁ = D₀ · A
[.5   .5   0   ]                                Ans*[A]
[.5   0    .5  ]                                  [[.80 .10 .10]]    ←—D₂ = D₀ · A²
                                                Ans*[A]
                                                  [[.90 .05 .05]]    ←—D₃ = D₀ · A³
```

↑ Probability of being in S_1

Notice that the first entry of these state distribution vectors increases from 0.6 to 0.8 and then to 0.9, showing that the probability of being in state S_1 increases with each transition. These probabilities will continue to increase, even if we begin with a different initial distribution D_0.

SECTION 7.2

Steady-State Distribution by Powers

To see whether the transition matrix for the truck rental company in Example 2 on page 260 has the property that the rows of higher powers "settle down" to the steady-state distribution, we may calculate as follows.

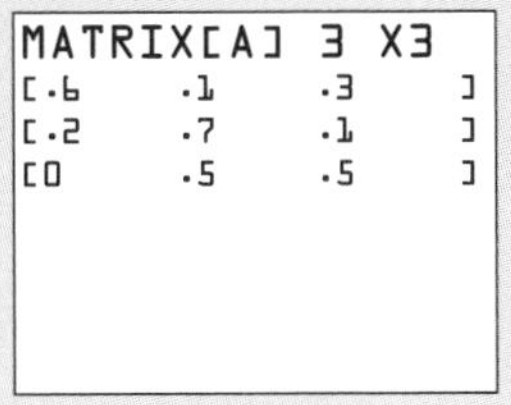

```
MATRIX[A] 3 X3
[.6   .1   .3   ]
[.2   .7   .1   ]
[0    .5   .5   ]
```

a. Enter the matrix $T = \begin{pmatrix} 0.6 & 0.1 & 0.3 \\ 0.2 & 0.7 & 0.1 \\ 0 & 0.5 & 0.5 \end{pmatrix}$ in matrix [A].

b. Then calculate T^k for $k = 2, 4, 26,$ and 52 (a year of weekly transitions):

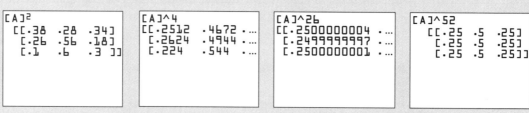

```
[A]²
[[.38  .28  .34]
 [.26  .56  .18]
 [.1   .6   .3 ]]
```

```
[A]^4
[[.2512  .4672 .…
 [.2624  .4944 .…
 [.224   .544  .…
```

```
[A]^26
[[.2500000004  .…
 [.2499999997  .…
 [.2500000001  .…
```

```
[A]^52
[[.25  .5  .25]
 [.25  .5  .25]
 [.25  .5  .25]]
```

Notice that the rows in successive screens all seem to be approaching the same distribution $\left(\frac{1}{4} \quad \frac{1}{2} \quad \frac{1}{4}\right)$, with the last screen (for T^{52}) showing exactly this steady-state distribution, the same one that we found in Example 2.

Steady-State Distribution Using MARKOV

We may solve Example 3 on pages 263–264 using a graphing calculator. The program MARKOV* displays powers of a transition matrix T, the augmented matrix $\left(\begin{array}{c|c} T^t - I & 0 \\ \hline 1 \cdots 1 & 1 \end{array}\right)$, and the row-reduced form showing the steady-state distribution. To find the steady-state distribution in Example 3, we proceed as follows:

a. Enter the T in matrix [A].

```
MATRIX[A] 3 X3
[.6   .1   .3   ]
[.2   .7   .1   ]
[0    .5   .5   ]
```

b. Run the program MARKOV and press ENTER several times to display T^k for $k = 1, 2, 3, 32,$ and 64, and then $\left(\begin{array}{c|c} T^t - I & 0 \\ \hline 1 \cdots 1 & 1 \end{array}\right)$ followed by its row-reduced form.

```
[A]
[[3/5 1/10 3/10…
 [1/5 7/10 1/10…
 [0   1/2  1/2 …
```

```
[A]^64
[[1/4 1/2  1/4]
 [1/4 1/2  1/4]
 [1/4 1/2  1/4]]
```

```
(Aᵀ-I)X=0
AND (1..1)X=1 IS
[[-2/5 1/5   0 …
 [1/10 -3/10 1/…
 [3/10 1/10  -1…
 [1    1    1 …
```

```
SOLUTION
[[1 0 0 1/4]
 [0 1 0 1/2]
 [0 0 1 1/4]
 [0 0 0 0]]
```

The steady state distribution $\left(\frac{1}{4} \quad \frac{1}{2} \quad \frac{1}{4}\right)$ shown in the right-hand column agrees with the answer found earlier.

SECTION 7.3

Finding an Absorption Probability

To find the probability that the absorbing Markov chain in Example 3 on pages 268–269 will be absorbed into state A_2 given that it started in state N_3, proceed as follows.

Enter the matrices Q and R from the solution of Example 3 into the calculator as [A] and [B], along with a 3×3 identity matrix in [I], as shown in the first three screens below.

```
MATRIX[A] 3 X3
[.8  .1  .05  ]
[.1  .2  .6   ]
[.05 .1  .8   ]

3,3=.8
```

```
MATRIX[B] 3 X2
[.03 .02  ]
[.07 .03  ]
[.01 .04  ]

3,2=.04
```

```
MATRIX[I] 3 X3
[1  0  0  ]
[0  1  0  ]
[0  0  1  ]

3,3=1
```

```
([I]-[A])⁻¹*[B]
[[.46  .54]
 [.43  .57]
 [.38  .62]]
```

Then calculate $F \cdot R$ (with $F = (I - Q)^{-1}$) as $([I] - [A])^{-1}*[B]$, as shown in the last screen above.

By part (3) of the result on page 270, the probability that the chain, beginning in state N_3, will be absorbed into state A_2 is the entry in row 3 and column 2 of $F \cdot R$, which is 0.62, in agreement with Example 5 on pages 271–272.

$$F \cdot R = \begin{array}{c} N_1 \\ N_2 \\ N_3 \end{array} \begin{pmatrix} \overset{A_1}{0.46} & \overset{A_2}{0.54} \\ 0.43 & 0.57 \\ 0.38 & 0.62 \end{pmatrix}$$

---

* Go to CourseMate for FINITE for graphing calculator programs, Excel spreadsheets, and technology guides. Access at login.cengagebrain.com.

Graphing Calculator Explorations

SECTION 8.2

Expected Value

The expected value found in Example 1 on pages 282–283 is easily calculated as a matrix product by entering the row probabilities in matrix [A], the payoff matrix in matrix [B], and the column probabilities in matrix [C], and then finding the product [A][B][C].

```
MATRIX[A] 1 x2      MATRIX[B] 2 x2      MATRIX[C] 2 x1       [A] [B] [C]
[.4    .6     ]     [-1    2      ]     [.5          ]                  [[.2]]
                    [2    -2      ]     [.5          ]
```

The answer, 0.2, agrees with the answer $\frac{1}{5}$ we found in Example 1.

Random Choices

A graphing calculator can help to make random choices for any strategy. For example, the strategy $r = (\frac{1}{4} \quad \frac{1}{2} \quad \frac{1}{4})$ separates the interval from 0 to 1 into three subintervals of lengths $\frac{1}{4}, \frac{1}{2}$, and $\frac{1}{4}$:

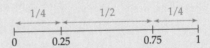

```
     1/4          1/2           1/4
  |-----|-------------------|--------|
  0    0.25               0.75      1
```

The random number command gives random numbers between 0 and 1.

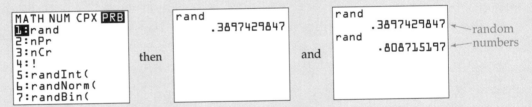

```
MATH NUM CPX PRB          rand                          rand
1:rand                          .3897429847                    .3897429847  ←random
2:nPr                                                    rand                numbers
3:nCr               then                          and          .8087151197  ←
4:!
5:randInt(
6:randNorm(
7:randBin(
```

The first random number, 0.3897, lies in the second of the above intervals, so you would play the *second* choice. The next random number, 0.8087, lies in third interval, so you would next play the *third* choice, and so on, using as many random numbers as you need. Of course, the row player and the column player should each pick his or her own random number to decide which move to make next.

SECTION 8.3

Solving a Game Using GameLP

The program GameLP* constructs and solves the simplex tableau for a game with all entries at least 1. To see that the particular constant k added to G in Step 1 of the method given on pages 292–293 does

not change the solution of the game, we use GameLP to solve $G + 4 = \begin{pmatrix} 5 & 2 & 5 \\ 6 & 3 & 2 \\ 3 & 6 & 5 \end{pmatrix}$ for the game

solved in Example 1 on pages 293–294 as follows.

---

* Go to CourseMate for FINITE for graphing calculator programs, Excel spreadsheets, and technology guides. Access at login.cengagebrain.com.

a. Enter the $G + k$ game as matrix [A] and then run the GameLP program:

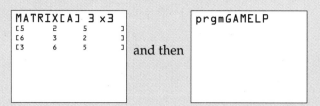

```
MATRIX[A] 3 x3
[5    2    5    ]
[6    3    2    ]
[3    6    5    ]
```
and then
```
prgmGAMELP
```

b. Press ENTER twice and select YES or NO depending on whether or not you wish to see all the intermediate tableaux calculations. Choosing NO, we arrive at the final tableau, and then ENTER for the final answer.

```
SHOW TABLEAUX
1:YES
2:NO
```

```
FINAL TABLEAU
[[0 0 1 9/34  -...
 [1 0 0 1/34  1...
 [0 1 0 -4/17 5...
 [0 0 0 1/17  1...
```

```
FINAL TABLEAU
... 1/34    1/17]
... -11/102 2/17]
... 10/51   1/17]
... 2/17    4/17]]
```

```
VALUE OF GAME IS
              17/4

R STRATEGY IS
  {1/4 1/4 1/2}
C STRATEGY IS
  {1/2 1/4 1/4}
```

Notice that despite the different value for k, the optimal strategy is still the same. Why is the value of the game now $4\frac{1}{4}$ instead of the $\frac{1}{4}$ found in Example 1?